PROGRESS IN CLINICAL AND BIOLOGICAL RESEARCH

Series Editors

Nathan Back Vincent P. Eijsvoogel Kurt Hirschhorn Sidney Udenfriend
George J. Brewer Robert Grover Seymour S. Kety Jonathan W. Uhr

RECENT TITLES

Vol 206: **Genetic Toxicology of the Diet,** Ib Knudsen, *Editor*

Vol 207: **Monitoring of Occupational Genotoxicants,** Marja Sorsa, Hannu Norppa, *Editors*

Vol 208: **Risk and Reason: Risk Assessment in Relation to Environmental Mutagens and Carcinogens,** Per Oftedal, Anton Brøgger, *Editors*

Vol 209: **Genetic Toxicology of Environmental Chemicals,** Claes Ramel, Bo Lambert, Jan Magnusson, *Editors*. Published in two volumes: Part A: *Basic Principles and Mechanisms of Action.* Part B: *Genetic Effects and Applied Mutagenesis*

Vol 210: **Ionic Currents in Development,** Richard Nuccitelli, *Editor*

Vol 211: **Transfusion Medicine: Recent Technological Advances,** Kris Murawski, Frans Peetoom, *Editors*

Vol 212: **Cancer Metastasis: Experimental and Clinical Strategies,** D.R. Welch, B.K. Bhuyan, L.A. Liotta, *Editors*

Vol 213: **Plant Flavonoids in Biology and Medicine: Biochemical, Pharmacological, and Structure–Activity Relationships,** Vivian Cody, Elliott Middleton, Jr., Jeffrey B. Harborne, *Editors*

Vol 214: **Ethnic Differences in Reactions to Drugs and Xenobiotics,** Werner Kalow, H. Werner Goedde, Dharam P. Agarwal, *Editors*

Vol 215: **Megakaryocyte Development and Function,** Richard F. Levine, Neil Williams, Jack Levin, Bruce L. Evatt, *Editors*

Vol 216: **Advances in Cancer Control: Health Care Financing and Research,** Lee E. Mortenson, Paul F. Engstrom, Paul N. Anderson, *Editors*

Vol 217: **Progress in Developmental Biology,** Harold C. Slavkin, *Editor.* Published in two volumes.

Vol 218: **Evolutionary Perspective and the New Genetics,** Henry Gershowitz, Donald L. Rucknagel, Richard E. Tashian, *Editors*

Vol 219: **Recent Advances in Arterial Diseases: Atherosclerosis, Hypertension, and Vasospasm,** Thomas N. Tulenko, Robert H. Cox, *Editors*

Vol 220: **Safety and Health Aspects of Organic Solvents,** Vesa Riihimäki, Ulf Ulfvarson, *Editors*

Vol 221: **Developments in Bladder Cancer,** Louis Denis, Tadao Niijima, George Prout, Jr., Fritz H. Schröder, *Editors*

Vol 222: **Dietary Fat and Cancer,** Clement Ip, Diane F. Birt, Adrianne E. Rogers, Curtis Mettlin, *Editors*

Vol 223: **Cancer Drug Resistance,** Thomas C. Hall, *Editor*

Vol 224: **Transplantation: Approaches to Graft Rejection,** Harold T. Meryman, *Editor*

Vol 225: **Gonadotropin Down-Regulation in Gynecological Practice,** Rune Rolland, Dev R. Chadha, Wim N.P. Willemsen, *Editors*

Vol 226: **Cellular Endocrinology: Hormonal Control of Embryonic and Cellular Differentiation,** Ginette Serrero, Jun Hayashi, *Editors*

Vol 227: **Advances in Chronobiology,** John E. Pauly, Lawrence E. Scheving, *Editors.* Published in two volumes.

Vol 228: **Environmental Toxicity and the Aging Processes,** Scott R. Baker, Marvin Rogul, *Editors*

Vol 229: **Animal Models: Assessing the Scope of Their Use in Biomedical Research,** Junichi Kawamata, Edward C. Melby, Jr., *Editors*

Vol 230: **Cardiac Electrophysiology and Pharmacology of Adenosine and ATP: Basic and Clinical Aspects,** Amir Pelleg, Eric L. Michelson, Leonard S. Dreifus, *Editors*

Vol 231: **Detection of Bacterial Endotoxins With the Limulus Amebocyte Lysate Test,** Stanley W. Watson, Jack Levin, Thomas J. Novitsky, *Editors*

Vol 232: **Enzymology and Molecular Biology of Carbonyl Metabolism: Aldehyde Dehydrogenase, Aldo-Keto Reductase, and Alcohol Dehydrogenase,** Henry Weiner, T. Geoffrey Flynn, *Editors*

Vol 233: **Developmental and Comparative Immunology,** Edwin L. Cooper, Claude Langlet, Jacques Bierne, *Editors*

Vol 234: **The Hepatitis Delta Virus and Its Infection,** Mario Rizzetto, John L. Gerin, Robert H. Purcell, *Editors*

Vol 235: **Preclinical Safety of Biotechnology Products Intended for Human Use,** Charles E. Graham, *Editor*

Vol 236: **First Vienna Shock Forum,** Günther Schlag, Heinz Redl, *Editors*. Published in two volumes: Part A: *Pathophysiological Role of Mediators and Mediator Inhibitors in Shock.* Part B: *Monitoring and Treatment of Shock.*

Vol 237: **The Use of Transrectal Ultrasound in the Diagnosis and Management of Prostate Cancer,** Fred Lee, Richard McLeary, *Editors*

Vol 238: **Avian Immunology,** W.T. Weber, D.L. Ewert, *Editors*

Vol 239: **Current Concepts and Approaches to the Study of Prostate Cancer,** Donald S. Coffey, Nicholas Bruchovsky, William A. Gardner, Jr., Martin I. Resnick, James P. Karr, *Editors*

Vol 240: **Pathophysiological Aspects of Sickle Cell Vaso-Occlusion,** Ronald L. Nagel, *Editor*

Vol 241: **Genetics and Alcoholism,** H. Werner Goedde, Dharam P. Agarwal, *Editors*

Vol 242: **Prostaglandins in Clinical Research,** Helmut Sinzinger, Karsten Schrör, *Editors*

Vol 243: **Prostate Cancer,** Gerald P. Murphy, Saad Khoury, Réne Küss, Christian Chatelain, Louis Denis, *Editors*. Published in two volumes: Part A: *Research, Endocrine Treatment, and Histopathology.* Part B: *Imaging Techniques, Radiotherapy, Chemotherapy, and Management Issues*

Vol 244: **Cellular Immunotherapy of Cancer,** Robert L. Truitt, Robert P. Gale, Mortimer M. Bortin, *Editors*

Vol 245: **Regulation and Contraction of Smooth Muscle,** Marion J. Siegman, Andrew P. Somlyo, Newman L. Stephens, *Editors*

Vol 246: **Oncology and Immunology of Down Syndrome,** Ernest E. McCoy, Charles J. Epstein, *Editors*

Vol 247: **Degenerative Retinal Disorders: Clinical and Laboratory Investigations,** Joe G. Hollyfield, Robert E. Anderson, Matthew M. LaVail, *Editors*

Please contact the publisher for information about previous titles in this series.

Regulation and Contraction of Smooth Muscle

REGULATION AND CONTRACTION OF SMOOTH MUSCLE

Proceedings of an International Union of Physiological Sciences Satellite Conference on Smooth Muscle Contraction held at Minaki, Ontario, Canada July 20–24, 1986

Editors

MARION J. SIEGMAN
Department of Physiology
Jefferson Medical College
Thomas Jefferson University
Philadelphia, Pennsylvania

ANDREW P. SOMLYO
Pennsylvania Muscle Institute
University of Pennsylvania
School of Medicine
Philadelphia, Pennsylvania

NEWMAN L. STEPHENS
Department of Physiology
University of Manitoba
Winnipeg, Manitoba, Canada

ALAN R. LISS, INC. • NEW YORK

Address all Inquiries to the Publisher
Alan R. Liss, Inc., 41 East 11th Street, New York, NY 10003

Copyright © 1987 Alan R. Liss, Inc.

Printed in the United States of America

Under the conditions stated below the owner of copyright for this book hereby grants permission to users to make photocopy reproductions of any part or all of its contents for personal or internal organizational use, or for personal or internal use of specific clients. This consent is given on the condition that the copier pay the stated per-copy fee through the Copyright Clearance Center, Incorporated, 27 Congress Street, Salem, MA 01970, as listed in the most current issue of "Permissions to Photocopy" (Publisher's Fee List, distributed by CCC, Inc.), for copying beyond that permitted by sections 107 or 108 of the US Copyright Law. This consent does not extend to other kinds of copying, such as copying for general distribution, for advertising or promotional purposes, for creating new collective works, or for resale.

Library of Congress Cataloging-in-Publication Data

Satellite Conference on Smooth Muscle Contraction
 (1986 : Minaki, Ont.)
 Regulation and contraction of smooth muscle.

 (Progress in clinical and biological research ; 245)
 Satellite conference of the International Union of Physiological Sciences 30th Congress, held in July 1986 at Vancouver, B.C.
 Includes bibliographies and index.
 1. Smooth muscle—Congresses. 2. Muscle contraction—Regulation—Congresses. I. Siegman, Marion J., 1933- . II. Somlyo, Andrew P. (Andrew Paul), 1930- . III. Stephens, Newman L., 1926- IV. International Union of Physiological Sciences. Congress (30th : 1986 : Vancouver, B.C.) V. Title. VI. Series: Progress in clinical and biological research ; v. 245. [DNLM: 1. Muscle Contraction—congresses. 2. Muscle, Smooth—physiology—congresses. W1 PR668E v.245 / WE 500 S253r 1986]
 QP321.S28 1986   599.01'852   87-16966
 ISBN 0-8451-5095-2

Contents

Contributors	xi
Preface	xix
Acknowledgments	xxi

ULTRASTRUCTURE AND CROSSBRIDGE KINETICS

Molecular Structure and Organization of Filaments in Single, Skinned Smooth Muscle Cells
P.H. Cooke, G. Kargacin, R. Craig, K. Fogarty, F.S. Fay, and S. Hagen . . . 1

Crossbridge Transients Initiated by Photolysis of Caged Nucleotides, and Crossbridge Structure in Smooth Muscle
Avril V. Somlyo, Yale Goldman, Taro Fujimori, Meredith Bond, David Trentham, and Andrew P. Somlyo . . . 27

Influence of ATP, ADP and AMPPNP on the Energetics of Contraction in Skinned Smooth Muscle
Anders Arner, Per Hellstrand, and J. Caspar Rüegg . . . 43

Comparative Studies on the Mechanism of Regulation of Smooth and Striated Muscle Actomyosin
Edwin W. Taylor . . . 59

BIOCHEMISTRY OF CONTRACTILE AND REGULATORY PROTEINS I

Isoforms of Myosin in Smooth Muscle
Malcolm P. Sparrow, Anders Arner, Mukhallad A. Mohammad, Per Hellstrand, and J. Caspar Rüegg . . . 67

Subunit Exchange Between Smooth Muscle Myosin Filaments
Kathleen M. Trybus and Susan Lowey . . . 81

Conformational Changes in Myosin and Heavy Meromyosin from Chicken Gizzard Associated with Phosphorylation
Sumitra Nag, Hiroshi Suzuki, Jan Sosinski, and John C. Seidel . . . 91

Ca^{2+} Regulation in Smooth Muscle; Dissociation of Myosin Light Chain Kinase Activity from Activation of Actin-Myosin Interaction
Setsuro Ebashi, Takashi Mikawa, Hideto Kuwayama, Masashi Suzuki, Hiroaki Ikemoto, Yasuki Ishizaki, and Ritsuko Koga . . . 109

Caldesmon, a Major Actin- and Calmodulin-Binding Protein of Smooth Muscle
Michael P. Walsh . . . 119

Modulation of Actomyosin ATPase by Thin Filament-Associated Proteins
Samuel Chacko, Hidetake Miyata, and Kurumi Y. Horiuchi   143

BIOCHEMISTRY OF CONTRACTILE AND REGULATORY PROTEINS II

MgATPase Activity of Vertebrate Smooth Muscle Actomyosin: Stimulation by Tropomyosin is Modified by Myosin Phosphorylation and Its Conformational State
Apolinary Sobieszek   159

Myosin Light Chain Kinases and Kinetics of Myosin Phosphorylation in Smooth Muscle Cells
Kristine E. Kamm, Sancy A. Leachman, Carolyn H. Michnoff, Mary H. Nunnally, Anthony Persechini, Andrea L. Richardson, and James T. Stull   183

Aortic Polycation-Modulable Protein Phosphatase(s): Structure and Function
Joseph Di Salvo   195

Characterization of the Smooth Muscle Phosphatases and Study of Their Function
Mary D. Pato and Ewa Kerc   207

Activation of Protein Kinase C and Contraction in Skinned Vascular Smooth Muscle
Meeta Chatterjee and Carolyn Foster   219

PHARMACOLOGICAL PROBES OF REGULATORY SYSTEMS

Cyclic AMP Dependent and Myosin Light Chain Kinase: Relationship to Altered Vascular Reactivity in Hypertension and Development of Direct Pharmacological Modulators
Paul J. Silver   233

Selective Inhibitors of Phosphorylation in Smooth Muscle
Masatoshi Hagiwara and Hiroyoshi Hidaka   251

Regulation of cAMP Content and cAMP-Dependent Protein Kinase Activity in Airway Smooth Muscle
Theodore J. Torphy, Miriam Burman, Lisa B.F. Huang, Stephen Horohonich, and Lenora B. Cieslinski   263

Molecular Mechanisms Underlying Increased Contractility to Norepinephrine Stimulation in SHR Vascular Smooth Muscle
R.C. Bhalla, R.V. Sharma, and M.B. Aqel   277

MECHANICS AND ENERGETICS

Slowing of Crossbridge Cycling Rate in Mammalian Smooth Muscle Occurs Without Evidence of an Increase in Internal Load
T.M. Butler, M.J. Siegman, and S.U. Mooers   289

Force: Velocity Relationship and Helical Shortening in Single Smooth Muscle Cells
David Warshaw, Whitney McBride, and Steven Work   303

The Effects of Calcium on Smooth Muscle Mechanics and Energetics
Richard J. Paul, John D. Strauss, and Joseph M. Krisanda   319

Stiffness and the Energetics of Active Shortening in Chemically Skinned Smooth Muscle
Per Hellstrand, Håkan Arheden, Lars Sjölin, and Anders Arner 333

Crossbridge Properties Studied During Forced Elongation of Active Smooth Muscle
Richard A. Meiss 347

REGULATORY MECHANISMS IN INTACT AND SKINNED MUSCLES

Smooth Muscle Contraction: Mechanisms of Crossbridge Slowing
Newman L. Stephens and C.Y. Seow 357

Are Contraction Kinetics Affected by the Activation Mode?
Ulrich Peiper, Brigitte M. Lobnig, and Bruno Zobel 377

High Myosin Light Chain Phosphatase Activity in Arterial Smooth Muscle: Can It Explain the Latch Phenomenon?
S.P. Driska 387

Dependence of Stress and Velocity on Ca^{2+} and Myosin Phosphorylation in the Skinned Swine Carotid Media
Meeta Chatterjee, Chi-Ming Hai, and R.A. Murphy 399

Determinants of the Latch State in Vascular Smooth Muscle
R.A. Murphy, P.H. Ratz, and C.M. Hai 411

Regulation of Carotid Artery Smooth Muscle Relaxation by Myosin Dephosphorylation
Joe R. Haeberle, Brett A. Trockman, and Anna A. Depaoli-Roach 415

Calcium/Calmodulin Activation of Gizzard Skinned Fibers at Low Levels of Myosin Phosphorylation
Jürgen Wagner, Gabriele Pfitzer, and J. Caspar Rüegg 427

Non-Ca^{2+}-Activated Contraction in Smooth Muscle
W. Glenn L. Kerrick and Phyllis E. Hoar 437

POSTER PRESENTATIONS

Synthesis of Inositol Phospholipids in Carbachol-Stimulated Canine Trachealis Muscle
Carl B. Baron and Ronald F. Coburn 449

Myosin Heavy Chain Isoforms in Human Smooth Muscle
N. DeMarzo, S. Sartore, L. Saggin, L. Fabbri, and S. Schiaffino 451

Contraction of Hog Carotid Arterial Smooth Muscle Cells Prepared by Digestion with Papain
S.P. Driska, M. Desilets, and C.M. Baumgarten 453

Guanylate-Cyclase-Dependent Gating of Receptor-Operated Calcium Channels in Vascular Smooth Muscle
T. Godfraind 455

Birefringence of Rat Anococcygeus: Correlation with the Density of Myosin Filaments
A. Godfraind-DeBecker, M.L. Cao, and J.M. Gillis 459

Mechanism of Shortening-Induced Depression of Contractility in Canine Tracheal Smooth Muscle
S.J. Gunst and J.Q. Stropp 461

Does a Limitation of Energy Supply to the Contractile Apparatus Underlie the Relaxation Induced by Hypoxia in Smooth Muscle?
Y. Ishida, M. Hashimoto, and R.J. Paul 463

Cultured Circular Smooth Muscle from the Rabbit Colon
H.W. Kao, S.E. Finn, A. Gown, J. Lechago, N. Lachant, and W.J. Snape, Jr. 465

A Geometric 3-Dimensional Thermodynamic Model for Crossbridge (XB) Attachment in Smooth Muscle (SM)
M. Li and D.M. Warshaw 467

Cytochalasin-Like Activity in Rat Aorta Smooth Muscle Cells
W. Magargal 469

Effects of Cyclic AMP-Dependent Protein Kinase in Skinned Coronary Artery
J.R. Miller and J.N. Wells 471

Latchbridges and Phosphorylated Crossbridges Are Activated Independently in Arterial Smooth Muscle
Robert S. Moreland and Suzanne Moreland 473

Inhibition of Cycling and Noncycling Crossbridges in Chemically Skinned Smooth Muscle by Vanadate
R.A. Nayler and M.P. Sparrow 475

The Effect of 2,3-Butanedione Monoxime (BDM) on Smooth Muscle Mechanical Properties
C.S. Packer, M.L. Kagan, S.A. Robertson, and N.L. Stephens 477

Effects of Antibodies to Turkey Gizzard (TG) Myosin Light Chain Kinase (MLCK) on Contraction and Myosin Phosphorylation (MLC-P_i) in Skinned Guinea Pig Taenia Coli (TC)
Richard J. Paul, John D. Strauss, and Primal de Lanerolle 479

Effects of Reduced Extracellular Calcium on Calcium Metabolism in Vascular Smooth Muscle
L.N. Russek and R.D. Phair 481

Stiffness of Tracheal Smooth Muscle During Active Shortening: Stiffness and Length Relationship
C.Y. Seow and N.L. Stephens 483

The Association of Intracellular Ca^{2+} Release with Contraction in Colonic Muscle
N. Sevy, H.W. Kao, and W.J. Snape, Jr. 485

Calcium, Magnesium and $MgATP^{2+}$ Dependence of Shortening in Skinned Single Smooth Muscle Cells
D.M. Warshaw and M.S. Hubbard 487

Index 489

Contributors

M.B. Aqel, Department of Anatomy, University of Iowa, Iowa City, IA 52242 **[277]**

Håkan Arheden, Department of Physiology and Biophysics, University of Lund, S-223 62 Lund, Sweden **[333]**

Anders Arner, Department of Physiology and Biophysics, University of Lund, S-223 62 Lund, Sweden **[43,67,333]**

Carl B. Baron, Department of Physiology, University of Pennsylvania, Philadelphia, PA 19104 **[449]**

C.M. Baumgarten, Department of Physiology and Biophysics, Medical College of Virginia, Richmond, VA 23298 **[453]**

R.C. Bhalla, Department of Anatomy, University of Iowa, Iowa City, IA 52242 **[277]**

Meredith Bond, Pennsylvania Muscle Institute, University of Pennsylvania School of Medicine, Philadelphia, PA 19104; present address: Department of Heart and Hypertension, Cleveland Clinic Research Institute, Cleveland, OH 44106 **[27]**

Miriam Burman, Department of Pharmacology, SmithKline and French Laboratories, Philadelphia, PA 19101 **[263]**

Thomas M. Butler, Department of Physiology, Jefferson Medical College, Thomas Jefferson University, Philadelphia, PA 19107 **[289]**

M.L. Cao, Physiologie des Muscles, Universite Catholique de Louvain, Brussels, Belgium **[459]**

Samuel Chacko, Department of Pathobiology, University of Pennsylvania, Philadelphia, PA 19104 **[143]**

Meeta Chatterjee, Department of Pharmacology, Schering Corporation, Bloomfield, NJ 07003 **[219,399]**

Lenora B. Cieslinski, Department of Pharmacology, SmithKline and French Laboratories, Philadelphia, PA 19101 **[263]**

Ronald F. Coburn, Department of Physiology, University of Pennsylvania, Philadelphia, PA 19104 **[449]**

Peter H. Cooke, Department of Physiology, University of Massachusetts Medical School, Worcester, MA 01605 **[1]**

The numbers in brackets are the opening page numbers of the contributors' articles.

Roger Craig, Department of Physiology, University of Massachusetts Medical School, Worcester, MA 01605 **[1]**

Primal de Lanerolle, Departments of Physiology and Biophysics, University of Chicago, Chicago, IL 60612 **[479]**

N. DeMarzo, Institute of Occupational Medicine, University of Padova, Padova, Italy **[451]**

Anna A. Depaoli-Roach, Department of Biochemistry, Indiana University School of Medicine, Indianapolis, IN 46223 **[415]**

M. Desilets, Department of Physiology and Biophysics, Medical College of Virginia, Richmond, VA 23298 **[453]**

Joseph Di Salvo, Department of Physiology and Biophysics, University of Cincinnati College of Medicine, Cincinnati, OH 45267-0576 **[195]**

Steven P. Driska, Department of Physiology and Biophysics, Medical College of Virginia, Richmond, VA 23298 **[387,453]**

Setsuro Ebashi, National Institute for Physiological Sciences, Myodaiji, Okazaki, 444 Japan **[109]**

L. Fabbri, Institute of Occupational Medicine, University of Padova, Padova, Italy **[451]**

Fredric S. Fay, Department of Physiology, University of Massachusetts Medical School, Worcester, MA 01605 **[1]**

S.E. Finn, Departments of Medicine and Pathology, Harbor-UCLA Medical Center, Torrance, CA and University of Washington, Seattle, WA **[465]**

Kevin E. Fogarty, Department of Physiology, University of Massachusetts Medical School, Worcester, MA 01605 **[1]**

Carolyn Foster, Department of Pharmacology, Schering Corporation, Bloomfield, NJ 07003 **[219]**

Taro Fujimori, Pennsylvania Muscle Institute, University of Pennsylvania School of Medicine, Philadelphia, PA 19104; present address: Brandeis University, Waltham, MA 02154 **[27]**

J.M. Gillis, Physiologie des Muscles, Universite Catholique de Louvain, Brussels, Belgium **[459]**

T. Godfraind, Laboràtoìre de Pharmacodynamie Génèralè et de Pharmacologie, Université Catholique de Louvain, Bruxelles, Belgique **[455]**

A. Godfraind-DeBecker, Physiologie des Muscles, Universite Catholique de Louvain, Brussels, Belgium **[459]**

Yale Goldman, Department of Physiology, University of Pennsylvania School of Medicine, Philadelphia, PA 19104 **[27]**

A. Gown, Departments of Medicine and Pathology, Harbor-UCLA Medical Center, Torrance, CA and University of Washington, Seattle, WA **[465]**

Susan J. Gunst, Mayo Clinic and Foundation, Rochester, MN 55905 **[461]**

Joe R. Haeberle, Department of Physiology and The Krannert Institute of Cardiology, University of Indiana School of Medicine, Indianapolis,IN 46223 **[415]**

Susan Hagen, Department of Medical Gastroenterology, Brigham and Women's Hospital, Boston, MA 02115 **[1]**

Masatoshi Hagiwara, Department of Molecular and Cellular Pharmacology, Mie University School of Medicine, Tsu, Mie 514, Japan **[251]**

Contributors / xiii

Chi-Ming Hai, Department of Physiology, School of Medicine, University of Virginia, Charlottesville, VA 22908 [399,411]

M. Hashimoto, Department of Physiology and Biophysics, University of Cincinnati College of Medicine, Cincinnati, OH 45267 [463]

Per Hellstrand, Department of Physiology and Biophysics, University of Lund, S-223 62 Lund, Sweden [43,67,333]

Hiroyoshi Hidaka, Department of Molecular and Cellular Pharmacology, Mie University School of Medicine, Tsu, Mie 514, Japan [251]

Phyllis E. Hoar, Departments of Physiology and Biophysics, University of Miami School of Medicine, Miami, FL 33101 [437]

Kurumi Y. Horiuchi, Department of Pathobiology, University of Pennsylvania, Philadelphia, PA 19104 [143]

Stephen Horohonich, Department of Pharmacology, SmithKline and French Laboratories, Philadelphia, PA 19101 [263]

Lisa B.F. Huang, Department of Pharmacology, SmithKline and French Laboratories, Philadelphia, PA 19101 [263]

M.S. Hubbard, Department of Physiology and Biophysics, University of Vermont, Burlington, VT 05405 [487]

Hiroaki Ikemoto, National Institute for Physiological Sciences, Myodaiji, Okazaki, 444 Japan [109]

Y. Ishida, Department of Physiology and Biophysics, University of Cincinnati College of Medicine, Cincinnati, OH 45267 [463]

Yasuki Ishizaki, National Institute for Physiological Sciences, Myodaiji, Okazaki, 444 Japan [109]

M.L. Kagan, Department of Physiology, University of Manitoba, Winnipeg, Manitoba, Canada R3E 0W3 [477]

Kristine E. Kamm, Department of Pharmacology and Moss Heart Center, University of Texas Health Science Center, Dallas, TX 75235; present address: Department of Physiology, University of Texas Health Science Center, Dallas, TX 75235 [183]

H.W. Kao, Departments of Medicine and Pathology, Harbor-UCLA Medical Center, Torrance, CA and University of Washington, Seattle, WA [465,485]

Gary J. Kargacin, Department of Physiology, University of Massachusetts Medical School, Worcester, MA 01605 [1]

Ewa Kerc, Department of Biochemistry, University of Saskatchewan, Saskatoon, Saskatchewan, Canada S7N 0W0 [207]

W. Glenn L. Kerrick, Departments of Physiology, Biophysics, and Pharmacology, University of Miami School of Medicine, Miami, FL 33101 [437]

Ritsuko Koga, National Institute for Physiological Sciences, Myodaiji, Okazaki, 444 Japan [109]

Joseph M. Krisanda, Department of Radiology, Brigham and Women's Hospital, Boston, MA 02115 [319]

Hideto Kuwayama, National Institute for Physiological Sciences, Myodaiji, Okazaki, 444 Japan [109]

N. Lachant, Departments of Medicine and Pathology, Harbor-UCLA Medical Center, Torrance, CA and University of Washington, Seattle, WA [465]

Sancy A. Leachman, Department of Pharmacology and Moss Heart Center, University of Texas Health Science Center, Dallas, TX 75235; present address: Department of Physiology, University of Texas Health Science Center, Dallas, TX 75235 [183]

J. Lechago, Departments of Medicine and Pathology, Harbor-UCLA Medical Center, Torrance, CA and University of Washington, Seattle, WA [465]

M. Li, Departments of Medical Biostatistics and Physiology and Biophysics, University of Vermont, Burlington, VT 05405 [467]

Bridgitte M. Lobnig, Institute of Physiology, University of Hamburg, Hamburg 20, Germany [377]

Susan Lowey, Rosenstiel Basic Medical Sciences Research Center, Brandeis University, Waltham, MA 02254 [81]

Wells W. Magargal, Hypertension Research, Division of Cardiology, Department of Medicine, University of Alabama at Birmingham, Birmingham, AL 35294 [469]

Whitney McBride, Department of Physiology and Biophysics, University of Vermont, Burlington, VT 05405 [303]

Richard A. Meiss, Departments of Physiology, Biophysics, and OB/GYN, Indiana University School of Medicine, Indianapolis, IN 46223 [347]

Carolyn H. Michnoff, Department of Pharmacology and Moss Heart Center, University of Texas Health Science Center, Dallas, TX 75235; present address: Department of Biochemistry, University of Texas Health Science Center, Dallas, TX 75235 [183]

Takashi Mikawa, National Institute for Physiological Sciences, Myodaiji, Okazaki, 444 Japan [109]

J.R. Miller, Department of Pharmacology, Vanderbilt University School of Medicine, Nashville, TN 37232 [471]

Hidetake Miyata, Department of Pathobiology, University of Pennsylvania, Philadelphia, PA 19104 [143]

Mukhallad A. Mohammad, Department of Physiology, University of Western Australia, Nedlands, Australia 6009 [67]

Susan U. Mooers, Department of Physiology, Jefferson Medical College, Thomas Jefferson University, Philadelphia, PA 19107 [289]

Robert S. Moreland, Bockus Research Institute, Graduate Hospital, Philadelphia, PA 19146 [473]

Suzanne Moreland, Department of Pharmacology, Squibb Institute for Medical Research, Princeton, NJ 08543 [473]

Richard A. Murphy, Department of Physiology, School of Medicine, University of Virginia, Charlottesville, VA 22908 [399,411]

Sumitra Nag, Department of Muscle Research, Boston Biomedical Research Institute, Boston, MA 02114 [91]

Ross A. Nayler, Department of Physiology, University of Western Australia, Nedlands, Western Australia, Australia [475]

Mary H. Nunnally, Department of Pharmacology and Moss Heart Center, University of Texas Health Science Center, Dallas, TX 75235; present address: Department of Physiology, University of Texas Health Science Center, Dallas, TX 75235 [183]

C.S. Packer, Department of Physiology, University of Manitoba, Winnipeg, Manitoba, Canada R3E OW3 **[477]**

Mary D. Pato, Department of Biochemistry, University of Saskatchewan, Saskatoon, Saskatchewan, Canada S7N 0W0 **[207]**

Richard J. Paul, Department of Physiology and Biophysics, University of Cincinnati College of Medicine, Cincinnati, OH 45267-0576 **[319,463,479]**

Ulrich Peiper, Institute of Physiology, University of Hamburg, Hamburg 20, Federal Republic of Germany **[377]**

Anthony Persechini, Department of Pharmacology and Moss Heart Center, University of Texas Health Science Center, Dallas, TX 75235; present address: Department of Biology, University of Virginia, Charlottesville, VA 22901 **[183]**

Gabriele Pfitzer, II. Physiologisches Institut, University of Heidelberg, D-6900 Heidelberg, Federal Republic of Germany **[427]**

R.D. Phair, Department of Biomedical Engineering, Johns Hopkins University School of Medicine, Baltimore, MD 21205 **[481]**

Paul H. Ratz, Department of Physiology, University of Virginia School of Medicine, Charlottesville, VA 22908; present address: Department of Pharmacology, Eastern Virginia Medical School, Norfolk, VA 23501 **[411]**

Andrea L. Richardson, Department of Pharmacology and Moss Heart Center, University of Texas Health Science Center, Dallas, TX 75235; present address: Department of Physiology, University of Texas Health Science Center, Dallas, TX 75235 **[183]**

S.A. Robertson, Department of Physiology, University of Manitoba, Winnipeg, Manitoba, Canada R3E OW3 **[477]**

Johann Caspar Rüegg, II. Physiologisches Institut, University of Heidelberg, D-6900 Heidelberg, Federal Republic of Germany **[43,67,427]**

L.N. Russek, Department of Biomedical Engineering, Johns Hopkins University School of Medicine, Baltimore, MD 21205 **[481]**

L. Saggin, Institute of General Pathology, University of Padova, Padova, Italy **[451]**

S. Sartore, Institute of General Pathology, University of Padova, Padova, Italy **[451]**

S. Schiaffino, Institute of General Pathology, University of Padova, Padova, Italy **[451]**

John C. Seidel, Department of Muscle Research, Boston Biomedical Research Institute, Boston, MA 02114 **[91]**

Chun Y. Seow, Department of Physiology, Faculty of Medicine, University of Manitoba, Winnipeg, Manitoba, Canada R3E OW3 **[357,483]**

N. Sevy, Harbor-UCLA Medical Center, Torrance, CA 90510 **[485]**

R.V. Sharma, Department of Anatomy, University of Iowa, Iowa City, IA 52242 **[277]**

Marion J. Siegman, Department of Physiology, Jefferson Medical College, Thomas Jefferson University, Philadelphia, PA 19107 **[289]**

Paul J. Silver, Division of Experimental Therapeutics, Wyeth Labs, Inc., Philadelphia, PA 19101; present address: Cardiopulmonary Section, Department of Pharmacology, Sterling-Winthrop Research Institute, Rensselaer, NY 12144 **[233]**

xvi / Contributors

Lars Sjölin, Department of Physiology and Biophysics, University of Lund, S-223 62 Lund, Sweden [333]

W.J. Snape, Jr., Departments of Medicine and Pathology, Harbor-UCLA Medical Center, Torrance, CA and University of Washington, Seattle, WA [465,485]

Apolinary Sobieszek, Institute of Molecular Biology, Austrian Academy of Sciences, A-5020 Salzburg, Austria [159]

Andrew P. Somlyo, Pennsylvania Muscle Institute and Department of Physiology, University of Pennsylvania School of Medicine, Philadelphia, PA 19104 [27]

Avril V. Somlyo, Pennsylvania Muscle Institute and Department of Physiology, University of Pennsylvania School of Medicine, Philadelphia, PA 19104 [27]

Jan Sosinski, Department of Muscle Research, Boston Biomedical Research Institute, Boston, MA 02114 [91]

Malcolm P. Sparrow, Department of Physiology, University of Western Australia, Nedlands, Australia 6009 [67,475]

Newman L. Stephens, Department of Physiology, Faculty of Medicine, University of Manitoba, Winnipeg, Manitoba, Canada R3E OW3 [357,477,483]

John D. Strauss, Department of Physiology and Biophysics, University of Cincinnati College of Medicine, Cincinnati, OH 45267-0576 [319,479]

John Q. Stropp, Mayo Clinic and Foundation, Rochester, MN 55905 [461]

James T. Stull, Department of Pharmacology and Moss Heart Center, University of Texas Health Science Center, Dallas, TX 75235; present address: Department of Physiology, University of Texas Health Science Center, Dallas, TX 75235 [183]

Hiroshi Suzuki, Department of Muscle Research, Boston Biomedical Research Institute, Boston, MA 02114 [91]

Masashi Suzuki, National Institute for Physiological Sciences, Myodaiji, Okazaki, 444 Japan [109]

Edwin W. Taylor, Department of Molecular Genetics and Cell Biology, University of Chicago, Chicago, IL 60637 [59]

Theodore J. Torphy, Department of Pharmacology, SmithKline and French Laboratories, Philadelphia, PA 19101 [263]

David Trentham, Medical Research Council, National Institute for Medical Research, London NW 7-1AA, England [27]

Brett A. Trockman, Department of Physiology, Indiana University School of Medicine, Indianapolis, IN 46223 [415]

Kathleen M. Trybus, Rosenstiel Basic Medical Sciences Research Center, Brandeis University, Waltham, MA 02254 [81]

Jürgen Wagner, II. Physiologisches Institut, University of Heidelberg, D-6900 Heidelberg, Federal Republic of Germany [427]

Michael P. Walsh, Department of Medical Biochemistry, University of Calgary, Calgary, Alberta, Canada T2N 4N1 [119]

David Warshaw, Department of Physiology and Biophysics, University of Vermont, Burlington, VT 05405 **[303,467,487]**

J.N. Wells, Department of Pharmacology, Vanderbilt University School of Medicine, Nashville, TN 37232 **[471]**

Steven Work, Department of Physiology and Biophysics, University of Vermont, Burlington, VT 05405 **[303]**

Bruno Zobel, Institute of Physiology, University of Hamburg, Hamburg 20, Federal Republic of Germany **[377]**

Preface

Symposia, like the Roman god Janus, have two faces, one surveying the past, the other looking into the future. They are instruments judging progress made, and crystallizing goals to be aimed for. The vitality of a field can be particularly well assessed through symposia that, like this one in Minaki, can be compared to their antecedents such as the 1975 Winnipeg meeting, organized by Newman Stephens, for the discussion of the biochemistry of smooth muscle.

It is gratifying to see that, whether we consider ultrastructure, energetics or regulation of smooth muscle contraction, major advances have been made in the past eleven years, and yet a sufficient number of important questions remain unanswered to assure continued growth. Thus, between the two meetings, the existence of myosin filaments in smooth muscle cells, resting or activated, has become generally accepted, and at Minaki we have seen the crossbridges attached in rigor configuration. These and other structural and physiological observations on single cells and skinned fibers have added strong evidence to what was largely taken on faith in 1975: the operation of a sliding filament mechanism of contraction in smooth muscle. Yet agreement, and more importantly, solid evidence, concerning the molecular packing of native myosin filaments is still lacking, as is a definition of the conformational changes of the crossbridge associated with the ATPase cycle. Phosphorylation of myosin light chains as the primary mechanism of contractile regulation, an idea barely hinted at in 1975, is now part of a generally accepted scheme that includes correlations between the conformation of isolated myosin molecules and their state of phosphorylation. There is also convincing evidence that the number of attached crossbridges that give rise to force production is regulated independently of crossbridge cycling rate. Force can be maintained at negligible energy costs, yet the mechanism(s) underlying such "catch-like state", or "latch" or resting links remains unsettled. The possibility of additional thin filament-mediated regulatory systems, such as caldesmon, actomyosin activating protein or kinase C, also experienced a vigorous revival at Minaki, and the visibility and contributions of "hard core" biochemists now allows physiologists to begin to relate the energetics and kinetics of crossbridge transients to the ATPase cycle. Another indication of the growth of smooth muscle physiology and biochemistry is that, unlike the 1975 meeting, major areas of smooth muscle physiology, such as excitation-contraction coupling could no longer be included. Thus, the excitement generated by the discovery of phosphatidyl inositol-biphosphate turnover and calcium mobilization by inositol trisphosphate could only be communicated during coffee breaks.

Visitors to the temple of Janus report that both faces of the god were smiling at Minaki, and his smile was particularly broad for the fools working on smooth muscle.

A.P. Somlyo
M.J. Siegman
Philadelphia, 1987

Acknowledgments

Support for this Conference was generously provided by the following organizations:

National Institutes of Health (Grant # HL/AM 35834)
Canadian Lung Association
Children's Hospital of Winnipeg Research Foundation, Inc.
Manitoba Heart Foundation
Medical Research Council of Canada

AB Hassle, Subsidiary of AB Astra
Ciba-Geigy Corporation
Schering Corporation
SmithKline and French Laboratories
The Upjohn Company

American Critical Care, American Hospital Supply Corp.
American Cyanamid Co., Medical Research Division
Ayerst Laboratories Research Inc.
Bayer AG/Miles Preclinical Pharmacology Group
Berlex Laboratories, Schering AG
Bristol-Myers Pharmaceutical Research & Development Div.
LC Services Corporation
McNeil Pharmaceutical Company
Merck, Sharpe & Dohme Research Laboratories
Searle Research and Development
Wyeth Laboratories, Inc.

Travel Support was provided, in part, for Dr. A. Sobieszek by The Society for Experimental Biology and Medicine

MOLECULAR STRUCTURE AND ORGANIZATION OF FILAMENTS IN
SINGLE, SKINNED SMOOTH MUSCLE CELLS

P.H. Cooke, G. Kargacin, R. Craig,
K. Fogarty, and F.S. Fay

Departments of Anatomy and Physiology,
University of Massachusetts
Medical School, Worcester, MA 01605, and

S. Hagen

Brigham and Women's Hospital, Boston, MA
02115

Introduction

   Gradual progress has been made towards understanding
the microscopic structure and organization of the
contractile apparatus in smooth muscle (Shoenberg and
Needham, 1976; Small and Sobieszek, 1980; Somlyo, 1980;
Cooke, 1983; Bagby, 1983), but the acquisition of new
structural information has not kept pace with recent
evidence indicating that a variety of enzymatic
mechanisms effect interactions between the major
contractile and cytoskeletal proteins, as described in
this Volume. The structure problem in smooth muscle is
attributable not only to special limitations of the
available methods (e.g. like the destructive artifacts
introduced by the preparative methods for electron
microscopy), but also to an apparently well-deserved
reputation of smooth muscle cells for revealing very
few quantitative structural features, on a sub-cellular
and molecular scale, that could lead to detailed
working models of the contractile apparatus. The
problem is not due to a lack of marked structural
change associated with contraction. Maximal reversible
shortening typically decreases the lengths of isolated
cells to less than 30 per cent of their resting length
(Fay and Delise, 1973). Active tension can be developed

by isolated cells and tissue strips over a broad range
of (extended) muscle lengths (Paul and Peterson, 1975;
Fay 1975; Siegman et al., 1976). For reasons that are
not yet apparent however, the essential structural
feature of the smooth muscle cell that allows for large
changes in length, but retains a capacity to develop
nearly optimal active force, has not been identified.
Our observations are aimed at resolving some of the
structural characteristics and organization of smooth
muscle cells that contribute to the remarkable
flexibility in the force generating mechanism.

Single, isolated cells
   A proven experimental model system for obtaining
direct insight into structural changes underlying
contraction and extension is the isolated amphibian
smooth muscle cell (Bagby, et al., 1971; Fay et al.,
1982). The isolated cells usually match the structure
and physiological properties of their counterparts
within tissue strips or intact stomach muscularis in
the giant toad, Bufo marinus (Linnaeus), despite the
loss of intercellular contacts and removal of most of
the extracellular matrix by enzymatic digestion (Fay
and Delise, 1973; Fay and Singer, 1977). The structure
of the contractile apparatus in the isolated amphibian
cells also compares closely with other vertebrate
smooth muscles, including mammalian types (Figure 1).
   A consistent, regular feature found in thin sections
of most fixed, embedded preparations of virtually all
types of vertebrate smooth muscle cells is an hexagonal
packing arrangement of actin filaments in small
bundles. The spaces around the bundles are occupied by
thick (myosin) filaments, with diameters around 25-30
nm in cross-sectional profile. The ratio of filament
profiles which can be identified as actin and myosin in

Figure 1. Comparison of an isolated toad smooth muscle
cell in cross-section (A) with a taenia coli smooth
muscle cell from guinea pig (B). Actin filament
profiles (small arrows) in lattice-like groups or
bundles. Myosin filaments (large arrows) in spaces
between bundles of actin filaments. Dense bodies (DB)
in toad cells appear to lack patent intermediate
filaments, which are abundant (IF) in the mammalian
smooth muscle. 66,000X.

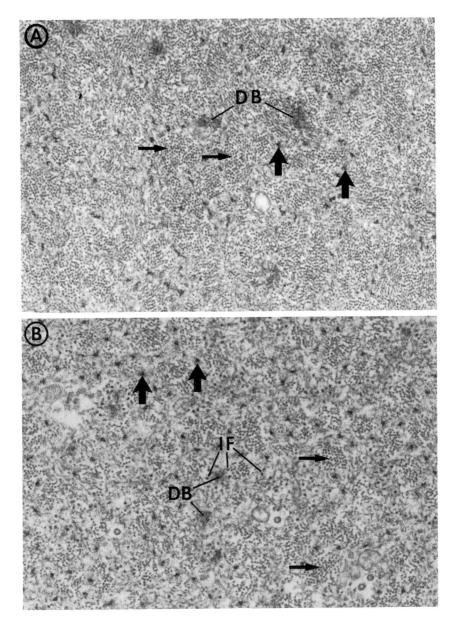

Figure 1

isolated amphibian smooth muscle cells is quite high, about 50:1, actin to myosin. An early estimate made from thin sections of mammalian smooth muscle is similar to this value (Rice, et al., 1970).

Fusiform dense bodies are another structure that is found consistently in the contractile apparatus of vertebrate smooth muscles. These objects contain amorphous electron-dense material(s) of unknown composition, and high concentrations of the protein alpha-actinin (Schollmeyer et al., 1976; Fay, et al., 1983). The core region of the dense bodies contains the alpha-actinin, and the bodies are linked, axially, to actin filaments, so they are thought to represent simply passive structures, like the Z discs in striated muscles. A special feature of the dense bodies in the amphibian cells is the absence of any obvious association with intermediate filaments (see Figure 1), like those located around isolated and sectioned dense bodies in avian and mammalian smooth muscles (Cooke 1976).

These three components, myosin and actin filaments, and fusiform dense bodies, are the most consistent features of the contractile apparatus seen by electron microscopy of vertebrate smooth muscles, and their basic arrangement, shown in Figure 1, conforms with the general model that is usually portrayed for cross-sections. Critical reviews of smooth muscle structure however, reveal many inconsistencies about the arrangement of the contractile elements, and little agreement on the details of organization is found (cf. Bagby, 1983). The models of structure are especially vague when it comes to explaining, from available direct evidence, how the filaments interact during contraction. Even when observations are focused on specific areas of isolated cells, localized shortening appears only to result in disorganization of actin and myosin filaments and axial disorientation of the cytoplasmic dense bodies (Figure 2). The gradual change in orientation of filaments and dense bodies in regions of transition between relaxed and contracted areas in partially contracted cells reveals nothing striking about the reorganization of filaments and dense bodies during the process of shortening. The transition from a uniform axial orientation in relaxed regions to bent and axially disoriented filaments in shortened regions only suggests that contraction does

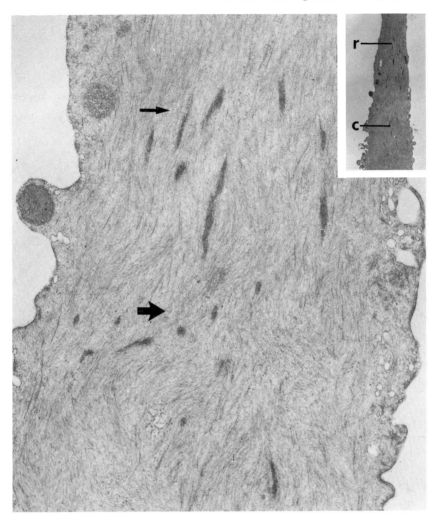

Figure 2. Single isolated smooth muscle cell fixed during a partial contraction. The inset shows the transitional region of the cell including the relaxed zone (r) in the upper half and contracted (c) zone in the lower half. At high magnification, the two zones are distinguished by axially oriented filaments and dense bodies in the relaxed zone (small arrow) and axially disoriented filaments and dense bodies in the contracted zone (large arrow). 10,000 X; inset 1,000 X.

not involve simply a uniform interdigitation of
filaments or a regular change of organization as would
be expected from a sarcomere-repeat, as found in
striated muscles. Even such negative information is
inconclusive, however, because so little of the
contracting structure is actually sampled in thin
sections. To compensate for this, three dimensional
views of whole cells might provide the information
necessary for some key insight into the longer range
order of the contractile elements in smooth muscle.

Three dimensional organization of the contractile
elements

A direct method of searching for long range order in
isolated muscle cells might be provided by optical
microscopy, but as implied by its name, smooth muscle
appears to be optically uniform, possibly because the
filaments and dense bodies are irregularly arranged and
densely packed, making the refractive indices of the
contractile elements match and refract light uniformly.

Immunofluorescent localization of antibodies
directed against isolated, purified components of the
contractile elements provides one way of successfully
modulating optical images of smooth muscle cells that
avoids their difficult optical properties. This method
also potentially reduces some of the complexity within
the contractile elements if one of the less abundant
antigens is selected for localization with
fluorescently labeled antibodies. This approach was
taken with tissue cells and single isolated cells using
fluorescently-labeled antibodies to alpha-actinin, a
major protein of the dense bodies (Schollmeyer et al.,
1976; Bagby, 1980; Fay et al., 1983). These structures
are nearly ideal to locate because they presumably
identify symmetrical attachment sites for actin
filaments in the cells, so their long range
organization should identify points of applied force
within the cells (Bond and Somlyo, 1982). The
organization of the dense bodies was visualized in
several ways: single and stereo-paired images of single
cells, and computed 3-D image reconstructions made from
multiple image planes of single cells (Fay et al.,
1983). The most regular detectable arrangements in
these images were short, string-like arrays of
fluorescent bodies, similar in size to the dense
bodies, with a mean interspace of 2.2 micrometers,

extending toward attachment plaques on the plasma membrane (Figure 3). In contracted cells, the mean interspace between the fluorescently labeled bodies decreased to about 1.4 micrometers, suggesting that shortening occurred in the spaces between these structures and was possibly transmitted through them,

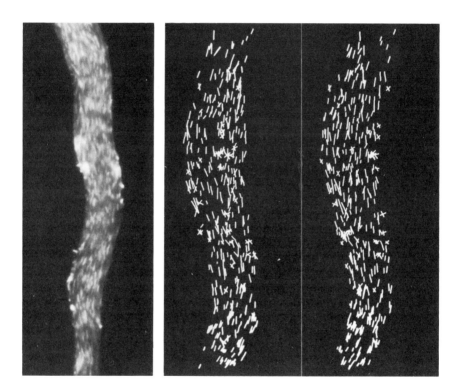

Figure 3. In left panel, the pattern of organization of fluorescent bodies containing alpha actinin in a single image plane of an isolated smooth muscle cell is shown. 2,000 X. On the right, a stereo-pair illustrating a reconstruction of the distribution of fluorescence computed from multiple image planes of a relaxed isolated cell.

as is the case with Z discs in striated muscles. These observations at "optical" resolution, provide some of the first clear indications for a functionally-related, internal structural change in smooth muscle during contraction. However, the string-like sequences represent only a portion of the total complement of fluorescent (dense) bodies, so even in complete 3-D views, the pattern of long range order including most of the dense bodies was not obvious.

Saponin skinned cells

The possibility that structural changes during contraction might reveal some crucial aspect of a regular organization has not been thoroughly tested in isolated smooth muscle cells. Two conditions for making such a study would be to control directly the functional state of the cells, chemically, and to follow the course of some relevant structure in the contractile apparatus, dynamically. These conditions were met recently with isolated cells made permeable by brief exposure to saponin (Kargacin and Fay, 1986). This treatment allowed the functional state of isolated, single cells to be regulated directly by [Ca++] and [ATP], and unlike removal of the membrane from isolated cells with Triton X-100 (Small, 1977), the procedure had the additional advantage of transforming the uniform optical density of the cells into regular structures that could be clearly resolved, optically, as elongated fibrils and fusiform objects resembling dense bodies (Figure 4). The functional properties of the skinned cells resembled native isolated cells: shortening was reversible; the pCa for half-maximal shortening was about 6.2; the rate of shortening for cells tethered to the substrate was between .05 and 0.1 cell lengths/sec., or 0.15 to 0.3 micrometers/sec. Under rigor conditions (no ATP), the fibrils observed in phase contrast images appeared to correlate with crossbridged groups of actin and myosin filaments in electron micrographs of thin sections. The (phase) dense bodies, located between the fibrils by optical microscopy, corresponded to the locations of fusiform dense bodies and associated actin filaments in electron micrographs of thin sections (Figure 5).

Structure and Organization of Filaments / 9

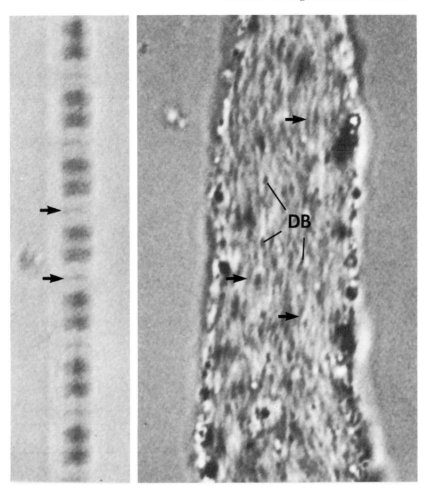

Figure 4. Fibrils (arrows) and dense bodies (DB) in an isolated, skinned smooth muscle cell resolved with a laser scanning microscope in the bright field mode (Model LSM-40, Carl Zeiss, Inc. Thornwood, NY). For comparison, a bright field image of a single myofibril, isolated from rabbit psoas muscle fibers, on the left, resolves the density of Z discs (arrows) to less than 100 nm thick 6,000 X.

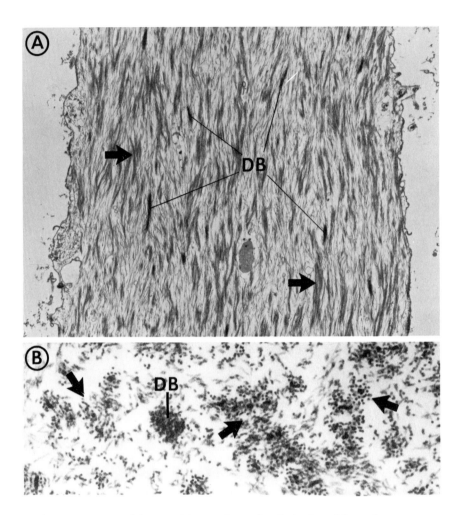

Figure 5. A. Thin section of an isolated, skinned smooth muscle cell showing the resolved structure of the fibrils (arrows) and dense bodies (DB) by electron microscopy. B. Cross-section showing clusters of myosin and actin filaments in fibrils (arrows) and a dense body (DB). A=10,000 X; B=75,000 X.

Structure of the contractile apparatus.

Thin filaments, as they appeared in negatively stained homogenates of isolated cells showed a periodic structure that was typical of F-actin. By optical diffraction from electron microscope images of paracrystalline bundles of the thin filaments, that appeared sporadically in negatively stained homogenates of skinned cells, prominent off-meridional reflections typical of F-actin were found at reciprocal spacings of 6 and 35-38 nm (Figure 6). Optical diffraction from the core region of isolated dense bodies showed the same evidence of subunit structure, albeit less ordered, that is characteristic of actin filaments (Figure 6). In the case of the dense bodies, the distribution of intensity is smeared along the 6th layer line, and a maximum occurs close to the meridian. This feature partially resembles calculated intensity distribution from model structures of F-actin bundles where the polarity of the filaments was randomly anti-parallel (O'Brien, et al., 1975). The similarities between the observed pattern from the dense bodies and the calculated intensity from the models might indicate that the core region of dense bodies contains an arrangement of anti-parallel F-actin filaments, as expected from the results of Bond and Somlyo, (1982).

Under identical conditions of isolation, numerous myosin filaments, obtained from the skinned cells, had an unusual subunit structure, when examined in negatively stained preparations, under relaxing conditions (pCa=8, 1 mM ATP). The filaments were between 0.8 and 3 micrometers long and 30-40 nm wide, but the major regular feature of the filaments was an axial repeat of 14.5 nm arising from crossbridge projections along the whole filament length. Two forms of this repeat were found along the filaments: the most common form was a regular transverse striation across the width of the filament shaft; the other form was a series of projections, 10-15 nm long, extending from the edges of the filament shaft, which was about 10-12 nm wide (Figure 7). On some filaments, both forms of crossbridge projection alternated along the length. This meant that either there were two forms of crossbridge projections or that one form assumed both appearances in projection due to twisting of the filament shafts. Since there was no evidence for a

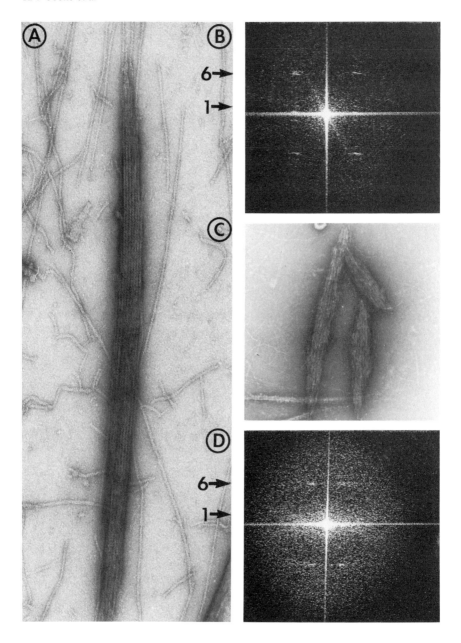

Figure 6

Figure 6. Electron micrographs of negatively stained filaments (A) and dense bodies (C) isolated from homogenates of skinned cells. Corresponding optical diffraction patterns showing reflections that are typical of F-actin at reciprocal spacings of about 35 nm (1) and 6 nm (6) for paracrystalline actin filaments (B), and dense bodies (D). A=115,000 X; C=69,000 X.

regular twist along the filament-shafts, a helix of crossbridge projections giving rise to the two forms seemed unlikely. The two different appearances could be straightforwardly interpreted, however, as orthogonal views (due to random twisting) of an arrangement of crossbridges projections emerging only from two, opposite sides of the filaments, as in the side polar filaments made from purified smooth muscle myosin (Craig and Megerman, 1977). To test this idea, selected filaments were tilted around their long axes in an electron microscope equipped with a eucentric goniometer stage. Figure 8 shows that the two different forms of crossbridge projections were interconverted by tilting. Regions of crossbridge projections in the form of striations or face views were converted into regions of lateral projections or edge views by tilting about 90 degrees. These edge views appear very similar to the side polar structure described by Craig and Megerman (1977), so the tilting operation suggests that the organization of myosin in the filaments isolated from these functionally active cells was essentially side-polar.

While these images of negatively stained homogenates resolve subunit repeats in isolated filaments and fusiform dense bodies that identify with typical axial spacings of myosin (14.5 nm) and F-actin (6 and 35-38 nm), when examined by optical diffraction, neither side-polar myosin filaments nor F-actin filaments in paracrystalline bundles are especially characteristic structures of most muscles. Side-polar myosin filaments and filaments with a closely related structure have only been observed from smooth muscles and certain non-muscle sources (Sobieszek, 1972; Small, 1977; Craig and Megerman, 1977; Hinssen, et al., 1978), and the lattice-like bundles of hexagonally packed thin filaments containing F-actin, apparently excluding

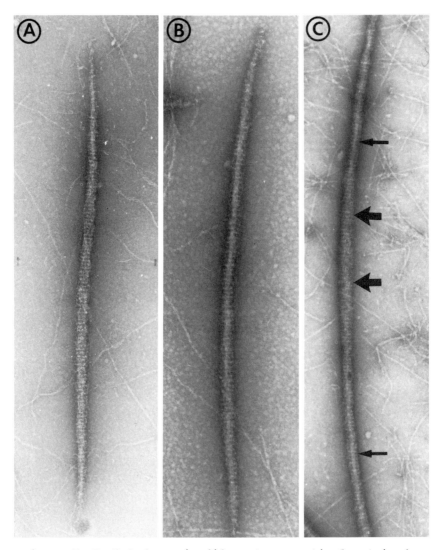

Figure 7. Isolated myosin filaments, negatively stained under relax conditions. A continuous 14.5 nm periodicity was found in two basic forms: all face view in (A), mostly in (B), and edge views in (C). Both forms were often found along adjacent regions of the same filament (C); face view (large arrows) and edge views (small arrows). A and C=69,000 X, B=82,000 X.

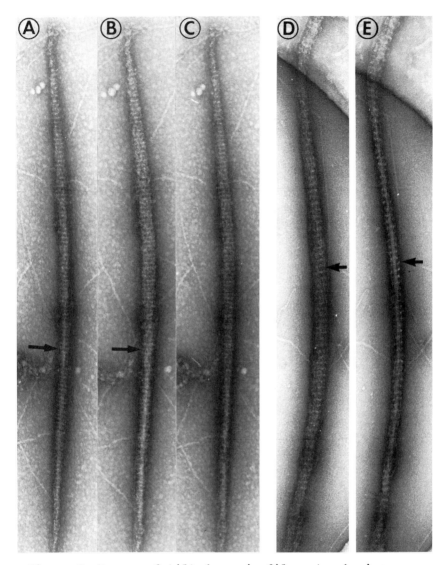

Figure 8. Images of tilted myosin filaments showing that face and edge views are different projections of a single, side-polar arrangement of crossbridges. On the left (A) rotation from −46 degrees through (B), 0 degrees, to (C) +38 degrees, changes some of the 14.5 periodicity from edge view to face view (arrows). In (D), a region of face view converts to edge view (E), at the arrow. 69,000 X.

myosin, are otherwise only common in molluscan smooth muscles containing paramyosin (Sobieszek, 1973), and in differentiated cytoskeletal structures, such as the filamentous cores in microvilli of intestinal epithelial cells (Hirokawa, et al., 1982). Despite the unusual feature of paracrystalline organization of actin and side-polar molecular arrangement of myosin in filaments of smooth muscle cells, the occurrence of typical subunit repeats nevertheless identifies these structures with the major contractile proteins. The fusiform dense bodies have been characterized by immunochemical labeling, providing evidence for their identification in optical images. Indirect labeling of dense bodies in thin sections with immunogold was used to locate antibodies bound to alpha-actinin exclusively over or within the core regions of these bodies, as shown in Figure 9, and described quantitatively by Hagen, et al.,(1985). Furthermore, the pattern of immunofluorescent labeling in single skinned cells was qualitatively similar to the distribution of (phase) dense bodies in optical images of the same cell (Figure 9). These results collectively support an interpretation of the structure in skinned cells which relates the fibrils and (phase) dense bodies in optical images, respectively to bundles of crossbridged myosin and actin filaments and fusiform dense bodies containing alpha actinin, as observed by electron microscopy. If this identification of contractile elements in optical images of skinned cells is accurate, it should be possible to resolve, dynamically, with an optical microscope, the locus of shortening in contracting skinned cells.

Figure 9. Immunochemical identification of alpha actinin in dense bodies. A. Location of colloidal gold-immunoglobin complexes over areas containing dense bodies (DB) on a thin section of an isolated smooth muscle cell embedded in Lowicryl. 45,000 X. Correlated optical images of a single skinned cell incubated with Rhodamine conjugated antibody to toad alpha actinin: (B) phase contrast image showing dense bodies (arrow), and fluorescent bodies (arrow), showing the distribution of antibody to alpha actinin, below. 2,000 X.

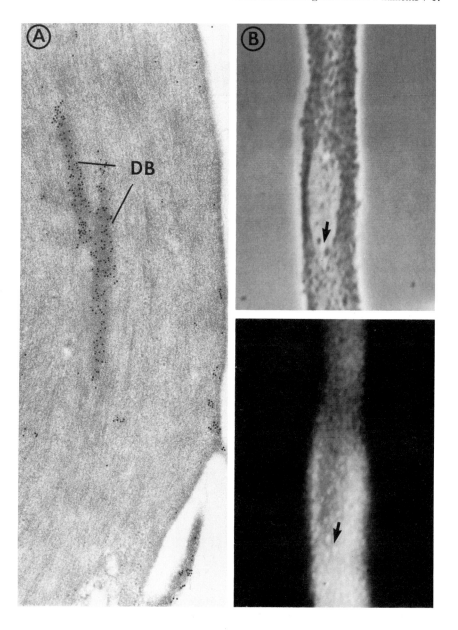

Figure 9

Structural changes during contraction

Suspensions of isolated, skinned cells were injected into a thin chamber between two coverglasses, that were coated with a layer of protamine sulfate. After individual, adherent cells were selected for microscopic observation, contraction was induced by pumping solutions containing 1-5 mM ATP and micromolar concentrations of free Ca++ through the chamber. A Nuvicon camera was used to record phase contrast or darkfield microscope images of contracting cells. Twenty to thirty digitized images of each cell were collected at 3 second intervals for analysis. The positions of (phase) dense bodies within the timed sequence of images were selected and marked as points using an interactive computer graphics program, and a moving coordinate system was plotted to compensate for overall cell movement during contraction. Coordinates of all the marked points in successive frames were found, and two dimensional trajectories of the points were plotted for the contraction sequence.

An example of two frames, one before contraction of a cell at rest and another near the completion of contraction, together with the corresponding sets of marked points, is shown in Figure 10. In Panel A, dense bodies and fibrils are resolved in the video image of the unshortened segment of a skinned cell under rigor conditions. Panel B illustrates the appearance of the dense bodies and fibrils in the same cellular segment after 28 seconds of contraction. Panels C and D contain the selected points marking the positions of dense bodies, sites on the cell membrane, and the reference marks for the coordinate system. As a result of cell movement during contraction, some points at the top of the early frame were lost from the contraction sequence and new points on the later frame appeared (see Figure 10). When plotted in relation to the moving reference point, the axial displacements of dense bodies were uniform when averaged over the whole contraction sequence (Figure 11), which took about one minute and shortened the cells about 20 percent. The rate of dense body movement, calculated to be about 0.3 um/sec during these contractions, was comparable to the maximum rate of movement of beads coated with phosphorylated smooth muscle myosin (0.4 um/sec) in an in vitro assay (Sellers, et al., 1985). Similar values were also found for directed movements of cytoplasmic

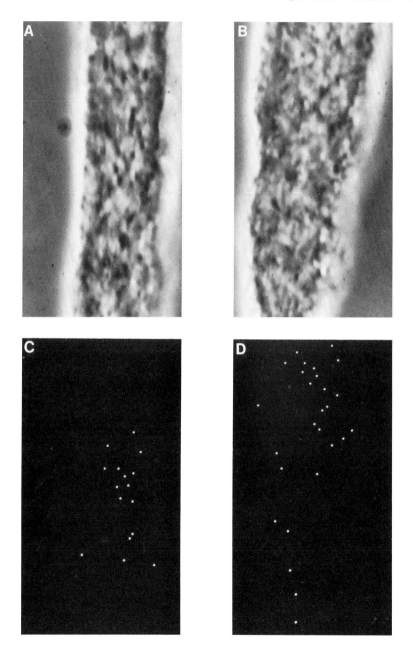

Figure 10

vesicles by myosin along bundles of actin filaments (Adams and Pollard, 1986). The radial displacements of dense bodies, averaged over the same sequence of frames, measured against the increase in cell radius, was less than would be expected if the bodies were moving passivly in the cell during shortening.

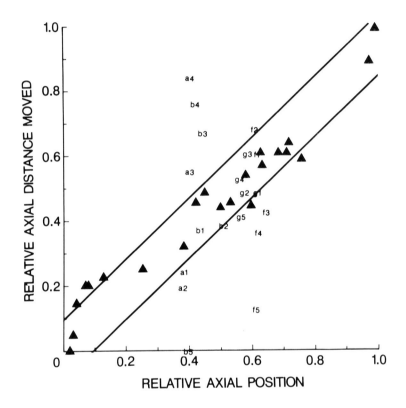

Figure 11. Plot of the relative axial displacement of marked points representing dense bodies (solid triangles) in relation to the center of the coordinate system, showing that movement was uniform when averaged over the whole contraction sequence. Alphanumeric symbols above and below the solid lines identify marked points and frame number of 3 to 6 sec departures from the average rate, indicated by values enclosed between the solid lines.

The starting points and trajectories of the dense bodies during contraction are illustrated in Figure 12. The points representing dense bodies, which remained in the plane of focus throughout contraction, were marked, frame by frame, and the arrangement of points was selected to mark oblique and axial arrangements of dense bodies for testing preferred orientations of movement during shortening. The trajectories of the points were in some cases coordinated, suggesting that groups of dense bodies were subject to the same forces, but in other cases, the directions of movement were independent, even when the bodies were closely spaced. These independent movements appear to indicate that some adjacent dense bodies were differentially influenced by transients in the contractile forces or by non-uniform spatial arrangements of the contractile elements.

Figure 12. A set of marked points related to dense bodies extending over a length of 20 micrometers in the cell, on the left, and trajectories of the points through the contraction sequence on the right. Some closely spaced points moved independently (large arrows), while others moved coordinately (small arrows).

Independent movements of the dense bodies were also indicated when the axial displacements of points marking the bodies were plotted over short time intervals (three to six seconds). Two closely spaced dense bodies, labeled a and b in Figure 11, together moved slower then faster than the average rate, outlined by the solid lines; whereas closely spaced bodies f and g moved independently. Dense body g moved at the average rate and f moved slower than the average rate of movement. These brief independent and coordinated movements suggest that separate contractile elements are located within small areas of the skinned cells, but the minimum size of a contractile unit has not yet been clearly resolved in these two dimensional images.

Summary

The contractile elements of single, skinned amphibian smooth muscle cells were resolved as fibrils containing F-actin and side-polar myosin filaments and adjacent cytoplasmic dense bodies, containing alpha-actinin and actin. The fibrils were found to be linked axially through actin filaments to the dense bodies, by electron microscopy. The arrangment of these two structures provides the essential features of a sarcomere to the contractile elements.

The effect of contraction upon the movement of the dense bodies was followed dynamically by optical microscopy as a method of analyzing the organization of the contractile apparatus and resolving contractile units. Averaged over long time periods (30 to 50 seconds), the axial movements of dense bodies was uniform. The dense bodies were not simply displaced passively during contraction because the extent of radial movements averaged over the same (long) time periods were less than would be expected for the observed increases in cell diameter. Measured over short time intervals (3 to 6 seconds), the axial movement of some closely spaced dense bodies was not uniform: individual and groups of bodies moved significantly faster and slower than the mean rate of axial translocation. These shifts in coordinated and independent movements might reveal separate anisometric contractile units, or they could represent a random temporal pattern of localized activation and deactivation of small contractile elements.

Acknowledgments
This work was supported by grants from the NIH to RC (AM34711) and FSF (HL14523), a post-doctoral fellowship to GK (AM07341), and grants from the MDAA, Inc. to RC and FSF. The authors thank L. Harris for preparing millions of isolated cells, S. Abramson for developing several computer programs, and R. Hutchinson for expert clerical assistance.

References

Adams, RJ, Pollard, T (1986). Propulsion of organelles isolated from Acanthamoeba along actin filaments myosin-1. Nature 322:754-756.

Bagby, R, Young, A, Dotson, R, Fisher, B, McKinnon, K (1971). Contraction of single smooth muscle cells from Bufo marinus stomach, Nature 234:351-352.

Bagby, R (1980). Double immunofluorescent staining of isolated smooth muscle cells. I. Preparation of anti-chicken gizzard alpha actinin and its use with anti-chicken gizzard myosin for co-localization of alpha actinin and myosin in chicken gizzard cells, Histochem 69:113-130.

Bagby, R (1983).
Organization of contractile/cytoskeletal elements. In Stephens, NL (ed): Biochemistry of Smooth Muscle, Vol 1: Boca Raton, FL, CRC Press, pp 1-84.

Bond, M, Somlyo, AV (1982). Dense bodies and actin polarity in vertebrate smooth muscle. J Cell Biol 95:403-413.

Cooke, P (1976). A filamentous cytoskeleton in vertebrate smooth muscle fibers. J Cell Biol 52:105-116.

Cooke, P (1983). Organization of contractile fibers in smooth muscle. In Dowben, RM, Shay, JW (eds): Cell and Muscle Motility, Vol 3, New York: Plenum Press, pp 57-77.

Craig, R, Megerman, J (1977). Assembly of smooth muscle myosin into side-polar filaments. J Cell Biol 75:990-996.

Fay, FS, Delise, CM (1973). Contraction of isolated smooth muscle cells- structural changes. Proc Nat Acad Sci (USA) 70:641-645.

Fay, FS (1975). Mechanical properties of single isolated smooth muscle cells. In Worcel, M, Vassort, G (eds): Smooth Muscle Pharmacology and Physiology, vol. 50, Paris, INSERM, pp 327-342.

Fay, FS Honeyman, R, Leclair, S, Merriam, P (1982).
Preparation of individual smooth muscle cells from
the stomach of Bufo marinus. In Frederickson, DW,
Cunningham, LW (eds): "Methods in Enzymology", New
York: Academic Press, pp 284-292.

Fay, F, Singer JJ (1977). Characteristics of response
of isolated smooth muscle cells to cholinergic drugs.
Amer J Physiol 232:C144-C154.

Hagen, S, Fujiwara, K, Cooke, P, Fay, FS (1985).
Localization of contractile proteins in isolated
smooth muscle cells by electron microscopy and
computer analysis: actin and alpha-actinin. J Cell
Biol 101:624a.

Hinssen, H, D'Haese, J, Small, JV, Sobieszek, A
(1978). Mode of filament assembly of myosins from
muscle and non-muscle cells. J Ultrastuct Res
64:282-302.

Hirokawa, N, Tilney, L, Fujiwara, K, Heuser, J (1982).
Organization of actin, myosin, and intermediate
filaments in brush border of intestinal epithelial
cells. J Cell Biol 94:425-443.

Kargacin, GJ, Fay, FS (1986). Physiological and
structural properties of saponin skinned single
smooth muscle cells, (submitted for publications).

O'Brien, EJ, Gillis, JM, Couch, J (1975). Symmetry and
molecular arrangement in paracrystals of
reconstituted muscle thin filaments. J Molec Biol
99:461-475.

Paul, RJ, Peterson, JW, (1975). Relation between
length, isometric force and O_2 consumption rate in
vascular smooth muscle. Amer J Physiol 228:915-922.

Rice, R, Moses, J, McManus, G, Brady, A, Blasik, L
(1970). The organization of contractile filaments in
a mammalian smooth muscle. J Cell Biol 47:183-196.

Schollmeyer, J, Furcht, L, Goll, D, Rolson, R, Stromer,
M (1976). Localization of contractile proteins in
smooth muscle cells and in normal and transformed
fibroblasts. In Goldman, R, Pollard, T, Rosenbaum, J
(eds): Cell Motility, Book A, NY, Cold Spring Harbor
Laboratory, pp 361-388.

Sellers, JR, Spudich, JA, Sheetz, MP (1985). Light
chain phosphorylation regulates the movement of
smooth muscle myosin on actin filaments. J Cell
Biol 101:1897-1902.

Shoenberg, CF, Needham, DM (1976). A study of the mechanism of contraction in vertebrate smooth muscle. Biol Rev 51:53-104.

Siegman, MJ, Butler, TM, Mooers, SU, Davies, RE, (1976). Calcium-dependent resistance to stretch and stress relaxation in resting smooth muscles. Amer J Physiol 231:1501-1508.

Small, JV (1977). Studies on isolated smooth muscle cells: The contractile apparatus. J Cell Sci 24:327-350.

Small, JV, Sobieszek, A (1980). The contractile apparatus of smooth muscle. Intl Rev Cytol 64:241-306.

Sobieszek, A (1972). Cross-bridges on self-assembled smooth muscle myosin filaments. J Mol Biol 70:741-744.

Sobieszek, A (1973). The fine structure of the contractile apparatus of the ABRM of mytilus edulis. J Ultrastruct Res 43:313-343.

Somlyo, AV (1980). Ultrastructure of vascular smooth muscle. In Bohr, DF, Somlyo, AP, Sparks, HV (eds): Handbook of Physiology: Sect. 2, The Cardiovascular System, Vol. 2, Vascular Smooth Muscle, Bethesda, MD: Amer Physiol Soc, pp 33-67.

Regulation and Contraction of Smooth Muscle, pages 27-41
© 1987 Alan R. Liss, Inc.

CROSSBRIDGE TRANSIENTS INITIATED BY PHOTOLYSIS OF CAGED NUCLEOTIDES AND CROSSBRIDGE STRUCTURE, IN SMOOTH MUSCLE

Avril V. Somlyo, Yale Goldman, Taro Fujimori, Meredith Bond, David Trentham and Andrew P. Somlyo

Pennsylvania Muscle Institute, University of Pennsylvania, School of Medicine, B42 Anatomy-Chemistry Bldg., Philadelphia, Pennsylvania 19104-6083 (A.V.S., Y.G., T.F., M.B., A.P.S) and Medical Research Council, National Institute for Medical Research, The Ridgeway, Mill Hill, London NW 7-1AA, England (D.T)

INTRODUCTION

The operation of a sliding filament mechanism of contraction in smooth muscle has been inferred from the presence of myosin filaments bearing crossbridges (Ashton et al., 1975; A.P Somlyo et al., 1973) and their association with actin filaments inserting into dense bodies (Ashton et al., 1975; Bond and Somlyo, 1982), as well as from the shape of the length-active tension curve (Gordon and Siegman, 1971; Herlihy and Murphy, 1973; Mulvany and Warshaw, 1979; Speden, 1960) and the length dependence of energy consumption (Hellstrand and Paul, 1982). Mechanical transients of isolated, single smooth muscle cells (Warshaw and Fay, 1983a, 1983b) are also indicative of crossbridge activity, and smooth muscle actomyosin ATPase, exhibits basic similarities to that of the striated muscle protein (Marston and Taylor, 1980; Rosenfeld and Taylor, 1984). Laser flash photolysis of chemically inert precursors of ATP ("caged ATP"; Kaplan et al., 1978) has been used successfuly to probe the mechanical transients accompanying the ATPase cycle (Goldman et al., 1982, 1984a, 1984b). We have used this approach (see also, Goldman et al., 1986; Somlyo, A.P. et al., 1986) for detecting crossbridge transients in smooth muscle in which we also obtained ultrastructural evidence of rigor bridges. Our major findings show 1) force development to be rate limited by light chain phosphorylation, 2) cooperative attachment of crossbridges not requiring LC phosphorylation and 3) force transients most probably ascribed to detachment of negatively strained crossbridges.

METHODS

Electronmicroscopy

Saponin (100µg/ml)-skinned strips of rabbit or guinea pig portal vein were incubated for 30 min in rigor solution containing 1µM leupeptin and 2mM dithiothreitol at room temperature. The muscles were fixed with a 2% glutaraldehyde solution containing 0.13M ammonium acetate, 0.065M sucrose, 0.0375M sodium cacodylate buffer for 2 hours at room temperature, followed by a rinse in buffer and a 30 min incubation in 0.1% tannic acid in 0.075M cacodylate buffer containing 4.5% sucrose. The strips were subsequently postfixed in 2% osmium, followed by en bloc staining in saturated uranyl acetate and embedment in Spurr's resin.

Crossbridge Transients

The laser flash photolysis method, the configuration of the optical components for the frequency-doubled ruby laser and the muscle trough arrangement with the solution exchange mechanism have been described previously (Goldman et al., 1984a). In-phase stiffness measurements were made by applying a 500 Hz, 1µm sinusoidal length oscillation.

P_3-1(2-nitrophenyl)ethyladenosine 5'-triphosphate (caged ATP) and P_3-1(2-nitrophenyl)ethylcytidine 5'-triphosphate (caged CTP) were synthesized by treating the parent nucleotide with 1(2-nitrophenyl) diazoethane in rapidly stirred $CHCl_3/H_2O$ for 18 hrs at 21°C and pH4-5. The diazoethane was prepared by MnO_2 oxidation of the hydrazone of 2-nitroacetophenone.

Smooth muscle strips were prepared from the portal anterior mesenteric vein of the guinea pig. Following removal of the adventitia, longitudinal strips 0.1-0.25mm wide and 2-3mm long were cut from the 60µm thick vessel wall. Freeze-glycerination (Peterson 1982) with or without prior exposure to (50µg/ml) saponin was used to permeabilize the cell membranes.

Low tension rigor was induced by transferring the muscle from a relaxing solution containing 2mM ATP to a rigor solution without ATP or calcium (20mM EGTA). High tension rigor was achieved by transferring the muscle to rigor solution following a maximal activation with calcium and ATP. The rigor force obtained in the high tension rigor state was approximately 50% of maximal force. In some preparations the myosin light chains were thiophosphorylated by incubation in 2mM ATPγS, in the presence of Ca^{2+}, for 10 mins. All solutions contained protease inhibitors, leupeptin 1μM, PMSF 1mM (phenylmethylsulphonyl fluoride) as protease inhibitors. Mitochondrial blockers, 1μM FCCP (carbonyl cyanide 4-(trifluoromethoxy phenylhydrazone) was present throughout, and 1μg/muscle strip oligomycin was added during the first exposure to relaxing solution following glycerination. Calmodulin, 5μM, was added to all Ca^{2+}-containing solutions. All experiments were carried out at 18-20°C at pH 7.1 and ionic strength of 0.2M.

RESULTS AND DISCUSSION

Crossbridge Structure

Longitudinal views of muscles in a rigor state were strikingly different from those of relaxed muscle, due to the presence of regular arrays of attached crossbridges. The crossbridges decorated the actin filaments, except in regions where the actin filaments were separated from the myosin filaments, such as in the upper right hand corner of Fig. 1. Occasionally one could find regions of alternating actin and myosin filaments (Fig. 2), where clearly visible chevron patterns of crossbridges were evident. In preliminary studies, measurements in stereo views of the most regular regions showed that the number of crossbridges attached to actin per 0.5μm length of thick filament was 18±1.7 S.D. (n=22) in low tension rigor and 18±3.6 S.D. (n=4) in rigor following thiophosphorylation. These values are in the same range as those observed (23±4 S.D. n=20) in rapidly frozen and freeze substituted rabbit psoas muscle fibers per 0.5μm thick filament in the rigor state (Tsukito and Yano, 1985) and in deep etched fish muscle (Varriano-Marston, et al., 1984). This similarity in the number of attached rigor bridges in, respectively, striated and smooth muscle is rather surprising, given the greater ratio of actin filaments to myosin filament in smooth than in striated muscle (13:1 vs 2:1). The result suggests that there are either more crossbridges per crown in the smooth muscle myosin filament or that not all of the surrounding actins interact with the crossbridges of a given myosin filament.

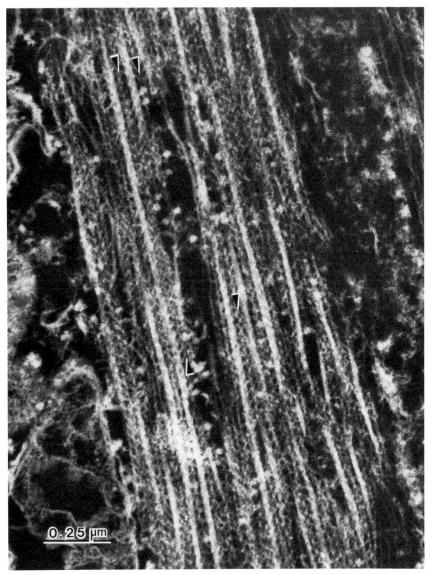

FIG. 1 Longitudinal view of actin and myosin filaments with rigor cross-bridges in a portal vein smooth muscle cell incubated in a rigor solution for 30 min. Arrowheads indicate regions where the chevron patterns of cross-bridges attached to actin are evident. The negative has been contrast reversed.

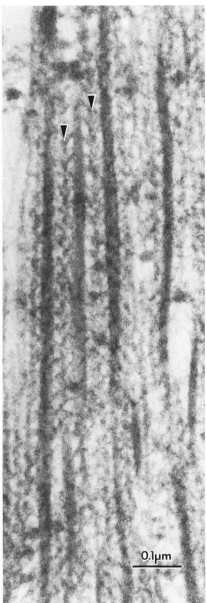

FIG. 2 Enlarged region taken from Fig. 1 showing alternating myosin and actin filaments with typical rigor chevron patterns of crossbridges (arrowheads). The surface densities of the first and third myosin filaments on the left may represent myosin heads projecting away from the filaments.

Further studies are required to resolve this issue. The length of the crossbridges measured (in these planar projections) from the backbone of the myosin filament to their attachment site on the actin filament was 27nm ± 3.5 S.D. (n=14), using the diameter of the intermediate filaments (10nm) as an internal calibration. Based on the length (16-19nm) of the head region (subfragment-1) of a myosin molecule (Elliot & Offer, 1978; Winkelman et al., 1985), the attached crossbridge in the rigor state in smooth muscle must include a portion of the subfragment-2 region of the myosin molecule positioned away from the shaft of the myosin filament.

These images provide structural support for the existence of a rigor state due to crossbridge attachment in smooth muscle, and are consistent with the increased in-phase stiffness during rigor in similar preparations described in the next section.

Crossbridge transients initiated by photolysis of caged ATP

Tension development from the rigor state in the presence of Ca^{2+} was significantly faster when initiated by photolysis of caged ATP than by the diffusion of ATP from the bath to the myofilament space in smooth, as in striated muscle (Goldman et al., 1984a). Phosphorylation of the 20,000 dalton myosin light chains (LC_{20}) is considered to be a necessary prerequisite for the physiological initiation of force development in a variety of smooth muscles (for review, see: Hartshorne, 1981; Kerrick et al., 1981; Kamm & Stull, 1985). ATPγS, a slowly hydrolyzed (Goody & Hoffman, 1980) analog of ATP can be used to thiophosphorylate the LC_{20} myosin light chains. Such thiophosphorylated light chains are insensitive to smooth muscle light chain phosphatases (Cassidy et al., 1979) and, following thiophosphorylation, the actomyosin ATPase of the isolated proteins or skinned fibers is Ca^{2+} independent, and can be activated by ATP without Ca^{2+} (Walsh et al., 1982). This enabled us to determine whether LC_{20} phosphorylation or the inherent properties of actomyosin are rate limiting in force development, because in muscles that are not phosphorylated, the contraction induced by ATP, in the presence of Ca^{2+} would have to be preceded by light chain phosphorylation. Tension development initiated by the release of 0.5mM to 1mM ATP from caged ATP was over an order of magnitude faster in muscles with thiophosphorylated light chains ($k=3.9$ s^{-1} ± 1.5 S.D. n=11) than in muscles that were not thiophosphorylated prior to activation ($k=0.3$ s^{-1} ± 0.12 S.D. n=29) (Fig. 3). In order to test whether force development was rate limited by the amount of light chain kinase present, skinned fibers were also incubated with 20μg/ml myosin light chain kinase (kindly supplied by Dr. David Hartshorne, University of Arizona); this did not alter the rate of tension development in nonthiophosphorylated muscles. In intact (nonpermeabilized) muscle strips the rate of force development following electrical (50ms DC pulse) stimulation ($k= 0.4$ sec^{-1} ± 0.10 S.D. n=4) or during high potassium contractures ($k= 0.4$ sec^{-1} ± 0.10 S.D. n=8) were similar to the rates found following caged ATP photolysis of nonthiophosphorylated skinned muscles.

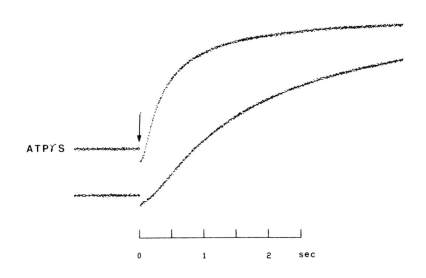

FIG. 3 Tension transients after liberation of ATP by photolysis of caged ATP (arrow) by a 50nsec laser pulse, in a permeabilized guinea pig portal vein. The protocol for the lower trace was to transfer the muscle strip from relaxing solution to a Ca^{2+}-free rigor solution for 10 min. followed by a 5 min incubation in a rigor solution containing Ca^{2+} and finally a 2.5 min. incubation in the photolysis solution containing caged ATP and Ca^{2+}. The protocol for the upper trace was identical except that the muscle was incubated for 10 min. in 2mM ATPγS in the presence of calcium prior to exposure to the rigor solution. The rate of force development is an order of magnitude faster in the thiophosphorylated muscle. The small drop in tension at the onset of the force transient is due to an artifact induced directly by the laser pulse and can be reduced by decreasing the laser energy or masking the hooks which attach to the muscle strip.

Therefore, we conclude that light chain phosphorylation is the rate limiting step in force development following liberation of ATP. This conclusion is consistent with the coincidence of the rise of stiffness and light chain phosphorylation in intact, electrically stimulated tracheal smooth muscle (Kamm & Stull, 1986). In non-permeabilized (intact) muscles, long latencies were observed following electrical stimulation (280msec $\pm$ 60 S.D. n=4) and during high K^+ depolarization (570msec $\pm$ 180 S.D. n=8), similar to latencies observed in other smooth muscles (Fay 1977; Kamm & Stull 1985), and unlike the short latency following photolysis of caged ATP (Fig. 3). Our results and those of Kamm and Stull's results suggest that at least one additional rate limiting process precedes light chain phosphorylation during activation of intact muscle.

The initiation of relaxation of the smooth muscle strips from rigor by photolysis of caged ATP_2 is illustrated in Fig. 4 (traces labelled "$-Ca^{2+}$"). This consisted of an initial rapid followed by a slow phase that generally did not follow a single exponential. A fall in stiffness accompanied the rapid detachment step, with approximately 90% of the total change in stiffness and approximately 30% of the total drop in tension occurring within this phase. In-phase stiffness is largely a measure of the number of attached crossbridges, thus the initial rapid fall in stiffness is evidence of this early phase being dominated by the detachment of rigor crossbridges. By fitting the tension trace during the fast phase to a single exponential, the rate of detachment was found to be dependent on the ATP concentration; at an ATP concentration of 700µM it was approximately $100s^{-1}$.

The shape of the tension trace depended on the strain imposed (by prestretches or prereleases) on the rigor crossbridges. When the muscles were prestretched prior to the laser pulse, the amplitude of the initial fast detachment phase was enhanced. This suggests that the initial tension change is modulated by the strain on the crossbridges.

The shape of the relaxation transients were similar when CTP, a nucleotide that is not a substrate for smooth muscle myosin light chain kinase (Cassidy et al., 1979), was released from caged CTP. Therefore, the slow relaxation following the initial fast component cannot represent reattachment of crossbridges due to LC_{20}

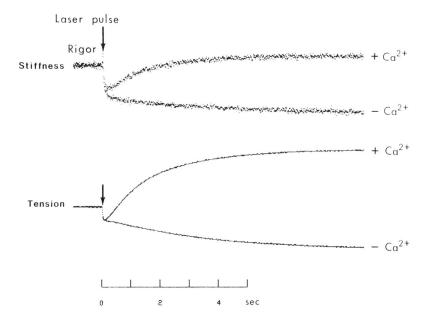

FIG. 4 Tension and in-phase stiffness transients initiated by photolysis of caged ATP in the presence and absence of calcium. The muscle strips were incubated in two 5 min changes of Ca^{2+}-free rigor solution followed by a 2.5 min. incubation in caged ATP rigor solution $\pm$ Ca^{2+}. The muscles were stretched by 1% l_o just prior to the laser pulse. The traces obtained in the presence and absence of calcium have been superimposed to illustrate the initial common rapid detachment step which is observed in both the stiffness and tension traces. The tension traces diverge by 50msec, indicating that cross-bridge attachment leading to positive force generation can occur within that time.

phosphorylation by a calcium independent kinase, such as the "protease clipped" kinase described by Walsh et al. (1982).

The shape of the slow phase of tension decay was complex; we interpret the early plateau and the slowness of final decay as being due to some crossbridges reattaching and producing force, following their detachment. The overall shape of the curve would then be due to the sum of the detachment and reattachment events.

The reattachment of detached crossbridges in the absence of calcium was also suggested by the following experiments. The release of small amounts (300µM-400µM) of ATP by photolysis of caged ATP in muscles in a rigor state, in the absence of calcium, resulted in a biphasic curve of force vs. ATP concentration, with peak force developing at approximately 25µM ATP. In phase stiffness fell at each increase of ATP concentration. Similar biphasic curves have been observed in striated muscle fibers (Reuben et al., 1971; Fabiato and Fabiato, 1975; Goldman et al., 1984a), and ascribed to thin filament cooperativity (Goldman et al., 1984a; Bremel et al., 1972; Bremel and Weber, 1972). At subsaturating ATP concentrations, some of the crossbridges would still be attached through rigor links, and this could cooperatively facilitate crossbridge cycling and force generation by the population of crossbridges with nucleotide bound. Thus, our experiments provide the first evidence of cooperative crossbridge attachment in a myosin regulated vertebrate muscle, although it remains to be determined whether this is mediated through the thin or the thick (myosin) filaments.

The tension transients following photolysis of caged ATP in the presence and absence of calcium diverged in less than 50msec (Fig. 4), indicating that crossbridge attachment leading to positive force generation can occur within that time. This experiment also shows that the rate of the initial rapid detachment is not affected by calcium.

Inorganic phosphate (P_i), one of the products of ATP hydrolysis, at millimolar concentrations attenuates tension both in activated skinned smooth and striated muscles (Schneider et al., 1981; Guth and Junge, 1982; Brandt et al., 1982; Hibberd et al., 1985; Arner et al., 1986). P_i greatly accelerated the rate of relaxation from

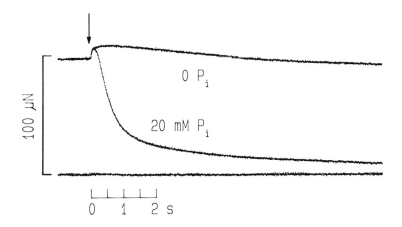

FIG. 5 The effect of 20mmol P_i on the tension transients initiated by the photolysis of caged ATP in the absence of Ca^{2+} in muscles preshortened by 1% l_o prior to the laser pulse. The initial rapid rise is independent of the P_i concentration and may represent the detachment of negatively strained cross-bridges. P_i markedly accelerated the subsequent rate of relaxation.

rigor, evoked by photolysis of caged ATP in smooth muscle in the absence of calcium (Fig. 5). Similar acceleration is seen in skeletal muscle (Hibberd et al., 1985). It has been suggested that in skeletal muscle an increase in $[P_i]$ drives the reaction back from a force generating $AM'.ADP$ state to $AM.ADP.P_i$ (weakly bound) and, $M.ADP.P_i$ state (Hibberd et al., 1985; Webb et al., 1986). Our results suggest that a similar mechanism might be functioning in smooth muscle. We interpret the marked enhancement of the rate of relaxation observed in the presence of 20mM P_i as showing that crossbridges are cycling during relaxation in the absence of calcium, consistent with the above arguments for the role of cooperativity during the slow phase of relaxation in smooth muscle.

In muscle strips allowed to shorten about 1% prior to photolysis of caged ATP in the absence of calcium, there was a rapid increase in force prior to relaxation, and this was not influenced by P_i (Fig. 5). We suggest that this increase in force is due to the detachment of negatively strained (A.F. Huxley, 1974) crossbridges, for the following reasons: 1. A fall in in-phase stiffness was coincident with the rise in tension, consistent with the detachment of crossbridges; 2. The rate of this force development was an order of magnitude higher than the rate of force development of prethiophosphorylated muscles initiated by photolysis of caged ATP (c.f. Figs 2 and 5) and; 3. P_i would not be expected to affect the detachment of the nucleotide free rigor state, and, indeed, had no effect on this force transient.

We conclude that: A. tension development from the relaxed state in smooth muscle is rate limited by light chain phosphorylation, but other steps, preceeding light chain phosphorylation, make a major contribution to the long latency between stimulus and the initiation of tension in intact smooth muscle. B. The complex shape of the slow relaxation transient is due to some cross-bridges reattaching and generating force without requiring myosin light chain phosphorylation. C. Release of micromolar ATP during rigor causes contraction (presumably without phosphorylation), in the absence of calcium, as the result of cooperative reattachment in a myosin regulated muscle. Crossbridges can bear negative as well as positive forces in smooth muscle, and the detachment of negatively strained crossbridges can be detected as positive force development.

REFERENCES

Arner A, Goody RS, Guth K, Rapp G, Ruegg JC (1986). Effects of phosphate and magnesium on rigor and ATP induced relaxation in chemically skinned guinea-pig taenia coli. J. Mus. Res. Cell Motil. 7: 381a.

Ashton FT, Somlyo AV, Somlyo AP (1975). The contractile apparatus of vascular smooth muscle: Intermediate high voltage stereo electron microscopy. J. Mol Biol 98: 17-29.

Bond M, Somlyo AV (1982). Dense bodies and actin polarity in vertebrate smooth muscle. J. Cell Biol 95: 403-413, 1982.

Brandt PW, Cox RN, Kawai M, Robinson T (1982) Regulation of tension in skinned muscle fibers: effect of crossbridge kinetics on apparent Ca^{2+} sensitivity. J. Gen Physiol. 79: 997-1016.

Bremel RD, Murray JM, Weber A (1972). Manifestations of cooperative behavior in the regulated actin filament during actin-activated ATP hydrolysis in the presence of calcium. Cold Spring Harbor Symp quant Biol 37: 267-275.

Bremel RD, Weber A (1972). Cooperative behavior within the functional unit of the actin filament in vertebrate skeletal muscle. Nature, New Biol 238: 97-101.

Cassidy P, Hoar PE, Kerrick WGL (1979). Irreversible thiophosphorylation and activation of tension in functionally skinned rabbit ieum strips by [^{35}S]ATPγS. J. Biol Chem 254: 11148-11153.

Dillon PF, Aksoy MO, Driska SP, Murphy RA (1981). Myosin phosphorylation and the cross-bridge cycle in arterial smooth muscle. Science 211: 495-497.

Elliot A, Offer G (1978). Shape and flexibility of the myosin molecule. J. Mol Biol 123: 505-519.

Fabiato A, Fabiato F (1975). Effects of magnesium on contractile activation of skinned cardiac cells. J. Physiol 249: 497-517.

Fay FS (1977) Isometric contractile properties of single isolated smooth muscle cells. Nature 265: 553-556.

Goldman YE, Hibberd MG, McCray JA, Trentham DR (1982). Relaxation of muscle fibres by photolysis of caged ATP. Nature 300: 701-705.

Goldman YE, Hibberd MG, Trentham DR (1984a). Relaxation of rabbit psoas muscle fibres from rigor by photochemical generation of adenosine-5'-trisphosphate. J. Physiol 354: 577-604.

Goldman YE, Hibberd MG, Trentham DR (1984b). Initiation of active contraction by photogeneration of adenosine-5'-triphosphate in rabbit psoas muscle fibres. J. Physiol 354: 605-624.

Goldman YE, Reid GP, Somlyo AP, Somlyo AV, Trentham DR, Walker JW (1986). Activation of skinned vascular smooth muscle by photolysis of 'caged inositol trisphosphate' to inositol 1,4,5-trisphosphate (InsP3). J. Physiol 377: 100P.

Goody RS, Hoffman W (1980). Stereochemical aspects of the interaction of myosin and actomyosin with nucleotides. J. Mus Res Cell Motil 1: 101-115.

Gordon AR, Siegman MJ (1971). Mechanical properties of smooth muscle. I. Length tension and force-velocity relations. Amer J. Physiol 221: 1243-1249.

Guth K, Junge J (1982). Low Ca2+ impedes cross-bridge detachment in chemically skinned taenia coli. Nature 300: 775-776.

Hartshorne DJ (1981). Biochemistry of the contractile process in smooth muscle. In Johnson LR (ed): Physiology of the Gastrointestinal Tract, New York: Raven Press, pp 243-267.

Hellstrand P, Paul RJ (1982). Vascular smooth muscle: Relations between energy metabolism and mechanics. In Crass MS III, Barnes CD (eds): "Vascular Smooth Muscle: Metabolic, Ionic and Contractile Mechanisms," New York: Academic Press, pp 1-35.

Herlihy JT, Murphy RA (1973). Length-tension relationship of smooth muscle of the hog carotid artery. Circ Res 33: 275-283.

Hibberd MG, Dantzig JA, Trentham DR, Goldman YA (1985). Phosphate release and force generation in skeletal muscle fibers. Science 228: 1317-1319.

Huxley AF (1974). Muscular contraction. J. Physiol 243:1-43.

Kamm KE, Stull JT (1985). The function of myosin and myosin light chain kinase phosphorylation in smooth muscle. Ann Rev Pharmacol Toxicol 25:593-620.

Kamm KE, Stull JT (1986). Activation of smooth muscle contraction: Relation between myosin phosphorylation and stiffness. Science 232: 80-82.

Kaplan JH, Forbush III B, Hoffman JF (1978). Rapid photolytic release of adenosine 5'-triphosphate from a protected analogue: utilization by the Na:K pump of human red blood cell ghosts. Biochemistry 17: 1929-1935.

Kerrick WGL, Hoar PE, Cassidy PS, Bridenbaugh RL (1981) Skinned muscle fibers: Funcational significance of phosphorylation and calcium-activated tension. Cold Spring Harbor Conferences on Cell Proliferation 8: 887-900.

Marsten SB, Taylor EW (1980). Comparison of the myosin and actomyosin ATPase mechanisms of the four types of vertebrate muscles. J. Mol Biol 139: 573-600.

Mulvany MJ, Warshaw DM (1979). The active tension-length curve of vascular smooth muscle related to its cellular components. J. Gen Physiol 74: 85-104.

Peterson JW (1982). Rate-limiting steps in the tension development of freeze-glycerinated vascular smooth muscle. J. Gen Physiol 79: 437-452.

Reuben JP, Brandt PW, Berman M, Grundfest, H (1971). Regulation of tension in the skinned crayfish muscle fiber. I. Contraction and relaxation in the absence of Ca (pCa>9). J. Gen Physiol 57: 385-407.

Rosenfeld SS, Taylor EW (1984). The ATPase mechanism of skeletal and smooth muscle acto-subfragment 1. J. Biol Chem 259: 11908-11191.
Schneider M, Sparrow M, Ruegg JC (1981). Inorganic phosphate promotes relaxation of chemically skinned smooth muscle of guinea-pig Taenia coli. Experientia 37: 980-982.
Somlyo AP, Devine CE, Somlyo AV, Rice RV (1973). Filaments organization in vertebrate smooth muscle. Phil Trans R. Soc Lond 265: 223-229.
Somlyo AP, Somlyo AV, Goldman YE, Fujimori T, Bond M, Trentham D (1986) Photolysis of caged nucleotides (ATP and CTP) for kinetic studies of vascular smooth muscle contraction. J. Mus Res Cell Motil 7: 380a.
Somlyo AV, Butler TM, Bond M, Somlyo AP (1981). Myosin filaments have nonphosphorylated light chains in relaxed smooth muscle. Nature 294: 567-570.
Somlyo AV, Somlyo AP (1986) Rigor crossbridges and other structural aspects of smooth muscle contractility. J. Mus Res Cell Motil 7: 377a.
Speden J (1960). The effect of initial strip length on the noradrenaline induced isometric contraction of arterial strips. J. Physiol Lond 154: 15-25.
Tsukita S, Yano M (1985). Actomyosin structure in contracting muscle detected by rapid freezing. Nature 317: 182-184.
Varriano-Marston E, Franzini-Armstrong C, Haselgrove JC (1984). The structure and disposition of crossbridges in deep-etched fish muscle. J. Mus Res Cell Motil 5: 363-386.
Walsh MP, Bridenbaugh R, Hartshorne DJ, Kerrick WGL (1982). Phosphorylation-dependent activated tension in skinned gizzard muscle fibers in the absence of Ca^{2+}. J. Biol Chem 257: 5987-5990.
Warshaw DM, Fay FS (1983a). Cross-bridge elasticity in single smooth muscle cells. J. Gen Physiol 82: 157-199.
Warshaw DM, Fay FS (1983b). Tension transients in single isolated smooth muscle cells. Science 219: 1438-1441.
Webb MR, Hibberd MG, Goldman YE, Trentham DR (1986). Oxygen exchange between Pi in the medium and water during ATP hydrolysis mediated by skinned fibers from rabbit skeletal muscle. J. Biol Chem 261: 15557-15564.
Winkelmann DA, Mekeel H, Rayment I (1985). Packing analysis of crystalline myosin subfragment-1. Implications for the size and shape of the myosin head. J. Mol Biol 181: 487-501.

INFLUENCE OF ATP, ADP AND AMPPNP ON THE ENERGETICS OF CONTRACTION IN SKINNED SMOOTH MUSCLE

Anders Arner, Per Hellstrand and J. Caspar Rüegg

Department of Physiology and Biophysics (A.A., P.H.), University of Lund, Sölvegatan 19, S-223 62 Lund, Sweden and II. Physiologisches Institut (J.C.R), University of Heidelberg, Im Neuenheimer Feld 326, D-6900 Heidelberg, FRG.

INTRODUCTION

Smooth muscle exerts a well regulated conversion of chemical energy, in the form of ATP, to a mechanical output: force and shortening. The basis of this process is thought to be a cyclic operation of cross-bridges between thick (myosin) and thin (actin) filaments. The biochemical pathways involved in the actin-myosin-nucleotide interaction, characterized for solubilized contractile proteins, have been found to be similar to those in striated muscle, although several of the reaction rates are slower (cf. Marston, 1983). Little is however known regarding the biochemical steps and their mechanical correlates in the structurally organized smooth muscle contractile system. Of interest is to determine the rate-limiting steps and their regulation under different contractile conditions such as isometric force maintenance or isotonic shortening. One approach is to investigate the effects of substrates and products of the cross-bridge cycle on muscle contraction. In chemically skinned preparations the cell membranes have been functionally removed, which enables a direct control of the ionic environment of the contractile proteins. In the study presented here we have investigated the effects of ATP, ADP and the nonhydrolyzable ATP-analogue AMPPNP (β-γ-imido-ATP, Yount et al., 1971) on the isometric force, force-velocity relation and rate of ATP-turnover in chemically skinned preparations of guinea pig taenia coli.

METHODS

Guinea pig taenia coli strips were chemically skinned with Triton X-100 as described by Sparrow et al. (1981). Experiments were performed at room temperature (21-23°C) in solutions containing (mM): EGTA 4, free-Mg^{2+} 2, DTE 0.5, NaN_3 1, TES-buffer 30 (pH 6.9) and calmodulin 0.0005. Ionic strength was 150 (adjusted with KCl) and MgATP 3.2 mM unless stated otherwise. The composition of the solutions was computed according to Fabiato and Fabiato (1979), Fabiato (1981) and Yount et al. (1971). Force-velocity relations were determined by the quick release method as described by Arner and Hellstrand (1985). A series of releases to different afterloads were applied at the plateau of active contractions. Shortening velocity at 100 ms after each release was determined as well as the corresponding afterload. The maximal shortening velocity (Vmax) was calculated by extrapolation to zero load of the Hill (1938) equation fitted to the data. ATP-turnover was measured by incubating the muscles in solutions containing an ATP-regenerating system consisting of phosphoenolpyruvate and pyruvate kinase, and then determining the amount of released pyruvate with a fluorimetric assay (Arner and Hellstrand, 1983). A rigor state was obtained as described by Arner and Rüegg (1985) by first activating the preparations with Ca^{2+} and ATP and then removing ATP at the plateau of contraction. The dynamic stiffness in rigor was determined from tension responses to quick (<0.8 ms) length steps with amplitudes less than 0.5% of muscle length. Caged-ATP experiments, investigating the ATP-induced relaxation from rigor, were performed as described by Arner et al. (1986a).

EFFECTS OF ADP ON Ca^{2+} ACTIVATED CONTRACTION

The addition of MgADP did not alter tension in the relaxing (pCa 9) solution in the skinned guinea pig taenia coli preparations as seen in Figure 1 (lower record). Ca^{2+}-independent contraction in skinned chicken gizzard elicited by ADP has previously been reported by Hoar and Kerrick (1985). This suggests a difference in the effects of ADP between the two types of smooth muscle. However, since ADP and ATP cannot be buffered simultaneously, the actual concentrations of these nucleotides at the level of the contractile proteins may be different due to differences in size or basal ATP-turnover between the preparations. Other factors, e.g. the ionic composition of the solutions, may also be influencing the ADP-induced responses. An increase in free-Mg^{2+} has been shown to induce a Ca^{2+}-insensitive contraction (Saida and Nonomura, 1978; Arner, 1983; Ikebe et al., 1984).

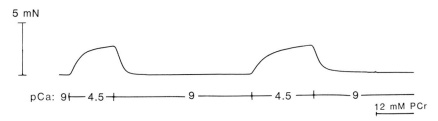

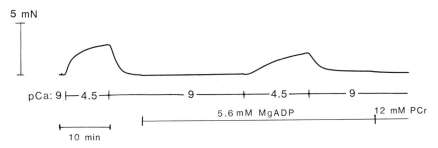

Figure 1. Original recordings of contraction-relaxation cycles from two skinned guinea pig taenia coli preparations. MgATP concentration was 3.2 mM.

The rate of tension increase following activation by Ca^{2+} (pCa 4.5) and the rate of relaxation were slower in the presence of ADP (Fig. 1). This may reflect an influence of ADP on the activation-deactivation systems or on the contractile system itself. The Ca^{2+} sensitivity was investigated in the experiments shown in Figure 2. Force was recorded at increasing concentrations of Ca^{2+} in 0 and 2.8 mM MgADP. The preparations were relaxed in pCa 9 and exhibited a slight shift in the Ca^{2+}-dependence towards higher levels in the presence of ADP. These results do not suggest a marked influence of ADP on the Ca^{2+}-sensitivity of contraction.

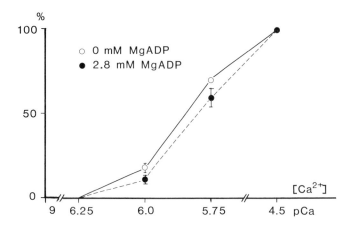

Figure 2. Ca^{2+} dependence of active force. In each preparation an initial active response was recorded (pCa 4.5 , 0 MgADP). Then, the fibres were relaxed and force was recorded for increasing Ca^{2+} levels (pCa 9 to pCa 4.5) at 0 or 2.8 mM MgADP. Force is expressed relative to the value at pCa 4.5. This force value was found to be approximately 15 % lower in the presence of MgADP when normalized to the initial active response. Data are shown as means ±SEM, n=6 in each group.

EFFECTS OF ADP ON THE FORCE-VELOCITY RELATION

The experiment in Fig. 3 show that a low concentration of MgADP (0.56 mM) causes a small reversible decrease in the maximal shortening velocity with unchanged isometric force in Ca^{2+}-activated contractions. In addition, the experiment show that at 3.2 mM MgATP (and no added ADP) the mechanics of the muscle is not influenced by any changes in the ADP/ATP-concentrations ocurring inside the fibre since the addition of the phosphocreatine backup system did not alter force or Vmax.

At a higher concentration of MgADP (2.8 mM) the influence on Vmax was more pronounced but still reversible (Fig. 4). A decrease in isometric force of about 15-20 % was observed at this MgADP concentration.

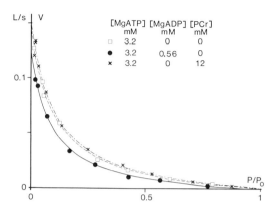

Figure 3. Force-velocity relations of a skinned fibre preparation activated by pCa 4.5 in successively 0, 0.56 and 0 mM MgADP. To ensure complete removal of ADP, an ADP-depleting system (phosphocreatine, creatine kinase) was included during the last period. The isometric force (P_o) was similar in all solutions. Shortening velocity is plotted against relative afterload (P/P_o).

In addition to the ATP-turnover associated with actin-myosin interaction, the contraction of smooth muscle also involves the ATP-dependent phosphorylation of the regulatory myosin light chains. An influence on myosin phosphorylation will affect the contractile system. A characterization of the effects of ADP and ATP requires a separation of these two processes. We have therefore investigated the effects of MgADP on force and Vmax of contractions elicited by thiophosphorylation as described by Hellstrand and Arner (1985). As seen in Figure 5, the addition of MgADP reduced force in thiophosphorylated fibres. This shows that ADP exerts an effect directly at the level of the cross-bridge interaction. The effect was however on the average slightly smaller than the depression of Ca^{2+}-activated force at a similar MgADP concentration. This suggests that a part of the decrease in tension in Ca^{2+}-activated tension, observed at high MgADP, is caused by an interference with the activation process.

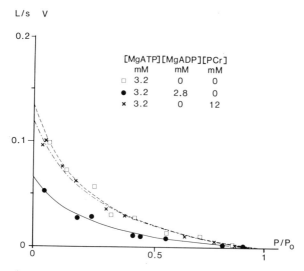

Figure 4. Force-velocity relation of a skinned fibre preparation activated by pCa 4.5. Same protocol as in Fig. 3, except that MgADP concentration was 2.8 mM.

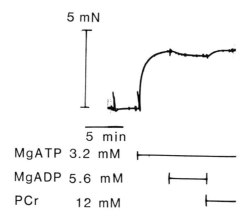

Figure 5. Effects of MgADP on force in a fibre activated by irreversible thiophosphorylation.

The maximal shortening velocity was determined at different MgADP concentrations in thiophosphorylated fibres. The results (Fig. 6) show that Vmax is reduced in a concentration-dependent way by MgADP. The experiment in Figure 7 shows that the reduction in Vmax induced by MgADP is dependent on the MgATP concentration; similar reduction is observed when the ratio of the MgADP and MgATP concentrations is held constant (0.28/1 and 0.90/3.2 mM/mM).

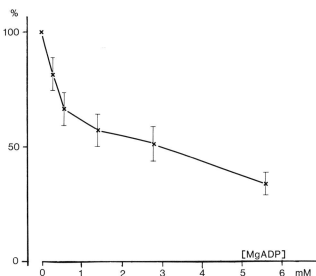

Figure 6. Effects of MgADP on the maximal shortening velocity (Vmax) in thiophosphorylated preparations at 3.2 mM MgATP. The data are expressed relative to Vmax in 0 MgADP . Means±SEM (n=6).

A direct quantitative analysis of the effects of MgADP on the basis of data from muscle fibres is difficult, since ADP and ATP cannot be buffered simultaneously in the preparations. The concentration gradients are dependent on factors such as shape, size and ATP-turnover of the muscle preparation. We have therefore chosen to investigate only high concentrations of ATP and ADP in order to minimize the gradients in the the preparation. Data from skeletal muscle suggest that MgADP acts as a competetive antagonist for MgATP with a K_i for inhibition of the shortening velocity of 0.2-0.3 mM (Cooke and Pate, 1985). The results in Figure 7 suggest competitive inhibition also in smooth muscle fibres and an analysis of our data using the concentrations of MgADP and MgATP in the solutions and assuming competitive inhibition give a K_i value for the inhibition of Vmax in thiophosphorylated fibres in the range 0.1-0.3 mM.

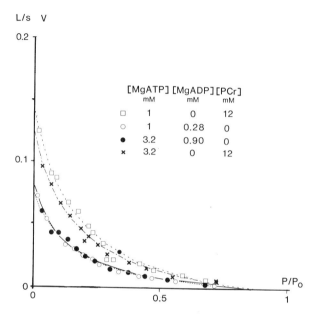

Figure 7. Effects of MgADP on the maximal shortening velocity (Vmax) in a thiophosphorylated fibre at different MgATP concentrations. Two MgADP concentrations were investigated. The MgADP/MgATP ratio was kept constant.

Results from studies of contractile proteins in solution suggest that the product ADP is released with myosin attached to actin (cf. review by Marston, 1983). The effects of MgADP in striated muscle have been interpreted to reflect an inhibition of cross-bridge detachment after the power stroke due to competition with MgATP-binding (Cooke and Pate, 1985). The effects of MgADP on shortening velocity in smooth muscle are qualitatively, and possibly also quantitatively, similar to those in striated muscle which may suggest a similar action. However, in skinned striated muscle fibres MgADP increases isometric tension (Cooke and Pate, 1985). This is interpreted as an effect of a decreased cross-bridge detachment. If cross-bridge detachment is inhibited also in smooth muscle then this effect is not directly coupled with an increased isometric force, since force rather tended to decrease in the presence of MgADP in concentrations which reduced Vmax (Fig. 5). In this context it is of interest to note that the rigor state in smooth muscle observed in the absence of ATP (Arner and Rüegg, 1985) where presumably the number of attached cross-bridges is high, is associated with a submaximal force. Thus an increased number of attached cross-bridges is not necessarily coupled with increased isometric force in smooth muscle.

EFFECTS OF ATP AND AMPPNP ON THE RATE OF ATP-TURNOVER

An important correlate to the cross-bridge cycling is the ATP-turnover (JATP) during muscle contraction. We have previously shown that the rate of this process in skinned smooth muscle can be influenced both absolutely and in relation to developed force (tension cost) by the concentration of free-Ca^{2+} (Arner and Hellstrand, 1983). Similarly, the Ca^{2+}-dependence of the maximal shortening velocity (Arner, 1983) suggests regulation by Ca^{2+} of the rate, and not only the extent, of cross-bridge turnover. When the concentration of substrate (MgATP) is reduced from the optimal 3.2 mM to 0.1 mM the JATP and Vmax of Ca^{2+}-activated contractions are decreased by about 75 and 50 % respectively (Arner and Hellstrand, 1985). The reduction in MgATP was associated with decreased levels of myosin light chain phosphorylation (Hellstrand and Arner, 1985) consistent with previous reports in the literature regarding the substrate dependence of the myosin light chain kinase. In thiophosphorylated, as compared to Ca^{2+} activated fibres, JATP and Vmax were accordingly less reduced in 0.1 mM MgATP, although still influenced (Hellstrand and Arner, 1985). In general, the MgATP-dependence of Vmax and JATP is similar to that in skeletal muscle (Cooke and Bialek, 1979; Kushmerick and Krasner, 1982).

In order to adequately measure the low rate of ATP turnover in skinned smooth muscle we have utilized the phosphoenolpyruvate backup system and determined the release of pyruvate by a fluorimetric assay. This technique cannot directly be applied to investigate the effects of ADP on JATP. We have therefore recently utilized the nonhydrolyzable ATP-analogue AMPPNP (Yount et al., 1971) to study the mechanical and metabolic effects of substances competing with ATP for binding to myosin. AMPPNP did not interfere with the assay for JATP. The addition of MgAMPPNP caused a reversible reduction of JATP. Preliminary determinations of the MgATP dependence of ATP-turnover in the presence of MgAMPPNP, suggest that MgAMPPNP acts as a competitive inhibitor for MgATP with a K_i value in the millimolar range. This is consistent with data regarding the effects on Vmax in striated muscle fibres where AMPPNP is a weak competitive inhibitor compared to ADP (Pate and Cooke, 1985). A further characterization of the effcts of AMPPNP and other ATP-analogues on ATP-turnover, force and Vmax in smooth muscle fibres will contribute to the linking of biochemical and physiological data on the kinetics of the cross-bridge interaction.

EFFECTS OF ADP ON THE ATP-INDUCED RELAXATION FROM RIGOR

Rigor offers the possibility to study the properties of attached cross-bridges in the muscle fibre. In contrast to skinned striated muscle fibres (White, 1970) skinned smooth muscle do not generate force of large amplitude when ATP in removed in the relaxed state. A rigor force can however be obtained by rapidly lowering the ATP concentration at the plateau of an active contraction as described by Arner and Rüegg (1985). The rigor force decreases with time to a plateau of about 30-40% of the maximal active force and is characterized by an increased stiffness/force ratio. When MgADP (5.6 mM) was added to the rigor solution, in the presence of a myokinase inhibitor (Feldhaus et al., 1975) no alteration in rigor force or stiffness could be observed.

MgATP causes relaxation from rigor in smooth muscle in micromolar concentrations (Arner and Rüegg, 1985). In the experiments shown in Figure 8, MgATP was allowed to diffuse into fibres in rigor. The ionic stength was kept constant at 90 mM. Under these conditions the rate of relaxation is slow, with a halftime of about 1 minute at 1 mM MgATP. No change in the rate could be observed with increased MgATP (4mM) or in the presence of 12 mM PCr suggesting that under these conditions the relaxation rate is not influenced by MgADP present in the fibre or by ATP limitation. When MgATP was added in the presence of MgADP the relaxation was markedly slower. The effect could be reversed by increasing the MgATP concentration, suggesting competition between MgATP and MgADP. This effect of MgADP was also observed at increased ionic strength (150 mM) where the rate of relaxation from rigor is higher (Arner and Rüegg, 1985; Arner et al., 1986b).

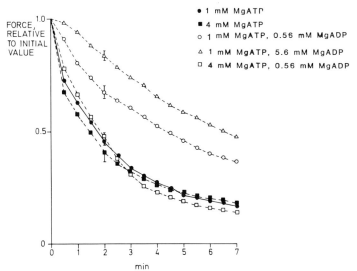

Figure 8. Effects of MgADP on the MgATP-induced relaxation from rigor. MgATP was allowed to diffuse into preparations in rigor. Mean values of 6 experiments. To increase clarity SEM is shown at 2 minutes only.

The rate of ATP-induced relaxation from rigor was investigated with the caged-ATP technique. Caged-ATP, which is biologically inert, is allowed to diffuse into fibres in rigor. Then, ATP can be rapidly released with a uv-light flash (Goldman et al., 1984). We have previously applied this technique to the study of smooth muscle (Arner et al., 1986a). In the experiment shown in Fig. 9 the fibre preparation in the left panel is illuminated (arrow) in rigor solution without MgADP and relaxes with a halftime of less than 1 s. The fibre in the right panel was investigated in the presence of MgADP which markedly slowed the rate of relaxation.

The relaxation from rigor in smooth muscle is complex and may involve several mechanical events in addition to the detachment of rigor cross-bridges. The relaxation from rigor following release of ATP from caged-ATP can be interpreted (Arner et al., 1986a) according to a model proposed for striated muscle (Goldman et al., 1984) where a rapid detachment of rigor cross-bridges (in the millisecond time scale) is followed by reattachment and a slower final relaxation. Thus the reduced rate of MgATP-induced relaxation in the presence of MgADP may be a direct measure of a slow rigor cross-bridge detachment or reflect a slow detachment of other cross-bridge states generated efter the detachment from rigor states. If cross-bridges reattach after the ATP-induced detachment from rigor, a slowing of the detachment by MgADP, as suggested for striated muscle by Dantzig et al. (1984), may cause a potentiation of cross-bridge reattchment and thereby slow relaxation.

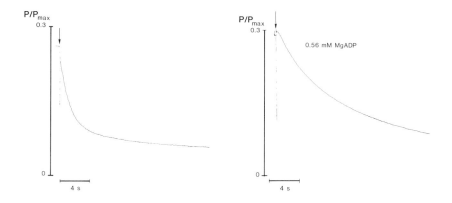

Figure 9. Relaxation from rigor following release of ATP from caged-ATP. Photolysis conditions: 12.5 mM caged-ATP was added to rigor solution containing (mM): DTE 12 , EGTA 4, Mg^{2+} 2, TES-buffer (pH 6.9) ionic strength 90. MgADP was 0 (left panel) or 0.56 mM (right panel). Illumination with uv-light from a xenon flash lamp (arrow) releases 1-2 mM MgATP. Force is expressed relative to the maximal active isometric value.

SUMMARY AND CONCLUSIONS

The contraction of smooth muscle is influenced by the substrate MgATP and the product MgADP. The effects on force, shortening velocity and ATP-turnover, are consistent with an influence on the kinetics of cross-bridge cycling.

Part of these effects are mediated via an influence on the regulation of contraction by myosin light chain phosphorylation. Results from preparations activated by thiophosphorylation, show that MgATP and MgADP also interact directly at the cross-bridge level, and are consistent with MgADP acting as a competitive ATP-analogue. The slow shortening velocity and decreased rate of ATP-induced relaxation from rigor in the presence of MgADP, suggest an inhibition of cross-bridge detachment. The rate of ATP-turnover was decreased in the presence of the nonhydrolyzable ATP-analogue AMPPNP. These results may contribute to the characterization of the biochemical reactions in the structurally organized smooth muscle contractile system. In addition, the influence of MgATP and MgADP on smooth muscle contraction suggest that the concentrations of substrate and products, at the level of the contractile proteins, may constitute important regulatory factors in vivo under conditions, such as hypoxia and ischemia, associated with impaired cellular energy supply.

ACKNOWLEDGEMENTS

We wish to thank Dr. Roger S. Goody, Max Planck Institut für Medizinische Forschung, Heidelberg, FRG for kindly giving us the caged-ATP. The study was supported by the Swedish Medical Research Council (14x-28), the Medical Faculty, University of Lund, Sweden and the Deutsche Forschungsgemeinschaft (SFB 90).

REFERENCES

Arner A (1983). Force-velocity relation in chemically skinned rat portal vein: Effects of Ca^{2+} and Mg^{2+}. Pflügers Arch 397:6-12.
Arner A, Goody RS, Güth K, Rapp G, Rüegg JC (1986a). Relaxation of chemically skinned guinea pig taenia coli smooth muscle from rigor by photolytic release of ATP. J Muscl Res Cell Motil 7:73.
Arner A, Goody RS, Güth K, Rapp G, Rüegg JC (1986b). Effects of phosphate and magnesium on rigor and ATP induced relaxation in chemically skinned guinea pig taenia coli. J Muscl Res Cell Motil 7:381.
Arner A, Hellstrand P (1983). Activation of contraction and

ATPase activity in intact and chemically skinned smooth muscle of rat portal vein. Circul Res 53:695-702.
Arner A, Hellstrand P (1985). Effects of calcium and substrate on force-velocity relation and energy turnover in skinned smooth muscle of the guinea pig. J Physiol (Lond) 360:347-365.
Arner A, Rüegg JC (1985). Cross-bridge properties in smooth muscle. Acta Physiol Scand (Suppl 542) 124: 206.
Cooke R, Bialek W (1979). Contraction of glycerinated muscle fibers as a function of the ATP concentration. Biophys J 28:241-258.
Cooke R, Pate E (1985). The effects of ADP and phosphate on the contraction of muscle fibers. Biophys J 48:789-798.
Dantzig JA, Hibberd MG, Goldman YE, Trentham DR (1984). ADP slows cross-bridge detachment rate induced by photolysis of caged ATP in rabbit psoas muscle fibers. Biophys J 45:8a.
Fabiato A (1981). Myoplasmic free calcium concentration reached during the twitch of an intact isolated cardiac cell and during calcium-induced release of calcium from the sarcoplasmic reticulum of a skinned cardiac cell from the adult rat or rabbit ventricle. J Gen Physiol 78:457-497.
Fabiato A, Fabiato F (1979). Calculator programs for computing the composition of the solutions containing multiple metals and ligands used for experiments in skinned muscle cells J Physiol (Paris) 75:463-505.
Feldhaus P, Flöhlich T, Goody RS, Isakov M, Schirmer RH (1975). Synthetic inhibitors of adenylate kinases in the assays for ATPases and phosphokinases. Eur J Biochem 57:197-204.
Goldman YE, Hibberd MG, Trentham DR (1984). Relaxation of rabbit psoas muscle fibres from rigor by photochemical generation of adenosine-5'-triphosphate. J Physiol (Lond) 354:577-604.
Hellstrand P, Arner A (1985). Myosin light chain phosphorylation and the cross-bridge cycle at low substrate concentration in chemically skinned guinea pig taenia coli. Pflügers Arch 405:323-328.
Hill AV (1938) The heat of shortening and the dynamic constants of muscle. Proc Roy Soc B 126:136-195.
Hoar PE, Kerrick WG (1985). Activation of smooth muscle contraction by MgCDP, MgADP and Mg^{2+} in the absence of myosin light chain phosphorylation. In Merlevede W, DiSalvo J (eds) "Advances in protein phosphatases", Vol 2, Leuven: Leuven University Press, p407.

Ikebe M, Barsotti RJ, Hinkins S, Hartshorne DJ (1984). Effects of magnesium chloride on smooth muscle actomyosin adenosine-5' triphosphatase activity, myosin conformation, and tension development in glycerinated smooth muscle fibers. Biochemistry 23:5062-5068.

Kushmerick MJ, Krasner B (1982). Force and ATPase rate in skinned skeletal muscle fibers. Fed Proc 41:2232-2237.

Marston SB (1983) Myosin and actomyosin ATPase: kinetics. In Stephens NL (ed) "Biochemistry of smooth muscle", Vol 1, Cht 4, Boca Raton: CRC Press, pp 167-191.

Pate E, Cooke R (1985). The inhibition of muscle contraction by adenosine 5' (β,γ-imido) triphosphate and by pyrophosphate. Biophys J 47:773-780.

Saida K, Nonomura Y (1978). Characteristics of Ca^{2+}- and Mg^{2+}-induced tension development in chemically skinned smooth muscle fibers. J Gen Physiol 72:1-14.

Sparrow MP, Mrwa U, Hofmann F, Rüegg JC (1981). Calmodulin is essential for smooth muscle contraction. FEBS Lett 125:141-145.

White DCS (1970). Rigor contraction and the effect of various phosphate compounds on glycerinated insect flight and vertebrate muscle. J Physiol (Lond) 208:583-605.

Yount RG, Babcock D, Ballantyne W, Ojala D (1971). Adenylyl imidodiphosphate, an adenosine triphosphate analog containing a P-N-P linkage. Biochemistry 10:2484-2489.

COMPARATIVE STUDIES ON THE MECHANISM OF REGULATION OF SMOOTH AND STRIATED MUSCLE ACTOMYOSIN

Edwin W. Taylor

Dept. of Molecular Genetics and Cell Biology

The University of Chicago
Chicago, Illinois 60637

The regulation of the activity of striated muscle actomyosin involves the binding of calcium to thin filaments and a very persuasive model was formulated (Huxley, 1972) in which tropomyosin blocks the binding of myosin to actin in the absence of calcium. However, recent studies with acto S1 failed to show an effect of calcium on the binding of S1 to regulated actin in the presence of ATP (Chalovich and Eisenberg, 1982). It was proposed that regulation is a kinetic effect and probably the rate of dissociation of phosphate from the acto S1.ADP.P complex is increased by calcium binding to the thin filament.

Smooth muscle actomyosin is activated by phosphorylation of the 20 kDa light chain of myosin. This mechanism at first appeared to be distinctly different from the steric blocking mechanism proposed for striated muscle actomyosin but the finding that product dissociation may be the step that is activated raised the possibility that the mechanisms may be similar for smooth and striated muscles. First we have to define what is meant by similar.

It is useful to distinguish between a kinetic mechanism of regulation and a structural mechanism. A detailed kinetic mechanism has been proposed for non-regulated acto S1 of striated and smooth muscle (Stein et al., 1984; Rosenfeld and Taylor, 1984). There is some disagreement on the details but the main features of the mechanism are described by the scheme:

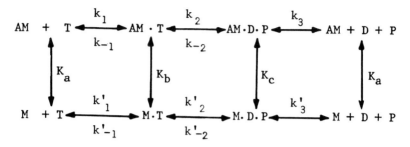

where T, D and P refer to ATP, ADP and phosphate and K_a, K_b and K_c are association constants of M, M.T and M.D.P with actin. The scheme is oversimplified in that ATP forms a collision complex with M or AM followed by a rapid conformational change to form M.T or AM.T. Also, dissociation of products involves at least two steps in which P and ADP dissociate sequentially and therefore an AM.D or a second AM.D.P state occurs. In this discussion these steps are represented by a single effective rate constant k_3.

The solution of the problem of the kinetic mechanism of regulation is obtained by measuring all of the rate and equilibrium constants of the scheme for the relaxed and activated system. The mechanism of regulation is specified by determining which steps have altered rate and equilibrium constants in the active versus the relaxed state. Smooth and striated muscle actomyosin have similar kinetic mechanisms of regulation if the same steps are altered in both cases.

Although a consideration of the kinetic mechanism can provide clues to the structural mechanism it does not explain how the rate or binding constant has been altered. The striated and smooth muscle systems could have the same kinetic mechanism but a different structural mechanism.

The kinetic mechanism for striated muscle acto S1 has been the subject of a detailed study (Rosenfeld and Taylor, 1986). The simple kinetic scheme has the advantage that the values of the rate and equilibrium constants can be measured.

Measurements of the apparent rate constant of the phosphate burst, the steady state rate and the degree of association as a function of actin concentration are sufficient to determine k_2, k_{-2} and k_3. Measurement of the average rate of dissociation of substrate and products by a competition experiment gives the value of k_{-1}. Measurements were made by means of single turnover experiments at low S1 to actin ratios to avoid activation by rigor complexes. For example, the dissociation of ATP and ADP.P was measured by forming the M.T and M.D.P states and then mixing with regulated actin and an excess of etheno-ATP. The dissociation of ATP and ADP.P was monitored by the increase in fluorescence for the binding of etheno-ATP which also dissociates rigor complexes.

The results of these experiments are summarized here. The rate constants for dissociation of substrate and products are increased 10 to 20 fold in the presence of calcium while there is little or no effect of calcium on the hydrolysis step. The binding constants K_b and K_c changed by less than 50%. K_a is 6 to 10 times larger in the presence of calcium as has been inferred previously from the co-operative binding of S1.ADP (Greene, 1982). A similar effect of calcium was obtained for the dissociation of ADP which is a simpler experiment. S1.ADP was mixed with actin plus a large excess of ATP to prevent activation. The rate constant extrapolated to infinite actin was increased twenty fold by calcium binding.

The step in the mechanism that is regulated is the transition between an acto S1 state in which the nucleotide is strongly bound to a state in which it is weakly bound. For ATP dissociation it is the transition from AM.T to the collision complex. For product dissociation it is probably a transition from AM.D.P to a state in which phosphate is weakly bound, possibly AM.D. It is not clear whether the complex formed from AM and ADP is the same AM.D state but if so then the next step, the dissociation of ADP, is also regulated by calcium.

The results confirm and extend the studies of Chalovich and Eisenberg (1982) that the primary effect is on a rate process rather than on the binding of myosin intermediates. The structural basis of regulation is not settled by these results. It can be concluded that the step (or steps) which is regulated is the same as the step which is activated by actin, namely the transition or transitions of the actin-S1-nucleotide complex to a state in which actin is strongly bound and nucleotide is weakly bound. This could be thought of as a transition which opens up the nucleotide binding site. The increased association constant for M or an M.D state could be the consequence of the change in first order rate constant of the transition between AM-nucleotide states. Thus, the increase in k_{-1} in the presence of calcium, if it is not compensated by an increase in k_1, would lead to a decrease in K_1. Therefore K_a must increase in the presence of calcium to satisfy the thermodynamic condition, $K_a K_1 = K_1' K_b$.

The results rule out a simple steric blocking mechanism. A partial steric mechanism could still be correct. The weakly bound complexes M.T and M.D.P could interact with a part of the binding site that is not occupied by tropomyosin in the relaxed state but the transition leading to nucleotide release and formation of a more strongly bound complex could require an interaction with an additional part of the binding site which is occupied by tropomyosin. This mechanism is consistent with the co-operative and calcium-dependent binding of M and M.D. The results could also be explained by a conformational mechanism. A change in the conformation of the thin filament accompanies the transition to the strongly bound state and the rate of the conformation change is effected by the interaction with tropomyosin. In this case the two positions occupied by tropomyosin need not overlap with the binding site. It is difficult to decide between these two structural mechanisms on the basis of the kinetic mechanism.

A similar analysis of the regulation of smooth muscle actomyosin ATPase is currently in progress but much less is known. A difficulty is that smooth muscle acto S1 is not regulated by phosphorylation and it is essentially in the active state (Ikebe and Hartshorne, 1985). A detailed study of regulation of striated muscle acto HMM has not been made although it appears that there is an effect of calcium in the binding of substrate and product intermediate states if LC-2 is intact (Wagner, 1984). With gizzard muscle acto HMM there is also a five fold increase in binding in the presence of calcium and ATP (Ikebe and Hartshorne, 1985). Two heads and a tail appear to be important for regulation by light chain phosphorylation. In addition smooth muscle myosin and HMM undergo a structural change which depends on ionic strength and phosphorylation (Ikebe et al., 1983; Suzuki et al., 1985). At low ionic strength the heads are folded back towards the tail for HMM and the tail is twisted around the heads for myosin (10S or folded form) while at high ionic strength the heads are extended (6S or extended form). Phosphorylation shifts the distribution toward the extended form.

The first problem is to determine the effects of phosphorylation and ionic strength on the kinetic mechanism of HMM and myosin in the absence of actin. The rate constant of substrate binding (k_1,) and the rate of the hydrolysis step (k_2, and k_{-2}) show only small effects of phosphorylation (Rosenfeld and Taylor, 1984a). The steady state rate of ATP hydrolysis increases with ionic strength and phosphorylation and the increase is correlated with the 10S to 6S transition (Ikebe et al., 1983). Thus k_3, is increased up to ten fold. Preliminary results on the rate constant of dissociation of εADP from HMM and S1 show a dependence on ionic strength and phosphorylation which is similar to the effects on the steady state ATPase. The rate constant decreases with ionic strength for S1 from 12 sec^{-1} in 10 mM KCl to 4 sec^{-1} in 300 mM KCl. The rate constant for dephosphorylated HMM increases with ionic strength from 0.5 sec^{-1} in 10 mM KCl to 4 sec^{-1} in 250 mM KCl and decreases slightly at higher ionic strength. Phosphorylation of the HMM increases the rate constant to 3.5 sec^{-1} at 10 mM KCl. The rate constant increases with ionic strength to a maximum of 5 sec^{-1} at 100 mM KCl and decreases at higher ionic strength. In contrast, the rate constant for εADP dissociation of striated muscle S1 or HMM is 3.5 sec^{-1} and is almost independent of ionic strength over the same range. Therefore the steps in dissociation of ADP.P and ADP are increased by phosphorylation at low ionic strength but the steps are almost unaffected at high ionic strength.

It has been shown that nucleotides which are strongly bound, ATP and AMP PNP, shift the equilibrium toward the folded form for myosin and HMM (Suzuki et al., 1985). Therefore, the nucleotide must be more strongly bound to the folded form which is consistent with a slower rate of dissociation of nucleotide from the folded form.

A tentative kinetic mechanism to be tested by experiments is:

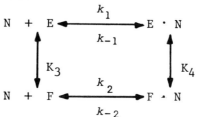

where N is a nucleotide (ADP, AMP PNP, ADP.P), K_3 = [E.N]/[F.N] and $K_3K_1 = K_2K_4$. The evidence suggest that $K_3 > K_4$ and $K_2 > K_1$. Increase in ionic strength or phosphorylation increases K_3 and thus shifts the equilibrium to the extended form. It is necessary to determine whether k_{-1} and k_{-2} are determined by the state (E or F) or also by phosphorylation.

The structural mechanism is not known. Apparently the regulatory light chain interacts with the heavy chain in the neck region or farther along S1 leading to a folding back of the head which directly or indirectly effects the interaction of nucleotide at the active site. Since nucleotide release involves some change in conformation to open the nucleotide pocket the rate of this step is reduced in the folded form. The folding is at least partly determined by a plus-minus charge interaction which is reduced by increasing ionic strength and by phosphorylation which adds two negative charges probably in the region of the junction with the coiled coil.

The effect of actin binding in every case is to increase the rate of the conformation change which opens the nucleotide pocket. Preliminary evidence showed little effect of actin binding on the hydrolysis step of dephosphorylated and phosphorylated HMM (Rosenfeld and Taylor, 1984a). Actin binding increases the rate constant for the dissociation of ADP from HMM or S1. However, the rate constants for phosphorylated and dephosphorylated acto HMM differed by only a factor of two. A much larger factor is required to explain the difference in V_{max} of phosphorylated and non-phosphorylated acto HMM ATPase because the difference in rates has to occur in the product release steps. The factor could be larger than 25 fold because of the presence of degraded HMM which is not inhibited in the dephosphorylated state (Sellers, 1985).

At present we do not have sufficient evidence to answer the question of whether the kinetic mechanisms of regulation are the same for smooth and striated muscle acto HMM. Studies on the ATPase mechanism are necessarily carried out at low ionic strength and phosphorylated HMM is partly in the F state. The increased affinity of phosphorylated HMM for actin may arise from stronger binding of the E form and a shift in the equilibrium to the E form as the degree of association increases at high actin concentrations. In the case of striated muscle we are able to separate kinetic effects from binding effects but for smooth muscle the F to E transformation may effect both binding and product release.

A further problem is that acto HMM may not be a satisfactory model of regulation in the muscle. The F to E transition for myosin occurs over a narrower range of ionic strength than for HMM and the two forms are not in rapid equilibrium since two peaks are observed in the ultracentrifuge. The F state is probably more stable for myosin than for HMM since a further interaction occurs which loops the tail around the head. In the muscle non-phosphorylated myosin may be present in the F state in thick filaments although reconstituted filaments are not stable in the F state in solution (Trybus and Lowey, 1985). F state filaments would probably not bind appreciably to actin filaments. Thus, phosphorylation could have a much larger effect on binding of myosin to actin in muscle than in solution studies with acto HMM.

In conclusion the preliminary results suggest that the product dissociation steps are increased in rate by phosphorylation while the hydrolysis step is unaffected and on this basis there is a similarity in the regulation mechanisms of smooth and striated actomyosin. However the F to E transition probably effects both binding to actin and product release. This transition apparently does not occur for striated muscle myosin and since the primary effect of phosphorylation may be on the F to E transition the structural mechanism of regulation may be quite different for the two muscles.

Acknowledgements

This work was supported by Program Project Grant HL 20592 from the Heart, Lung, and Blood Institute of the National Institutes of Health and by the Muscular Dystrophy Association of America.

REFERENCES

Chalovich, J.M. and Eisenberg, E. (1982). Inhibition of actomyosin ATPase. Activity by troponin-tropomyosin without blocking the binding of myosin to actin. J. Biol. Chem. 257:2432-2437.

Greene, L. (1982). Effect of nucleotide on the binding of myosin S1 to regulated actin. J. Biol. Chem. 257: 13993-13999.

Huxley, H.E. (1972). Structural changes in the actin and myosin-containing filaments during contraction. Cold Spring Harbor Symp. Quant. Biol. 37:361-376.

Ikebe, M. and Hartshorne, D.J. (1985). Proteolysis of smooth muscle myosin. Biochemistry 24:2380-2387.

Ikebe, M., Hinkins, S. and Hartshorne, D.J. (1983). Enzymatic properties and conformation of smooth muscle myosin. Biochemistry 22:4580-4587.

Rosenfeld, S.S. and Taylor, E.W. (1984). Mechanism of smooth muscle actomyosin ATPase. In Stephens, N.L. (ed): "Smooth Muscle Contraction", New York and Basel: M. Dekker pp. 175-187.

Rosenfled, S.S. and Taylor, E.W. (1986). Mechanism of regulation of acto S1 ATPase. Submitted by J. Biol. Chem.

Sellers, J.R. (1985). Mechanism of phosphorylation-dependent regulation of smooth muscle HMM. J. Biol. Chem. 260:15815-15819.

Stein, L.A., Chock, P.B. and Eisenberg, E. (1984). The rate limiting step of the actomyosin ATPase cycle. Biochemistry 23:1555-1563.

Suzuki, H., Stafford, III, W.F., Slayter, H.S. and Seidel, J.C. (1985). Conformational transition of gizzard HMM. J. Biol. Chem. 260:14810-14817.

Trybus, K.M. and Lowey, S. (1985). Mechanism of smooth muscle myosin phosphorylation. J. Biol. Chem. 260: 15988-15995.

Wagner, P.D. (1984). Effect of skeletal muscle myosin light chain two on Ca sensitive interaction of myosin and heavy meromyosin with actin. Biochemistry 22:5950-5956.

ISOFORMS OF MYOSIN IN SMOOTH MUSCLE

Malcolm P. Sparrow, Anders Arner, Mukhallad A. Mohammad, Per Hellstrand and J. Caspar Rüegg

II. Physiologisches Institut (J.C.R.) Univ. of Heidelberg, 6900 Heidelberg, Fed. Rep. of Germany, Dept. of Physiology (M.P.S., M.A.M.) Univ. of Western Australia Nedlands, Australia 6009, and Dept. of Physiology and Biophysics (A.A., P.H.) Univ. of Lund, 22362 Lund, Sweden

The existence of isoforms of myosin in vertebrate smooth muscle has been sought as a means of explaining its functional diversity. The evidence currently available arises from (1) multiple bands on pyrophosphate gels when tissue extracts containing native myosin are electrophoresed under non-denaturing conditions, (2) two heavy chains of myosin of different molecular weight on sodium dodecylsulphate (SDS) acrylamide gels of high porosity when electrophoresed under denaturing conditions and (3) heterogeneity of myosin antigenic expression in vascular smooth muscles.

The first approach, namely the separation of native myosins on pyrophosphate gel electrophoresis seemed to indicate that isoforms of myosin exist in smooth muscle (Hoh and Yeoh, 1977; Takano-Ohmuro et al., 1983; Pagani et al., 1985; Morano et al., 1986a). Beckers-Bleukx and Maréchal (1985) ascribed the difference in the electrophoretic mobility of the two "native myosin" bands to the presence of two heavy chains of myosin of 200 and 230 kDa. The appearance of vascular smooth muscle myosin bands separated by native (pyrophosphate) gel electrophoresis remained unchanged in the spontaneously hypertensive rat when compared with normotensive age-matched controls (Morano et al., 1986b). While the pyrophosphate gel electrophoretic system of Hoh et al. (1978) has universal acceptance for the separation of isoforms of myosin in skeletal and cardiac muscle, the application of this technique to smooth muscle has not been rigorously tested. Recently Sparrow et al. (1986) have reported that the slower migrating "myosin" band may be predominantly filamin or may comprise both myosin and filamin. Also multiple bands may arise from phosphorylation of the myosin (Persechini et al., 1986).

FIGURE 1

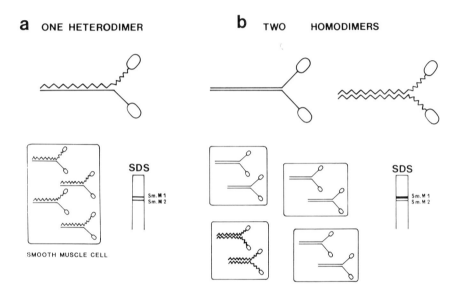

Fig. 1a. Scheme proposed by Rovner et al. (1986) for the stoichiometry of the putative myosin heavy chains in smooth muscle. The two different heavy chains which occur in equal proportions are organized as a heterodimer in the smooth muscle cell. To the right is a 4% acrylamide SDS gel showing two bands which are the myosin heavy chains designated Sm M1 and Sm M2.

Fig. 1b. Alternative scheme in which the myosin heavy chains occur as homodimers. This scheme is compatible with varying proportions of heavy chains.

More widely accepted is the existence of two different heavy chains of myosin in a number of vascular and non-vascular smooth muscle tissues (Schildmeyer and Seidel, 1986; Cavaille, 1986; Rovner et al., 1986). These differ in molecular weight by approximately 3-4 kDa, as shown on SDS acrylamide gels of low porosity. Rovner et al. (1986) found that the two heavy chains occurred in virtually equal proportions in the four tissues they studied, viz porcine carotid artery, aorta and stomach and the taenia coli of guinea pig. They proposed that the heavy chains were present as a heterodimer in the native myosin molecule, thereby inferring that only one isoform or isoenzyme of myosin occurred (Fig. 1a).

We have also examined the stoichiometry of the heavy chains of myosin in a variety of smooth muscle tissues and report here that their proportions can vary widely thus leading us to propose that two native isoforms occur which are composed of heavy chain homodimers (Fig. 1b). Nonetheless, it has not been possible as yet to show unequivocally that these putative isoforms of native myosin can be separated on pyrophosphate gels.

METHODS

The smooth muscle tissues were freed of mucosa, submucosa and serosal connective tissue: e.g. in uterus the endometrium was teased off completely, leaving virtually only longitudinal smooth muscle. All procedures were done at 4°C in a cold room, then the smooth muscle was frozen in liquid N_2, crushed and immediately homogenized in 1% SDS buffer to solubilize the proteins into their component polypeptides (see Rovner et al., 1986). The SDS extracts were then electrophoresed on 4% acrylamide SDS gels stained with Coomassie Blue, and the protein bands were scanned with a densitometer. Tissues from at least 3 animals were used.

For pyrophosphate (native myosin) electrophoresis the frozen tissue was crushed at -170°C and extracted at 0°C with a Guba-Straub type salt solution at pH 6.7 containing 0.3 M NaCl, 0.1 M NaH_2PO_4, 0.05 M Na_2HPO_2, 1 mM $MgCl_2$, 10 mM EDTA, 10 mM sodium pyrophosphate, 0.1 mM NaN_3 and 0.1% 2-mercaptoethanol. Alternatively the myosin was solubilized with an ATP salt solution at pH 6.7 using freshly dissected tissue at 4°C after washing first with a low ionic strength solution to remove soluble cytoplasmic proteins (Hoh and Yeoh, 1977). In general, both methods yielded the same profile of bands on the pyrophosphate gels although it seemed that the latter procedure produced a clearer gel (i.e. less background stain) with possibly better separation in some instances.

RESULTS

Two myosin heavy chains (MHC) of different M.W. were found in the smooth muscle layers of all the tissues listed below, guinea pig myometrium excepted, where three putative MHCs were detected. Fig. 2 shows that the proportions of the MHCs varied considerably: from the extreme in rat uterus to virtually equal proportions in rat bladder. In general the slower migrating band, designated Sm M1 was present in the greater proportion in most tissues, but in rat portal vein the pattern was reversed. Porcine trachealis from adult pig showed nearly equal proportions (see Fig. 6) and, in confirmation of Rovner et al. (1986), we found that porcine carotid artery media contained equal proportions, similar to rat bladder.

FIGURE 2

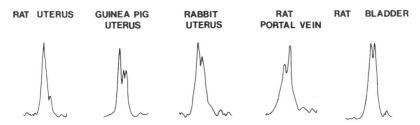

Fig. 2. Densitometric scans of the myosin heavy chains of SDS extracts of smooth muscle run on 4% acrylamide SDS slab gels. Origin is on the left.

In rat myometrium whether from pregnant or non-pregnant animals, Sm M1 was always predominant. Myosin from the circular smooth muscle of guinea pig ileum had the same profile. In guinea pig myometrium 3 bands were present with the highest M.W. band predominant. Where the proportion of MHCs was sufficiently unequal, only one band was seen at very low loadings on the gel, when the amount of protein in the minor band fell below the threshold of detection by the Coomassie Blue stain. In rat uterus, pregnant or non-pregnant and in guinea pig ileum this occurred. As the amount of protein loaded on the gel was increased the bands broadened, and Sm M1 approached saturation with the dye with Sm M2 merging with Sm M1 towards its apex.

Schildmeyer and Seidel (1986) report only one MHC in rat uterus. To further check this, we excised the "native myosin" band from a pyrophosphate gel and electrophoresed it on the 4% SDS acrylamide. Fig. 3 clearly shows that two MHCs are present in rat myometrium in unequal proportions.

Since these MHCs differ in M.W. by about 3-4 kDa (Rovner et al., 1986) it would seem likely that the amino acid substitutions in the native myosin molecules would give rise to a charge difference and thus should be separable on pyrophosphate gels, as do isoforms of myosin from skeletal and heart muscle.

FIGURE 3

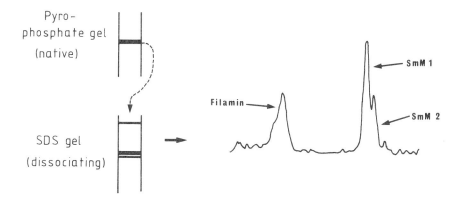

Fig. 3 Composition of "native myosin" on pyrophosphate gel from late pregnant rat. The band was excised, and the protein extracted and run on 4% acrylamide SDS gel. The densitometric trace shows that the band comprised a ~ 250 kDa-protein (filamin) and two heavy chains of myosin (M.W. 200 kDa) in which Sm M1 is predominant.

Unfortunately, in the case of smooth muscle the pyrophosphate technique produces a variety of band patterns which are dependent on the electrophoresis running conditions and the extraction of myosin from the tissue (e.g. washing first with low ionic salt solution before extracting the myosin with ATP). One band obtained on one run may appear as two bands on a subsequent run. Where three bands appear routinely, it is sometimes possible to see one split into two to give 4 bands when the resolution is improved. The particular conditions which promote this degree of resolution are difficult to pinpoint: polymerization of the gels in the cold room, the use of the extraction procedure of Hoh and Yeoh (1977) and using fresh tissue may be some of the factors involved in obtaining well separated multiple bands. Cardiac or skeletal muscle myosins are commonly used to provide a standard marker of isoforms in order to show that the gel is resolving isoforms well, but it is not a positive indicator that the gel is capable of separating the "myosin" components of smooth muscle extracts as is supposed. It is not surprising therefore, that the number of bands reported for smooth muscle extracts run on pyrophosphate gels varies greatly, but most frequently is only one (Rovner et al., 1986; Cavaillé, 1986).

FIGURE 4

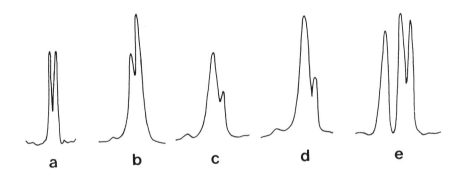

Fig. 4 Densitometric scans of "native myosin" band obtained on pyrophosphate gels. Origin is at the left. (a) rabbit myometrium, (b) rat myometrium, (c) pig carotid artery media, (d) pig ileum, (e) toad stomach circular muscle.

Fig. 4 shows some band patterns that we have obtained on pyrophosphate gels. They are summarized in Table 1. It should be realized from the foregoing that the pattern and proportion of the bands reported is not invariant, particularly where 3 bands occur, so that the table has been made to provide a qualitative guide to the usual band patterns obtained. Band patterns were unaffected by the location of sampling: e.g. in pig trachea they were identical in samples taken from the proximal, mid and distal regions of the trachea. In toad stomach, the same multiple band pattern was present from the cardiac to the pyloric end.

The effect of hypertrophy of uterine smooth muscle on the band pattern seen on pyrophosphate gels was investigated. The non-gravid myometrium of rabbits exhibited two widely spaced, well defined bands in equal proportions and these were unchanged in early and late pregnancy.

TABLE 1

Band patterns on pyrophosphate gels

species	bands	proportion	separation	number of animals
pig:				
carotid artery	2	slower predominant	wide	7
trachealis (adult)	2	about equal	wide	3
trachealis (young)	3	about equal	intermediate	7
jejunum	3	about equal	intermediate	2
ileum	2-3	about equal	intermediate	3
guinea pig ileum	3	variable	wide	6
rabbit myometrium	2	equal	wide	4
rat myometrium	2	faster predominant	close	6
toad stomach	2-3	slower predominant	wide	5

In the rat the two bands seen in the non-gravid myometrium persisted until late pregnancy at which stage only one band was found (Fig. 5).

FIGURE 5

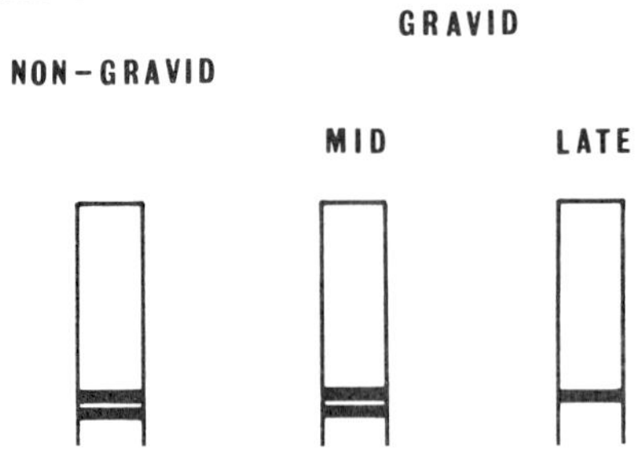

                GRAVID
NON-GRAVID
              MID   LATE

Fig. 5. Diagram of the band patterns seen on pyrophosphate gels of myometrium from non-pregnant and pregnant rats.

To find out what proteins the bands comprised we ran standards of filamin and myosin from chicken gizzard alongside the samples, and in addition, co-migrated these proteins in the samples applied to the gels. In rat and rabbit myometrium, filamin migrated closely with the slower migrating band, and in guinea pig ileum with the slowest migrating band. Filamin also migrated closely with the slower band in both rabbit and rat myometrium but in the latter where the bands do not separate as widely as in the rabbit, the migration of the filamin was very load dependent. When very small amounts of filamin were added to the rat myometrial sample, the filamin band was seen to run just above the upper band of the sample. At higher loadings the filamin band fused into the upper sample band (Morano, personal communication). Myosin from chicken gizzard co-migrated with the faster band of the myometrial samples from rabbit uterus but in addition, seemed to affect the migration and proportion of the slower band. A mixture of filamin and myosin could reproduce the two bands of rat uterus very closely, but again, separation was very load dependent. In view of the sensitivity of migration of the samples bands to the load of myosin or filamin added during co-migration, this procedure was abandoned in favour of cutting out the bands and determining their protein composition on acrylamide SDS gels. This could be undertaken readily in rabbit myometrium where the separation is wide. The gels were stained for 3 min using Coomassie Blue G 250 in 3% perchloric acid to visualize the bands and the protein from the separated upper and lower slices of 9 gels extracted into SDS buffer and applied to a 6% acrylamide SDS gel. The upper band seemed to contain almost exclusively filamin (a trace of myosin could be detected); the lower band comprised mostly myosin but a significant amount of a 230 kDa-protein (filamin) was also present. In the case of myometrium from the late gravid uterus of rat where only one band was present (Fig. 5), analysis of this band after excision from the gel using a 4% acrylamide SDS gel showed it to contain a substantial proportion of a 230 kDa-protein (filamin) in addition to the myosin (Fig. 3). What is even more significant is that the two MHCs are present in quite different proportions, a basis for suggesting that two homodimer species should have been present - yet they are not separating native myosins on the basis of a charge difference. Why this does not happen is not clear. It is possible, albeit unlikely, that in spite of a M.W. difference they just by chance have very little difference in charge, depending on which amino acids were substituted. A second possibility is that the presence of filamin co-migrating with the myosins modifies the charge or interacts with the myosins, perhaps by changing the conformation of the myosin between the 10S and 6S form. Clearly, the solution is to first eliminate the filamin which is present in abundance in the smooth muscle Guba-Straub extracts that are applied to the pyrophosphate gels.

FIGURE 6

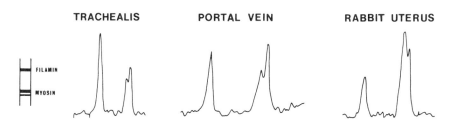

Fig. 6. Densitometric scans of 4% acrylamide SDS gels illustrating the filamin and myosin content of Guba-Straub extracts of native myosin. Origin is at the left. The first peak is filamin and the second peak shows the heavy chains of myosin.

Fig. 6 shows the content of filamin and myosin in Guba-Straub extracts before application to the gels. It seems that the Guba-Straub solution used for extracting myosin is more effective at extracting the filamin so that the filamin/myosin ratio is high in tracheal smooth muscle, is about equal in portal vein extract and low in rabbit myometrium. The filamin/myosin ratio is lower in the SDS extract of the total protein obtained by homogenizing the tissue in SDS (Fig. 7).

A consistent feature seen with Guba-Straub extracts is that MHCs are better resolved on 4% acrylamide SDS gels than with the SDS extract of tissue.

FIGURE 7

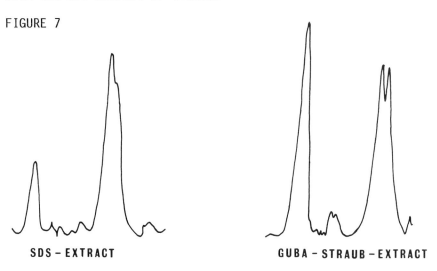

SDS – EXTRACT   GUBA – STRAUB – EXTRACT

Fig. 7. Filamin and myosin content in an SDS extract of rat bladder smooth muscle compared with a Guba-Straub extract of native myosin. The extracts were run on 4% acrylamide SDS gel and scanned. Origin is at the left. The first peak is filamin and the second peak comprises the myosin heavy chains.

Besides the problem encountered with the presence of filamin in the myosin bands, phosphorylation of the myosin also results in the appearance of multiple bands, as Persechini et al. (1986) have reported: a non-phosphorylated, a mono-phosphorylated, and a di-phosphorylated band, in order of increasing mobility. We have similarly observed that an additional faster migrating band appears after phosphorylating myosin from chicken gizzard and from rabbit myometrium. We cannot exclude the possibility that the faster migrating bands seen in some of the tissues listed in Table 1 may arise from different levels of phosphorylation.

If pyrophosphate gels are continued to be used for the purpose of separating native myosin, then unphosphorylated myosin which is free of filamin should be used. In conclusion, the only evidence currently available to indicate that isoforms of myosin exist in smooth muscle is the variable proportion of the two different heavy chains which occur in smooth muscle tissues. This leads us to propose a scheme of two native myosin molecules comprised of homodimer heavy chains (Fig. 1b). Whether one smooth muscle cell expresses one homodimer and another cell expresses the other or that they are co-expressed by one cell must await immunocytochemical verification. But even this model would be inadequate to explain the situation where three MHCs occur, such as in guinea pig myometrium (Fig. 2) or in human smooth muscle tissues where De Marzo et al. (1986) have presented immunochemical evidence for a third heavy chain. The proportions of these heavy chains varied among tissues. Whether a third band of lower M.W. may arise due to splitting of part of the head or rod region (as readily occurs with non-muscle myosin) remains a possibility which can only be tested when antibodies selective for a variety of sites on the head and rod section become available. Nonetheless, the evidence favouring a minimum of two heavy chains is strengthening. Rovner et al. (1986) have shown that both MHCs gave positive immunoblots with a smooth muscle myosin antibody. The possibility of a band or bands arising from non-muscle myosin cannot be excluded in some tissues. Immunocytochemical evidence for the presence of smooth muscle and non-muscle myosin in the media of arteries has been reported (Larson et al., 1984).

The possibility of isoforms arising from molecular variation in the light chains requires greater scrutiny. Cavaillé (1986) has reported that the 17 kDa light chain, which exhibits two forms on isoelectric focusing, undergoes a shift in proportions during pregnancy in human myometrium. From a functional standpoint, the range of shortening velocities in smooth muscles varies over a 40 fold range (0.017 to 0.74 L_o/sec for V_{max}, Hellstrand and Paul, 1983) yet it does not seem possible to utilize this information to select appropriate tissues for characterizing myosin isoforms until a valid and reliable method is available.

REFERENCES

Beckers-Bleukx G, Maréchal G (1985). Detection and

distribution of myosin isozymes in vertebrate smooth muscle. Eur J Biochem 152:207-211

Cavaillé F (1986). Study of myosin from gravid and non gravid human, monkey and rat uteri. J Muscle Res Cell Mot, in press

De Marzo N, Fabbri L, Schiaffino S (1986). Proceeding of IUPS Satellite Symposium on "Smooth Muscle Contraction", Minaki, Canada

Hellstrand P, Paul RJ (1982). Vascular smooth muscle: relations between energy metabolism and mechanics. In: Vascular smooth muscle: metabolic, ionic, and contractile mechanisms. Academic Press, Inc

Hoh JFY, McGrath PA, Hale PT (1978). Electrophoretic analysis of multiple forms of rat cardiac myosin: Effects of hypophysectomy and thyroxine replacement. J Mol Cell Cardiol 10:1053-1076

Hoh JFY, Yeoh GPS (1979). Rabbit skeletal myosin isoenzymes from fetal, fast-twitch and slow-twitch muscles. Nature (London) 280:321-323

Larson DM, Fujiwara K, Alexander RW, Gimbrone MA jr (1984). Heterogeneity of myosin antigenic expression in vascular smooth muscle in vivo. Lab Invest 50:401-407

Morano I, Gagelmann M, Sparrow M, Arner A, Rüegg JC (1986a). Smooth muscle myosin of the rat: a native gelelectrophoretic study. Pflügers Arch 406:R36

Morano I, Gagelmann M, Arner A, Ganten U, Rüegg JC (1986b). Myosin isoenzymes of vascular smooth and cardiac muscle in the spontaneously hypertensive and normotensive male and female rat: a comparative study. Circ Res, in press

Pagani F, Faris R, Shemin R, Julian FJ (1985). Evidence for smooth muscle myosin isozymes. Biophys J 47:301a

Persechini A, Kamm KE, Stull JT (1986). Different phosphorylated forms of myosin in contracting tracheal smooth muscle. J Biol Chem 216:6293-6299

Rovner AS, Thompson MM, Murphy RA (1986). Two different heavy chains are found in smooth muscle myosin. Am J Physiol 250 (Cell Physiol 19):C861-C870

Schildmeyer LA, Seidel CL (1986). Differences in native myosins from rat smooth muscles. Biophys J 49:70a

Sparrow MP, Morano I, Gagelmann M, Rüegg JC (1986). "Isoforms of myosin" in uterine smooth muscle comprise both filamin and myosin. J Muscle Res Cell Mot, in press

Takano-Ohmuro H, Obinata T, Mikawa T, Masaki T (1983). Changes in myosin isozymes during development of chicken gizzard muscle. J Biochem 93:903-908

NOTE ADDED IN PROOF:

Identification of the heavy chain shown in Fig. 2 as myosin was accomplished by immunoblotting myosin antibodies to smooth muscle myosin and to non-muscle myosin. In guinea pig myometrium, the third band at 200 kD was not a myosin heavy chain.

SUBUNIT EXCHANGE BETWEEN SMOOTH MUSCLE MYOSIN FILAMENTS

Kathleen M. Trybus & Susan Lowey

Rosenstiel Basic Medical Sciences
Research Center, Brandeis University
Waltham, Massachusetts   02254

ABSTRACT

Myosin filaments are in equilibrium with a "critical concentration" of monomer (Josephs & Harrington, 1966). Our recent studies with smooth muscle myosin minifilaments and larger synthetic filaments suggest that this monomer pool undergoes more extensive exchange with the polymer than would occur if only polymer ends were involved. This observation provides a possible explanation for the behavior of copolymers of dephosphorylated and phosphorylated myosin in the presence of nucleotide. Upon addition of 1 mM MgATP, essentially all of the dephosphorylated myosin disassembled to the folded monomeric conformation, despite the presence of the more stable phosphorylated molecules in the same polymer. If molecules exchange freely in and out of the filament, phosphorylated myosin would not be expected to exert a significant stabilizing influence on neighboring dephosphorylated molecules.

INTRODUCTION

There has been considerable interest over the past decade regarding the question of how a monomer pool of subunits interacts with the polymer that it forms. Among the most well-studied polymers are actin and microtubules, where the various mechanisms proposed include end exchange of subunits, directional exchange or treadmilling of monomers (Wegner, 1976), and the most recently formulated dynamic instability model of Mitchison & Kirschner (1984), which suggests that individual microtubules can grow while others shrink. The assembly of myosin into filaments differs from the polymerization of actin and tubulin in that nucleotide binding is not required for assembly. Nevertheless, it does appear that myosin from the monomer pool can undergo a quite rapid and extensive exchange with subunits along the length of both smooth and skeletal (Saad et al., 1986) myosin filaments. The idea of subunit exchange in smooth muscle myosin resulted from experiments designed to investigate the properties of copolymers of dephosphorylated and phosphorylated myosin.

RESULTS

It has been known for some time that the stability of synthetic smooth muscle myosin filaments in the presence of MgATP depends on the state of phosphorylation of the regulatory light chain (Suzuki et al., 1978). This differential stability is retained by "minifilaments,"

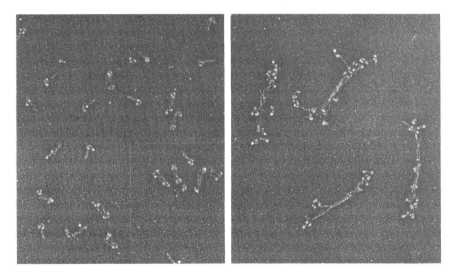

Figure 1. Dephosphorylated turkey gizzard myosin minifilaments are dissociated to a mixture of folded monomers and dimers in the presence of MgATP (left), while phosphorylated minifilaments remain assembled (right). Magnification, X 80,000.

small bipolar aggregates composed of 10-14 molecules (5 mM citrate/22 mM tris). As illustrated in Figure 1, addition of 1 mM MgATP disassembles dephosphorylated minifilaments to a 15S species, which is a mixture of folded monomers and antiparallel folded dimers, while phosphorylated minifilaments remain assembled. Although this property of homopolymers is well established, the stability of copolymers containing both phosphorylated and dephosphorylated myosin is not known. Can dephosphorylated myosin be stabilized in the filamentous form by phosphorylated myosin when nucleotide is present?

To answer this question, varying ratios of phosphorylated and dephosphorylated myosin were mixed in high salt, and dialyzed into citrate/tris for minifilament formation. Upon addition of 1 mM MgATP to these copolymers, both the folded and the filamentous form of myosin were observed by sedimentation velocity. Centrifugation could not be used to quantitate the relative amounts of the 22S minifilament and the 15S folded myosin because of the similarity of their sedimentation coefficients. Since the shapes of the two species are so

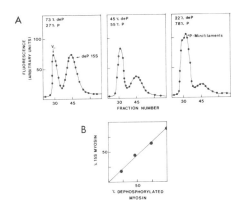

Figure 2. (A) Copolymers were formed from mixtures containing approximately 25, 50 or 75% phosphorylated myosin. One millimolar MgATP was added, and minifilaments (void volume, first peak) were separated from the 15S species by gel filtration on Sepharose 4B. (B) The percent disassembled myosin is plotted as a function of the percent dephosphorylated myosin in the copolymer.

different, however, gel filtration could be used to separate filamentous myosin from folded myosin (Figure 2). The data show that essentially all the dephosphorylated myosin in the copolymer disassembled to the 15S folded form, even when the copolymer contained more than 50% phosphorylated myosin. A possible explanation for this observation was that the two myosins did not copolymerize. Since the same minifilament is formed in the absence of MgATP regardless of the state of light chain phosphorylation, there is no way to demonstrate copolymer formation.

The ability of myosin species to copolymerize was therefore investigated by using myosin and rod, the proteolytic fragment of myosin. Rod forms a minifilament type of structure in 5 mM citrate/22 mM tris that, like phosphorylated myosin, remains assembled in the presence of MgATP. The sedimentation coefficient of rod minifilaments ($s°_{20,w}$=13S), however, is approximately half that of myosin minifilaments ($s°_{20,w}$=22S), consistent with the rod having approximately half the molecular weight of whole myosin. Copolymer formation between rod and myosin can therefore be detected by a change in sedimentation rate from that of the homopolymers, as shown in Figure 3 (A-C). A single boundary of intermediate mobility (14S) was obtained when an equimolar mixture of rod and myosin was dialyzed from high salt into citrate/tris. When MgATP was added to this known copolymer, dephosphorylated myosin depolymerized as usual to its favored folded conformation, as evidenced by gel filtration. Thus even the more stable rod molecules do not cause dephosphorylated myosin to remain filamentous in the presence of nucleotide.

A possible explanation for this lack of stabilization was provided by the following experiments. When rod minifilaments and myosin minifilaments were sedimented in the same cell in the analytical ultracentrifuge, only a single boundary, and not the expected two boundaries were observed. The single boundary (Figure 3D) sedimented at a rate similar to that observed when copolymers were formed from monomeric mixtures in high salt (Figure 3C). One way two populations of homopolymers can form copolymers is for molecules to dissociate from the filament, enter the monomer pool, and then randomly reassociate with either homopolymer until all filaments contain both types of molecules. A surprising feature of the single boundary

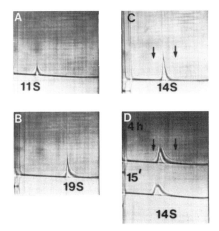

Figure 3. Sedimentation patterns of (A) rod minifilaments, (B) myosin minifilaments, and (C) the copolymer formed from an equimolar mixture of rod and myosin. (D) The patterns obtained 4 hours (upper) or 15 min (lower) after mixing rod minifilaments with myosin minifilaments.

was that it formed within 15 min of mixing the homopolymers.

The time course of copolymer formation was followed more precisely by a pelleting assay. Rod minifilaments were incubated with myosin filaments for 2-40 minutes; actin was then added and the mixture pelleted in the airfuge. Myosin binds to actin and is pelleted regardless of the time of incubation, but rod will be found in the pellet only if it has been incorporated into a myosin containing filament. Figure 4A shows that approximately half of the rod molecules were recovered in the pellet after a two min incubation. After 40 min essentially no rod remained in the supernatant. This assay confirmed the observation that copolymers are rapidly formed simply by mixing populations of filaments. Metal-shadowed images of the polymers obtained after a 15 min incubation of rod and myosin minifilaments also showed the presence of both rod and myosin in the same filament (Figure 4B).

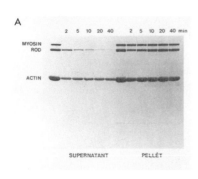
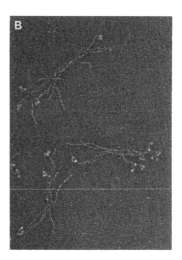

Figure 4. (A) The rate of copolymer formation was determined by incubating rod and myosin minifilaments for the indicated length of time, pelleting the mixture with actin, and following the appearance of rod in the pelleted fraction. (B) Filaments obtained after a 15 min incubation of rod and myosin minifilaments. Magnification, X 80,000.

In order to determine if rapid exchange of molecules also occurs in larger synthetic filaments, minifilaments were dialyzed into citrate/tris that also contained 0.1 M KCl and 7 mM $MgCl_2$, conditions which cause the filaments to grow in both length and width. Because these larger synthetic filaments sediment at a much faster rate than the folded conformation, the polymer could be pelleted without added actin, and the amount of myosin remaining in the supernatant quantitated. In the absence of MgATP, less than 5% of the myosin was recovered in the supernatant (Figure 5, open circles). Addition of 1 mM MgATP to copolymers containing the indicated fraction of dephosphorylated myosin solubilized most of the dephosphorylated myosin (Figure 5, solid circles). The deviation of the points from the dashed line indicates that a small amount of dephosphorylated myosin was stabilized in the filamentous form. Nevertheless, the majority of the dephosphorylated myosin disassembled even when present in a copolymer containing an excess of phosphorylated myosin, in agreement with the results obtained with the smaller minifilaments.

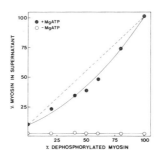

Figure 5. The amount of disassembled myosin in the presence (filled circles) or the absence (open circles) of 1 mM MgATP is plotted as a function of the percent dephosphorylated myosin in the copolymer. The synthetic filaments used in this experiment were formed in 5 mM citrate/16 mM tris, 0.1 M KCl, 7 mM $MgCl_2$.

These results suggested that if exchange of molecules is the mechanism whereby the dephosphorylated molecules initially dissociate from the filament, then this process must occur along the length of the filament. If molecules exchanged with the monomer pool primarily from filament ends, only a small percentage of the dephosphorylated myosin would be recovered in the supernatant, assuming a random distribution of phosphorylated and dephosphorylated myosin throughout the filament.

Electron microscopy was another technique used to show that molecules can be incorporated along the filament length. The first approach was based on the exchange of skeletal myosin into smooth muscle myosin filaments. The skeletal myosin was detected by incubating the filaments with a primary antibody specific for skeletal muscle myosin, followed by a secondary antibody (goat anti-rabbit) that was conjugated to gold. Figure 6A shows that the smooth muscle myosin filaments are not labeled with gold, whereas skeletal myosin filaments are labeled with gold along their entire length (Figure 6B).

When the critical concentration of monomer in equilibrium
with the skeletal muscle myosin filaments was incubated
with the smooth muscle myosin filaments, gold particles
appeared along the length of the smooth muscle myosin
filament, indicating that skeletal muscle myosin had been
incorporated (Figure 6C).

Another approach was to follow the exchange of
biotinylated smooth muscle myosin into unmodified smooth
muscle myosin filaments. Biotinylated myosin was detected
by the binding of streptavidin-gold. Figure 7 (A, B)
shows that the control smooth muscle myosin filaments do
not bind gold, while the filaments formed from
biotinylated myosin, although considerably smaller, are
heavily labeled with gold. After a two min incubation of
control and biotinylated filaments, gold was detected on
most filaments. After a 60 min incubation, the filaments
no longer resembled the control or the biotinylated
filaments, in itself evidence of exchange. Moreover, by
this time, all filaments in the field contain biotinylated
myosin, as indicated by the bound gold particles (Figure
7C, D).

DISCUSSION

The stability of copolymers composed of
dephosphorylated and phosphorylated smooth muscle myosin
was examined in the presence of MgATP. It might have been
expected that the degree to which dephosphorylated myosin
disassembled would depend on the amount of phosphorylated
myosin in the copolymer. Surprisingly, essentially all of
the dephosphorylated myosin was able to disassemble to the
folded conformation upon nucleotide addition, even when
phosphorylated myosin was present in excess. This result
held true for both minifilaments and larger synthetic
filaments.

One explanation for the lack of stabilization in the
copolymer, which cannot be excluded at present, is that
the entire copolymer dissolves upon MgATP addition,
followed by reassembly of the phosphorylated molecules.
Another plausible explanation, provided by the experiments
described here, is that molecules along the length of the
filament can rapidly exchange with the monomer pool of
myosin. If this is the case, neighboring molecules would

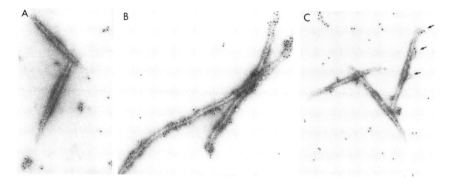

Figure 6. (A) Smooth and (B) skeletal muscle myosin filaments reacted with antibody specific for skeletal myosin, followed by goat anti-rabbit IgG conjugated to gold. (C) Incorporation of skeletal myosin into smooth muscle myosin filaments, as detected by the binding of gold. Magnification, X 35,000.

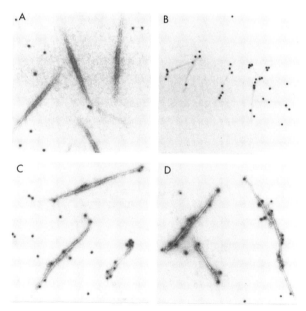

Figure 7. (A) Control and (B) biotinylated filaments reacted with streptavidin gold. (C) The two populations of filaments were then incubated for two min and (D) 60 min. Magnification, X 32,000.

not be expected to exert a significant stabilizing influence on each other. Moreover, phosphorylated myosin could leave and reenter the polymer in the presence of MgATP, but dephosphorylated myosin, once dissociated from the filament, would assume the folded conformation and remain monomeric. This process would lead to the observed disassembly of the dephosphorylated myosin in MgATP.

The exchange of molecules between monomer pool and filament is not unique to smooth muscle myosin. Saad et al. (1986) showed a similar phenomenon in skeletal muscle myosin filaments. They suggested that exchange could be a mechanism to account for the turnover of myosin in the myofibril, although exchange <u>in vivo</u> has yet to be established. Other contractile proteins, such as alpha-actinin, have been shown to be incorporated into cellular structures when injected into living cells (McKenna et al., 1985). Thus structures which have been thought of as relatively static could turn out to be undergoing constant exchange, as appears to be the case with actin filaments and microtubules.

REFERENCES

Josephs R, Harrington W (1966). Studies on the formation and physical chemical properties of synthetic myosin filaments. Biochemistry 5:3474-3487.
McKenna NM, Meigs JB, Wang Y (1985). Exchangeability of alpha-actinin in living cardiac fibroblasts and muscle cells. J Cell Biology 101:2223-2232.
Mitchison T, Kirschner M (1984). Dynamic instability of microtubule growth. Nature 312:237-242.
Saad AD, Fischman DA, Pardee JD (1986). Fluorescence energy transfer studies of myosin thick filament assembly. Biophys J 49:140-142.
Suzuki H, Onishi H, Takahashi K, Watanabe S (1978). Structure and function of chicken gizzard myosin. J Biochem (Tokyo) 84:1529-1542.
Wegner A (1976). Head to tail polymerization of actin. J Mol Biol 108:139-150.

(Supported by NIH grant R23 HL34791 to KT and grant R01 AM17350 to SL.)

CONFORMATIONAL CHANGES IN MYOSIN AND HEAVY MEROMYOSIN FROM CHICKEN GIZZARD ASSOCIATED WITH PHOSPHORYLATION.

Sumitra Nag, Hiroshi Suzuki, Jan Sosinski and John C. Seidel.

Department of Muscle Research, Boston Biomedical Research Institute, Boston, MA 02114.

INTRODUCTION

Smooth muscle myosin undergoes a conformational transition induced by addition of ATP in the presence of Mg^{2+}, that is accompanied by an increase in the sedimentation coefficient from 6S to 10S (Suzuki, et al., 1978, 1982; Onishi et al., 1983; Ikebe et al., 1983). The tail of the molecule folds at sites approximately 50 and 100 nm from the head-tail junction (Trybus et al., 1982; Onishi and Wakabayashi, 1982) and the heads, which extend away from the tail in the 6S form, reorient and project back toward the tail in the 10S state (Onishi and Wakabayashi, 1982). At ionic strengths below 0.3 M, ATP increases the concentration of myosin in the 10S conformation and phosphorylation of the 20 kDa light chain favors the 6S form, while at higher ionic strength myosin remains in the 6S form regardless of the state of phosphorylation or the presence of ATP (Suzuki et al., 1978, 1982; Trybus et al., 1982).

The head-tail junction has been implicated in myosin-linked regulation of actomyosin ATPase activity in molluscan muscle, by the observations that the activities of HMM and single-headed myosin are regulated by Ca^{2+}, while the activity of S1 is not (Szent-Gyorgyi et al., 1973; Stafford et al., 1979). On the basis of crosslinking studies of scallop myosin light chains, Hardwicke et al. (1983) have proposed a model in which the N-terminal region of the regulatory light chain overlaps the S2 region and movement of the light chains is associated with regulation.

Functional similarities between myosins from molluscan muscle and vertebrate smooth muscle, including cooperative interactions between myosin heads (Chantler et al., 1980; Persechini and Hartshorne, 1981; Sellers et al., 1983) and the loss of regulation on formation of S1 (Szent-Gyorgyi et al., 1973; Seidel, 1978) suggest that similar molecular changes might be involved in regulation of actomyosin

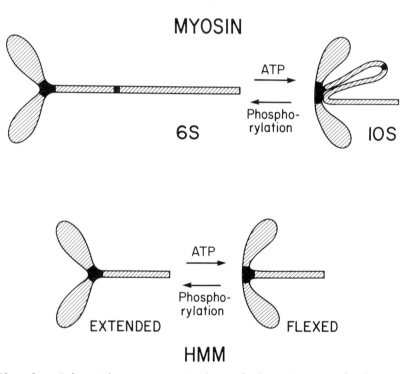

Fig. 1. Schematic representation of the structural changes in myosin and heavy meromyosin related to the changes in sedimentation velocity induced by MgATP and by phosphorylation of the 20 kDa light chain. Myosin exists in folded and extended forms having sedimentation velocities of 10S or 6S, respectively, phosphorylation favoring the 6S form and ATP favoring the 10S. Two changes in the molecular structure are seen on addition of ATP to myosin, the tail bends apparently interacting with the heads (Onishi and Wakabayashi 1982; Trybus et al., 1982) and the heads reorient so that they point back towards the tail (Onishi and Wakabayashi, 1982). HMM undergoes similar changes in structure and sedimentation velocity,

sedimenting as a single peak whose sedimentation velocity is determined by the fraction of molecules in a flexed form sedimenting at 9S and an extended form sedimenting at 7.5S. The 9S form is flexed at the head-tail junction but no folding of the shortened tail and no change in the shape of the heads is observed (Suzuki et al., 1985). The fraction of HMM in the 9S form is increased by ATP and the fraction in the 7.5S form is increased at high ionic strengths or upon phosphorylation at low ionic strength.

ATPase activity in vertebrate smooth muscle. An important role for the head-tail junction in smooth muscle is suggested by the effects of phosphorylation on proteolytic digestion of myosin and HMM at this site (Kumon et al., 1984; Onishi and Wakabayshi, 1984; Ikebe and Hartshorne, 1984), and by the electron microscopic appearance of HMM in the 7.5 and 9S conformations that suggests flexing at the head-tail junction is associated with the conformational transition (Suzuki et al., 1985).

Proteolytic Digestion of Smooth Muscle Myosin and HMM.

The resistance of unphosphorylated myosin to degradation by papain seen in the presence of ATP has been attributed to conversion of the myosin from the 6S form to the papain-resistant 10S conformation (Onishi and Watanabe, 1984) and at slightly lower ionic strengths to dissociation of papain-sensitive myosin filaments forming resistant 10S monomers (Kumon et al., 1984). The observation that MgATP does not inhibit digestion of phosphorylated myosin was attributed to conversion of myosin by phosphorylation to the papain-sensitive 6S form (Kumon et al., 1984; Onishi and Watanabe, 1984; Ikebe and Hartshorne, 1984). Raising the ionic strength also increases digestibility by papain, converting myosin to the 6S form. The proteolytic digestion of HMM by papain follows a pattern similar to that seen with myosin. The degradation rate of HMM, is also decreased by MgATP and the decrease is partially reversed by phosphorylation or by increasing the ionic strength (Ikebe and Hartshorne, 1984; 1985; Suzuki et al., 1986).

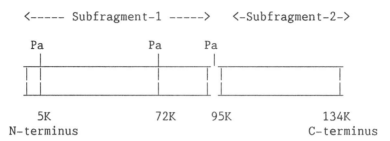

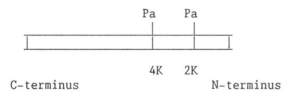

Fig. 2 Location of the major papain cleavage sites of the heavy and light chains of myosin and HMM. The cleavage of either myosin or HMM occurs initially at the same three sites 5, 72 and 95 kDa from the N-terminus (Nath et al., 1982; Kumon et al., 1984; Ikebe and Hartshorne, 1984; Onishi and Watanabe, 1984; Suzuki et al., 1986). Chymotrypsin cleaves at the 5 and 72 kDa sites, but not at the 95 kDa site (Seidel, 1978, 1980; Okamoto et al., 1980). ATP alters the rate of proteolysis at each of the three sites, enhancing cleavage at the 5 kDa site (Okamoto et al., 1980; Okamoto and Sekine, 1981; Onishi and Watanabe, 1984) and inhibiting cleavage at the 72 and 95 kDa sites (Kumon et al., 1984; Ikebe and Hartshorne, 1984; Onishi and Watanabe, 1984; Suzuki et al., 1986). Phosphorylation of the 20 kDa light chain reverses or partially reverses the effects of MgATP at the 5 kDa and 95 kDa sites (Kumon et al., 1984; Ikebe and Hartshorne, 1984; Onishi and Watanabe, 1984; Suzuki et al., 1986). The 20 kDa light chain of myosin or HMM is cleaved by papain or chymotrypsin at sites 2 and 4 kDa from the N-terminus (Sellers et al. 1981; Ikebe and Hartshorne, 1984). Cleavage occurs at the 4 kDa site with dephosphorylated HMM and at the 2 kDa site with

phosphorylated HMM.

The relationship of the digestibility of HMM to the sedimentation coefficient ($s_{20,w}$) was investigated by determining the degradation rates of the heavy and light chains with increasing concentrations of NaCl. Papain cleaves the heavy chain at the head-tail junction, 95 kDa from the N-terminus of the subfragment-1 region of myosin (Fig. 2), thus the results were expected to provide further insight into the molecular changes associated with flexing at the head-tail junction of the HMM and myosin molecules. At low ionic strength where the 9S form of HMM predominates (Suzuki et al., 1985), papain degrades the heavy chain rapidly in the absence of ATP. Addition of ATP reduces the overall cleavage rate of dephosphorylated HMM by an order of magnitude and that of phosphorylated HMM by a factor of 3 (Table 1). Phosphorylation increases the degradation rate in the presence and decreases it in the absence of ATP. At 0.4 M NaCl, HMM is almost entirely in the 7.5S form and the degradation rates are not influenced by ATP or by phosphorylation. These results indicate the presence of three forms of HMM differing in susceptibility to papain.

Table 1

Rate of Proteolysis of Heavy Chains of Gizzard HMM with Papain

HMM	ATP	NaCl, mM	
		23	500
	$k \times 10^3$, min^{-1}		
dephosphorylated	–	124	50
phosphorylated	–	81	49
phosphorylated	+	28	42
dephosphorylated	+	10	45

The degradation rates were determined by digestion of HMM,

0.81 mg/ml, at 20°C in 2 mM $MgCl_2$, 20 mM Tris-HCl, pH 7.5, 1 to 5 ug/ml papain with or without 1 mM ATP. Digested HMM was subjected to polyacrylamide gel electrophoresis in sodium dodecyl sulfate. The stained gels were scanned with a soft laser scanning densitometer emitting at 630 nm and the total area under the peaks of the intact heavy chain (134 kDa and 129 kDa peptides) were determined by integration. Pseudo first order rate constants were estimated from plots of log of the area under the scanning peak against time of incubation with papain (Ueno and Harrington, 1984). The rates represent the overall rate of degradation at the sites 72 and 95 kDa from the N-terminus but are not significantly influenced by cleavage at the 5 kDa site.

When the ionic strength is varied, sedimentation velocity and ATPase activity of HMM measured in the absence of actin change in parallel (Suzuki et al., 1985), suggesting that changes in ATPase activity reflect changes in the fraction of HMM in the rapidly and slowly sedimenting forms. Parallel changes in $s_{20,w}$ and the papain degradation rates of the HMM heavy chain (Suzuki, et al., 1986) can be interpreted in the same way. The observation that values of the heavy chain degradation rates for phosphorylated and dephosphorylated HMM fall on the same straight line when plotted against $s_{20,w}$ suggest that phosphorylation alters the degradation rates by changing the fractions of HMM in the rapidly and slowly sedimenting forms.

During ATP hydrolysis catalyzed by smooth muscle S1, HMM, or the 6S and 10S forms of myosin, the predominant intermediate is a complex of the enzyme with the products of hydrolysis (Takeuchi and Tonomura, 1980; Martson and Taylor, 1980; Onishi, 1982; Sellers, 1985). Direct measurements of the dissociation of ADP and P_i from HMM suggest that the product-complex contains both ADP and P_i (Sellers, 1985). The linear relationship between the degradation rate of the heavy chain and $s_{20,w}$ observed in the presence of ATP suggests that the fraction of HMM in the rapidly and slowly sedimenting product-complexes determines the proteolytic cleavage rate during ATP hydrolysis (Suzuki et al., 1986). The fraction of HMM present as the 9S-product-complex is undoubtedly influenced by the rate constants associated with ATP hydrolysis as well as those associated with the conformational transition between the flexed and extended forms, though little information is available on rate constants for these two forms of HMM. On the other hand, the effects of phosphorylation can be attributed largely to shifts in conformational transition, since phosphorylation produces parallel changes in $s_{20,w}$, ATPase activity and proteolytic degradation rates, which can be attributed to changes in the orientation of HMM heads (Suzuki et al., 1985). Although we refer to the two rapidly sedimenting forms seen in the presence and absence of ATP as 9S forms and both have sedimentation coefficients greater than 8.5 (Suzuki et al., 1985), there is no direct evidence that they have the same sedimentation velocities. The simplest interpretation of the proteolytic degradation rates and the structural evidence obtained by electron microscopy is that both forms are flexed and have essentially the same sedimentation velocities.

$$\begin{array}{ccc} M_{7S}^{*} + ATP & \longrightarrow & M_{7S}^{*}\text{-Pr} \longrightarrow \\ \big\Updownarrow K & & \big\Updownarrow K' \\ M_{9S}^{**} + ATP & \longrightarrow & M_{9S}\text{-Pr} \longrightarrow \end{array}$$

Fig. 3. The relationship of the proteolytic susceptibility of HMM to the enzymatic hydrolysis of ATP and to the transitions between rapidly and slowly sedimenting conformational forms. The relative degradation rates increase in the order $M_{9S}\text{-Pr} < M_{7S}^{*}\text{-Pr} = M_{7S}^{*} < M_{9S}^{**}$. Hydrolysis of ATP by the 9S form is accompanied by a decrease in the rate of proteolysis attributable to formation of the HMM-ADP-P_i complex, while no change in digestibility is seen when HMM is in the 7.5S form. Vertical transitions reflect changes sedimentation velocity that in turn involve changes in orientation or distribution of HMM heads relative to the tail (Suzuki et al., 1985).

As indicated above, ATP and phosphorylation affect proteolysis of the heavy chain only at low ionic strengths when HMM is in the flexed form; the effects of ATP and those of phosphorylation decrease in magnitude with increasing ionic strength and disappear as the sedimentation coefficient approaches 7.5S. This result suggests that the mechanism of ATP hydrolysis by the rapidly and slowly sedimenting forms differs with respect to the conformational transitions occurring at the head-tail junction, viz. the 9S nucleotide-free HMM differs in susceptibility to papain from the ADP-Pi complex, while no difference is observed in the corresponding 7.5S forms of HMM.

Degradation of the 20 kDa Light Chain by Papain

Proteolytic cleavage of 20 kDa light chain of dephosphorylated HMM occurs at a site 4 kDa from the N-terminus. Two changes occur on phosphorylation; the rate of cleavage at the 4 kDa site is strongly suppressed and the site of cleavage shifts to a site 2 kDa from the N-terminus (Sellers et al., 1981; Ikebe and Hartshorne, 1984). Since this shift occurs not only with HMM but also with S1, whose sedimentation rate and ATPase activity are not changed by phosphorylation, it appears to be a direct effect of phosphorylation on the structure of the light chain (Suzuki et al., 1985, 1986) and not related to the 7.5S-9S transition (see also Ikebe and Hartshorne, 1984). On digestion of dephosphorylated HMM the degradation rate of the 20 kDa light chain at the 4 kDa site is dependent on ionic strength and ATP and it correlates closely with $s_{20,w}$. The dependence of light chain degradation on sedimentation velocity suggests that the structure of the light chain at the 4 kDa site is substantially altered by interactions with the heavy chain, probably at the head-tail junction. Phosphorylation of HMM results in a loss of the dependence of light chain cleavage on $s_{20,w}$ that reflects at least in part the shift in the cleavage site from the 4 kDa to the 2 kDa site.

Relationship of Flexing of HMM to Structure and Function

The heads of myosin or HMM possess rotational motion that is independent of the molecule as a whole and thought in part to reflect flexibility at the head-tail junction (Mendelson et al., 1973; Thomas et al., 1975), perhaps allowing the detached crossbridge to find a binding site on actin and to change its orientation once attached. Therefore, HMM is not likely to have a completely static structure with the heads assuming discrete angles relative to the tail. The presence of such motion in gizzard HMM is supported by the variety of structures seen in the electron microscope, including molecules having two heads forward, two heads back, or one forward and the other back (Suzuki et al., 1985). Nevertheless, there must be some restrictions superimposed on the motion of the heads of smooth muscle HMM that may not occur with HMM from skeletal muscle.

The flexing of HMM at the head-tail junction appears to be responsible at least in part for the changes in $s_{20,w}$ based on a correlation between the fraction of flexed HMM observed by electron microscopy and $s_{20,w}$ determined under various ionic conditions, and by computer modeling which shows that rotation of the heads by 90° from a flexed to an extended position predicts the observed changes in sedimentation coefficient (Suzuki et al., 1985). Comparision of myosin, HMM and S1 in terms of the structural changes and regulation of ATPase activity by phosphorylation provides some insight into possible regulatory roles for the conformational changes that occur at the head-tail junction. The observation that the actin-activated ATPase activities of both myosin and HMM are regulated by phosphorylation (Sellers et al., 1981), together with the flexing of both molecules at the head-tail junction, points to the redistribution of heads that accompanies this flexing as being potentially important in regulation of activity. The observation that the tail of HMM does not fold (Suzuki et al., 1985) indicates that interactions between the LMM region and the head that might occur in the myosin filament, cannot account by themselves for regulation of activity. The observation that subfragment-1, which undergoes no changes in conformation or sedimentation velocity (Suzuki et al., 1985), has actin-activated ATPase activities that is not regulated by phosphorylation (Seidel, 1978) is consistent with the view that changes in structure or flexibility at the head-tail junction play an important role in regulation of activity.

Location of the Reactive Thiols of Smooth Myosin Myosin.

The enhancement of Mg^{2+}-activated ATPase activity of smooth muscle myosin by thiol directed reagents is very similar to that produced by enzymatic phosphorylation of the 20 kDa light chain (Seidel, 1979). Thiol reagents also inhibit the formation of the folded, 10S conformation (Chandra et al., 1985) and stabilize the myosin filaments (Chandra et al., 1985; Onishi, 1985), effects that are also produced by light chain phosphorylation (Suzuki et al., 1978, 1982). The pattern of incorporation of thiol reagents into gizzard myosin indicates the presence of at least four reactive sites, SH-A through SH-D: in the myosin rod, the 17 kDa light chain, and the 23 and 72 kDa papain fragments of the S1 heavy chain (Okamoto and Sekine, 1980; Bailin and Lopez, 1981; Chandra et al., 1985; Onishi, 1985; Nath et al., 1986).

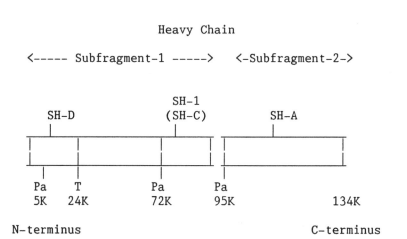

Fig. 4 Location of the reactive thiols in the myosin heavy chain. SH-A, the most reactive thiol of the heavy chain with MalNEt is in subfragment-2, but its position in the amino acid sequence is not known (Chandra et al., 1985; Nath et al., 1986). A thiol of the 17 kDa light chain (Okamoto and Sekine, 1980; Bailin and Lopez, 1981), SH-B, has about equal reactivity toward MalNEt (Chandra et al., 1985). SH-C, located in the C-terminal, 23 kDa papain fragment of S1, reacts slowly with thiol blockers - ATP accelerates its reaction with MalNEt or IAEDANS in the 6S

form and inhibits in the 10S form (Onishi, 1985; Nath et al., 1986). From the amino acid sequence of the C-terminal region of gizzard S1, SH-C corresponds to SH-1 of skeletal muscle myosin[1]. SH-D is located in the 24 kDa, N-terminal peptide of S-1 produced by tryptic digestion (Onishi and Watanabe, 1985). It also reacts slowly with thiol blockers but its reactivity is enhanced by ATP in both the 6S and 10S forms. Pa, papain; T, trypsin.

Stimulation of ATPase Activity of Myosin by MalNEt

The rates of MalNEt-induced changes in Mg^{2+}-activated ATPase activity and the rates of incorporation of ^{14}C-MalNEt into the four reactive thiols of gizzard myosin were used to determine which of the four are involved in the activation of ATPase activity and in the loss of the ability of myosin to assume the 10S form. When myosin is in the 6S form at high ionic strength, ATP increases the rate of incorporation of MalNEt into SH-C and SH-D, by factors of 10 and 3, respectively, but does not affect incorporation into other thiols of the molecule (Nath et al., 1986). The rates of the changes in ATPase activity induced by MalNEt both in the presence and absence of actin agree most closely with the rates of incorporation of MalNEt into SH-C (Nath et al., 1986). ATP also accelerates the increase in actin-activated ATPase activity of unphosphorylated myosin produced by MalNEt. The actin-activated activity of phosphorylated myosin decreases slightly on reaction with MalNEt but not until the activity of unphosphorylated myosin has risen to the value seen with the phosphorylated form.

Conversion of myosin to the 10S form by MgATP enhances incorporation of IAEDANS or MalNEt into SH-D but blocks incorporation into SH-C (Onishi, 1985; Nath et al., 1986). Under these conditions where only SH-D is blocked with MalNEt, the Mg^{2+}-activated ATPase activity of myosin with or without actin is essentially uneffected by either reagent.

Footnote 1. Onishi, H., Maita, T., Miyanishi, T., Matsuda, G., and Watanabe, S. (1986), personal communication.

A similar set of experiments demonstrated that blocking of SH-C with MalNEt prevents myosin from assuming the 10S conformation. After a short incubation of myosin with MalNEt in 0.5 M NaCl with or without ATP, the sedimentation behavior of myosin was studied by analytical ultracentrifugation. Under conditions where unmodified myosin sedimented as a single peak with a sedimentation velocity of 10S, myosin modified with MalNEt in the absence of ATP exhibited both 6S and 10S peaks, while myosin modified with MalNEt in the presence of ATP showed only a single peak sedimenting at 6S (Nath et al., 1986).

The linear relationship of the changes in sedimentation velocity to those in ATPase activity, or to the rates of degradation by papain, are consistent with the view that phosphorylation shifts an equilibrium between rapidly and slowly sedimenting forms of HMM toward the slowly sedimenting form. These relationships and the effects of phosphorylation on $s_{20,w}$ and ATPase activity suggest a possible role for the flexing of HMM at the head-tail junction in regulation of actomyosin ATPase activity. The possibility that the 6S-10S transition may play a role in the regulation of ATPase activity accompanying phosphorylation of other sites in the 20 kDa light chain has been raised by the observations that phosphorylation of a site near the N-terminus of the light chain by protein kinase C decreases actin-activated ATPase activity (Nishikawa et al., 1983) and promotes formation of 10S myosin (Umekawa et al., 1985), while phosphorylation of threonine 18 by myosin light chain kinase increases actin-activated ATPase activity (Ikebe et al., 1986a) and inhibits formation of the 10S conformation (Ikebe et al., 1986b).

The observation that reaction of SH-1 with thiol reagents enhances actomyosin ATPase activity[1] (Onishi, 1985; Nath et al., 1986) and prevents myosin from assuming the 10S conformation (Chandra et al., 1985; Nath et al., 1986) provides a model system, that may be useful in understanding the molecular events associated with activation of actomyosin ATPase activity. In skeletal muscle myosin, SH-1 has been estimated to be 30-40A from the ATP binding site (Tao and Lamkin, 1981; Perkins et al., 1984), suggesting that upon blocking SH-1, the activating signal may be transmitted for 30A along or across the head.

On the other hand, SH-1 and the ATP binding site of the skeletal muscle myosin head are 130 to 140 A from the head-tail junction (Sutoh et al., 1984, 1986). An understanding of the spatial relationships among these sites in smooth muscle myosin and the means by which information is transferred among them, should provide further insight into the role of the conformational transitions in myosin produced on activation.

SUMMARY

Heavy meromyosin (HMM) undergoes a conformational transition between a rapidly and a slowly sedimenting form, during which it sediments as a single peak in the ultracentrifuge with sedimentation coefficients between 7.5 and 9S. Changes in sedimentation velocity and ATPase activity produced by changes in ionic strength, phosphorylation of HMM or addition of MgATP are interpreted in terms of equilibria between the rapidly and slowly sedimenting forms, the observed values of activity and sedimentation velocity being determined by the fraction of HMM in each form. Phosphorylation of the 20 kDa light chain or raising the ionic strength decrease the sedimentation velocity, by decreasing the fraction of HMM in the rapidly sedimenting form, while addition of ATP increases sedimentation velocity upon forming a 9S HMM-ADP-Pi complex. Electron microscopic studies support this interpretation showing the presence of two distinct conformations of HMM - extended and flexed, which correspond to the 7.5S and 9S forms, respectively (Suzuki et al., 1985). In samples prepared at high ionic strengths, the heads extend away from the tail in a more or less random orientation, while at low ionic strength, the molecule is flexed at the head-tail junction assuming a more compact structure, that appears to account for its more rapid sedimentation rate.

The degradation rates of the heavy chain and the 20 kDa light chain of HMM on digestion with papain indicate the presence of three forms of HMM differing in their susceptibility to papain. At 25 mM NaCl, HMM is rapidly digested in the absence of ATP, while addition of ATP decreases digestibility by a factor of ten, upon formation of a complex of HMM with the products of ATP hydrolysis. Above 0.4 M NaCl, HMM is degraded at an intermediate rate that is not affected by ATP. When the ionic strength is varied, the rate of disappearance of the heavy chain depends linearly on the sedimentation velocity in both the phosphorylated and dephosphorylated states, indicating that the rate of proteolysis is determined primarily by the fraction of HMM in the rapidly and slowly sedimenting forms. The same pattern is seen in the disappearance of the 20 kDa light chain of dephosphorylated HMM on cleavage at a site 4 kDa from the N-terminus, indicating that the cleavage of the light chain also depends on the fraction of HMM in the rapidly and slowly sedimenting forms. Phosphorylation of HMM inhibits the rate of the light chain cleavage at the 4 kDa site with loss of the dependence of cleavage on ionic strength and MgATP.

A stimulation of actin-activated ATPase activity similar to that produced by phosphorylation, accompanies reaction of myosin with the thiol reagent, MalNEt. This increase in activity and a parallel loss of the ability of myosin to assume the 10S form, can be attributed to reaction of MalNEt with SH-1, as indicated by parallel changes in ATPase activity and the rate of incorporation of MalNEt into the C-terminal papain fragment of subfragment-1. The interactions between SH-1 and the catalytic site are reciprocal; blocking of SH-1 with MalNEt alters ATPase activity and ATP alters the reactivity of the thiol. Reciprocal interactions between SH-1 and the head-tail junction are suggested by the MalNEt-induced loss of the 10S form and the inability of SH-1 to react with MalNEt in the 10S form.

Abbreviations: HMM, heavy meromyosin; IAEDANS, N-iodoacetyl-N'-(5'-sulfo-1-naphthyl) ethylenediamine; MalNEt, N-ethyl maleimide; S1, subfragment-1; $s_{20,w}$, sedimentation coefficient.

REFERENCES

Bailin, G. and Lopez, F. (1981) Dinitrophenylation of chicken gizzard myosin. Reactivity of the 17,000-dalton light chain. Biochim. Biophys. Acta, 668, 46-56.

Chandra, T.S., Nath, N., Suzuki, H. and Seidel, J.C. (1985) Modification of Thiols of Gizzard Myosin Alters ATPase Activity, Stability of Myosin Filaments and the 6S-10S Conformational Transition. J. Biol. Chem. 260,

202-207.
Chantler, P.D. and Szent-Gyorgyi, A.G. (1980) Regulatory light chains and scallop myosin: full dissociation, reversibility and co-operative effects. J. Mol. Biol. 138, 473-492.
Hardwicke, P.M.D., Wallimann, T. and Szent-Gyorgyi, A.G. (1983) Light chain movement and regulation in scallop myosin. Nature 301, 478-481.
Ikebe, M. Hinkins, S. and Hartshorne, D.J. (1983) Correlation of enzymic properties and conformation of smooth muscle myosin. Biochemistry 22, 4580-4586.
Ikebe, M. and Hartshorne, D.J. (1984) Conformation -dependent proteolysis of smooth muscle myosin. J. Biol. Chem. 259, 11639-11642.
Ikebe, M. and Hartshorne, D.J. (1985) Proteolysis of smooth muscle myosin by staphyloccus aureus protease. Preparation of heavy meromyosin and subfragment 1 with intact 20,000-Dalton light chains. Biochemistry 24, 2380-2386.
Ikebe, M., Hartshorne, D.J. and Elzinga, M. (1986a) Identification, phosphorylation, and dephosphorylation of a second site for myosin light chain kinase on the 20,000-dalton light chain of smooth muscle myosin, J. Biol. Chem. 261, 36-39.
Ikebe, M., Hartshorne, D.J. and Elzinga, M. (1986b) Effects of phosphorylation at a second light chain site on conformation and enzymatic activity. Biophys. J. 49, 390a.
Kumon, A., Yasuda, S., Murakami, N. and Matsumura, S. (1984) Discrimination of assembled and disassembled forms of gizzard myosin by papain. Eur. J. Biochem. 140, 265-271.
Mendelson, R.A., Morales, M.F. and Botts, J. (1973) Segmental flexibility of the S-1 moiety of myosin. Biochemistry 12, 2250-2255.
Marston, S.B. and Taylor, E.W. (1980) Comparison of the myosin and actomyosin ATPase mechanisms of the four types of vertebrate muscles. J. Mol. Biol. 139, 573-600.
Nath, N., Chandra, T.S., Suzuki, H., Carlos, A. and Seidel, J.C. (1982) Digestion of Chicken Gizzard Myosin with Papain, Biophys. J. 37, 47a.
Nath, N., Nag, S. and Seidel, J.C. (1986) Location of the sites of reaction of MalNEt in papain and chymotryptic fragments of the gizzard myosin heavy chain. Biochemistry, in press.
Nishikawa, M., Hidaka, H. and Adelstein, R.S. (1983)

Phosphorylation of Smooth Muscle Heavy Meromyosin by Calcium-activated, Phospholipid-dependent Protein Kinase. Effect on Actin-activated MgATPase Activity. J. Biol. Chem. 258, 14069-14072.

Okamoto, Y., Okamoto, M. and Sekine, T. (1980) Two opposite effects of ATP on the chymotryptic cleavages in smooth muscle myosin head determination of cleavage points and their characterization. J. Biochem. 88, 361-371.

Okamoto, Y. and Sekine, T. (1981) Light chain phosphorylation alters the N-terminal structure of gizzard myosin heavy chain in an ATP dependent manner. J. Biochem. 89, 697-700.

Onishi, H. (1982) Possible role of ATP in dimerization of myosin molecules from chicken gizzard muscle. J. Biochem. 91, 157-166.

Onishi, H. and Wakabayashi, T. (1982) Electron microscopic studies of myosin molecules from chicken gizzard muscle (I). J. Biochem. 92, 871-879.

Onishi, H., Wakabayashi, T., Kamata, T. and Watanabe, S. (1983) Electron microscopic studies on myosin molecules from chicken gizzard muscle (II). J. Biochem. 94, 1147-1154.

Onishi, H. and Watanabe, S. (1984) Correlation between the papain digestibility and the conformation of 10S-myosin from chicken gizzard. J. Biochem. 95, 899-902.

Onishi, H. (1985) N-iodoacetyl-N'-(5'-sulfo-1-naphthyl) ethylenediamine modification of myosin from chicken gizzard. J. Biochem. 98, 81-86.

Perkins, W.J., Weiel, J., Grammer, J. and Yount, R.G. (1984) Introduction of a donor-acceptor pair by a single protein modification. J. Biol. Chem.. 259, 8786-8793.

Persechini, A. and Hartshorne, D.J. (1981) Phosphorylation of smooth muscle myosin: evidence for cooperativity between the myosin heads. Science 213, 1383-1385.

Seidel, J.C. (1978) Chymotryptic Heavy Meromyosin from Gizzard Myosin: A Proteolytic Fragment with the Regulatory Properties of the Intact Myosin. Biochem. Biophys. Res. Commun. 85, 107-113.

Seidel, J.C. (1979) Activation by Actin of ATPase Activity of Chemically Modified Gizzard Myosin Without Phosphorylation. Biochem. Biophys. Res. Commun. 89, 958-964.

Seidel, J.C. (1980) Fragmentation of Gizzard Myosin by

alpha Chymotrypsin and Papain and the Effects on ATPase Activity and the Interaction with Actin. J. Biol. Chem. 255, 4355-4361.

Sellers, J.R., Pato, M.D. and Adelstein, R.S. (1981) Reversible phosphorylation of smooth muscle myosin, heavy meromyosin, and platelet myosin. J. Biol. Chem. 256, 13137-13158.

Sellers, J.R., Eisenberg, E. and Adelstein, R.S. (1982) The binding of smooth muscle heavy meromyosin to actin in the presence of ATP. J. Biol. Chem. 257, 13880-13883.

Sellers, J.R. (1985) Mechanism of the phosphorylation-dependent regulation of smooth muscle heavy meromyosin. J. Biol. Chem. 260, 15815-15819.

Sellers, J.R., Chock, P.B. and Adelstein, R.S. (1983) The apparently negatively cooperative phosphorylation of smooth muscle myosin at low ionic strength is related to its filamentous state. J. Biol. Chem., 89, 14181-14188.

Sobieszek, A., and Small, J.V. (1976) Myosin-linked calcium regulation in vertebrate smooth muscle. J. Mol. Biol. 102, 75-92.

Stafford, W.F., Szentkiralyi, E.M. and Szent-Gyorgyi, A.G. (1979) Muscle heavy meromyosin regulatory properties of single-headed fragments of scallop myosin. Biochemistry 18, 5273-5280.

Sutoh, K., Yamamoto, K. and Wakabayashi, T. (1984) Electron microscopic visualization of the SH-1 thiol of myosin by the use of an avidin-biotin system. J. Mol. Biol. 178, 323-340.

Sutoh, K., Yamamoto, K. and Wakabayashi, T. (1986) Electron microscopic visualization of the ATPase site of myosin by photoaffinity labeling with a biotinylated photoreactive ADP analog. Proc. Natl. Acad. Sci., U.S.A. 83, 212-216.

Suzuki, H., Onishi, H., Takahashi, K. and Watanabe, S. (1978) Structure and function of chicken gizzard myosin. J. Biochem. 84, 1529-1542.

Suzuki, H., Kamota, T., Onishi, H. and Watanabe, S. (1982) Adenosine triphosphate-induced reversible change in the conformation of chicken gizzard myosin and heavy meromyosin. J. Biochem. 91, 1699-1706.

Suzuki, H., Stafford, W.F., Slayter, H.S and Seidel, J.C. (1985) A Conformational Transition in Gizzard Heavy Meromyosin Involving the Head-Tail Junction, Resulting in Changes in Sedimentation Coefficient, ATPase Activity and Orientation of Heads. J. Biol. Chem. 260, 14810-14817.

Suzuki, H., Kondo, Y., Carlos, A.D., and Seidel, J.C. (1986) A Conformational Transition in Gizzard Heavy Meromyosin Involving the Head-tail Junction Studied by Rates of Digestion of Heavy and Light Chains by Papain. Biophys. J. 49, 184a.

Szent-Gyorgyi, A.G., Szentkiralyi, E.M. and Kendrick-Jones, J. (1973) The light chains of scallop myosin as regulatory subunits. J. Mol. Biol. 74, 179-203.

Takeuchi, K. and Tonomura, Y. (1980) Comparison of kinetic properties of the ATPase reaction of arterial smooth muscle myosin with skeletal muscle myosin. J. Biochem. 88, 1693-1702.

Tao, T. and Lamkin, M. (1981) Excitation energy transfer studies on the proximity between SH1 and the adenosinetriphosphase site in myosin subfragment-1. Biochemistry 20, 5051-5055.

Thomas, D.D., Seidel, J.C., Hyde, J.S. and Gergely, J. (1975) Motion of subfragment 1 in myosin and its supramolecular complex: saturation transfer electron paramagnetic resonance. Proc. Nat. Acad. Sci. U.S.A. 72, 1729-1733.

Trybus, K.M., Huiatt, T.W. and Lowey, S. (1982) A bent monomeric conformation of myosin from smooth muscle. Proc. Natl. Acad. Sci., U.S.A. 79, 6151-6155.

Ueno, H. and Harrington, W.F. (1984) An enzyme-probe method to detect structural changes in the myosin rod. J. Mol. Biol. 173, 35-61.

Umekawa, H., Naka, M., Inagaki, M., Onishi, H., Wakabayashi, T. and Hidaka, H. (1985) Conformational Studies of Myosin Phosphorylated by Protein Kinase C. J. Biol. Chem. 260, 9833-9837.

Ca^{2+} REGULATION IN SMOOTH MUSCLE; DISSOCIATION OF MYOSIN LIGHT CHAIN KINASE ACTIVITY FROM ACTIVATION OF ACTIN-MYOSIN INTERACTION

Setsuro Ebashi, Takashi Mikawa, Hideto Kuwayama, Masashi Suzuki, Hiroaki Ikemoto, Yasuki Ishizaki, Ritsuko Koga
National Institute for Physiological Sciences

Myodaiji, Okazaki, 444 JAPAN.

INTRODUCTION

We have proposed that Ca^{2+} regulation in vertebrate smooth muscle is carried out by an actin-linked factor, called "leiotonin" (Ebashi, 1977). It was once reported that the factor was obtained in a homogeneous state, having a molecular weight of about 80,000 (Mikawa et al., 1977), but later investigation showed that it was a proteolytic product. Search for the original leiotonin led us to a protein very similar to myosin light chain kinase (MLCK) or inseparable from it (Ebashi and Nonomura, 1983). Occasionally a fraction without kinase activity could be obtained, but it was not always reproducible.

On the other hand, the idea that phosphorylation of myosin light chain by Ca^{2+} dependent kinase (MLCK) and its dephosphorylation is the mechanism underlying the contraction-relaxation cycle of smooth muscle has been supported by a number of smooth muscle researchers (Sobieszek and Small, 1976; Aksoy et al., 1976; Chacko et al., 1977; Hartshorne and Siemankowski, 1981). Now it appears that this idea is almost established.

In the meantime we have been presenting a plenty of observations that cannot be explained by the MLCK concept; there is a way of activating the contractile system of smooth muscle other than that via MLCK (Ebashi, 1980; Ebashi et al., 1981).

There is no doubt that the fraction prepared as MLCK

has an activating effect on the actin-myosin-ATP interaction of smooth muscle. In this article, we will show that the activating effect of the fraction, obtained by following the principles of preparing MLCK, is not mediated by its MLCK activity.

EXPERIMENTAL

Bovine stomach was most frequently used as the material of MLCK or leiotonin, but in some cases, chicken gizzard and bovine aorta were also used. Bovine aorta was a good material for its containing little intrinsic protease. As the source of natural actomyosin, or myosin B, bovine stomach and chicken gizzard were both used as well.

In most experiments desensitized actomyosin prepared by the routine procedure in our laboratory was used. Briefly, myosin B is subjected to precipitation at an ionic strength a little lower than the critical concentration to keep actomyosin in solubilized state, and the centrifuged pellet was solubilized in 0.4M KCl (or NaCl). This procedure is repeated until the preparation becomes insensitive to Ca^{2+}.

Superprecipitation was mainly used as the criterion for assessing the <u>in vitro</u> contraction. Kohama et al. (1986) have shown that superprecipitation is an <u>in vitro</u> model for contraction more akin to the tension development of glycerinated fibers of skeletal muscle than is the ATPase activity. Detailed studies on the superprecipitation of smooth muscle system have not yet been made, but the preliminary experiment has shown that the relationship between superprecipitation and the ATPase activity is in principle not different from skeletal muscle system. Furthermore, this method is simple and reproducible, and easy to handle.

RESULTS

Separation of 130K Protein from 155K Protein

The extract of bovine stomach was subjected to routine procedures to purify MLCK and the eluates from DEAE chromatography were checked for both activities, viz., MLCK activity (MA) and actomyosin-activating or leiotonin activity (LA) (Kuwayama et al., to be published). MA exhibited a rather wide plateau, whereas main part of LA appeared in early eluates (Fig. 1). If we plot the ratio of

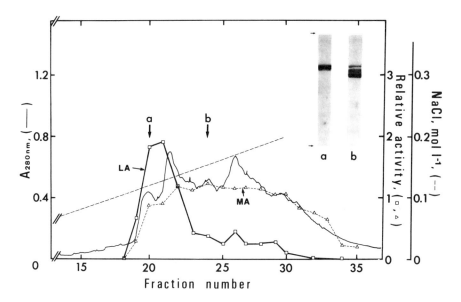

Fig. 1. Elution pattern of DEAE cellulose (DE 32) chromatography of bovine stomach extract.
MA : Myosin light chain kinase activity. LA : Actomyosin-activation assessed by superprecipitation (quoted from Kuwayama et al., submitted).

LA to MA, a sharp peak could be demonstrated. The ratio of the peak, which is composed mainly of 155K protein, is ten times higher than those of later eluates (Fig. 2). The fractions showing lower ratios were dominated by 130K protein, but not so homogeneous as is 155K protein in the peak. From bovine aorta, 130K protein can be obtained in a more homogeneous state (Kuwayama, Suzuki, Koga and Ebashi, to be published; cf. Ebashi, in press). In the case of stomach, it cannot be denied that 130K protein is a proteolytic product of 155K protein. Since aorta contains only a little intrinsic protease, 130K protein does not seem a proteolytic product of 155K protein but a protein entity, independent of the latter.

In the following experiments, ammonium sulfate

fractions before subjecting to DEAE chromatography was also used.

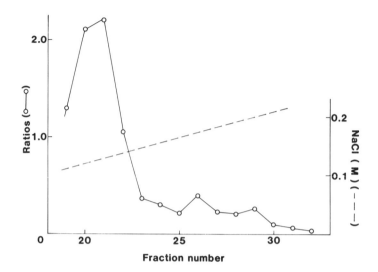

Fig. 2. LA/MA ratio of each fraction shown in Fig. 1.

Procedures to Remove LA without Affecting MA

The above-mentioned fact itself, i.e., the 155K protein shows higher LA/MA activity more than ten times than that of the 130K protein, clearly indicates the dissociation of LA from MA. Essentially the same observations could be made with several other procedures:

a) One of the most remarkable features of LA is its susceptivity to proteolytic digestion, whereas MA is slightly less sensitive to it. In Fig. 3 an example is exhibited; some other proteolytic enzymes such as V_8, Ca dependent neutral protease and thermolysin behave like trypsin, whereas chymotrypsin affects both MA and LA almost to the same extent. Generally speaking, MA of gizzard preparation is more sensitive to proteolysis than is MA of bovine stomach; the profile shown in Fig. 3 can be seen only with crude preparation of gizzard, not with purified 130K protein.

b) Various chemical modifications repress LA preferentially, but so far no agent comparable to some proteases has not been found.

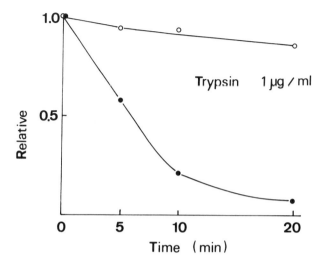

Fig. 3. Time course of LA (●) and MA (○) upon treatment with trypsin of 155K protein of bovine stomach (quoted from Ebashi et al., submitted).

c) Polyclonal antibodies against 130K protein of gizzard, reactive with 155K proteins of bovine stomach and aorta as well as the antigen, may be classified into three groups: i) represses both MA and LA; ii) represses only LA; iii) does not repress both activities. No antibody which represses only MA or preferentially MA has not yet been obtained. A monoclonal antibody which behaves like the group (ii) has been obtained.

Procedures or Conditions to Suppress MA without Affecting LA

There has been found no definite procedure to act just in an opposite direction to those mentioned above, though some experimental conditions definitely favor LA as shown below, (b) and (c):
a) It was reported that N-ethylmaleimide (NEM) exerts its deteriorating effect more intensely on MA of gizzard crude preparation (Ebashi and Nakasone, 1981); this is well demonstrated with bovine stomach preparation (Fig. 4).
b) The pH-activity relationship is quite distinct between MA and LA; while MA peak is attained at pH 6.5-6.8 and falls to a low level at 8.0, LA shows a wide plateau from

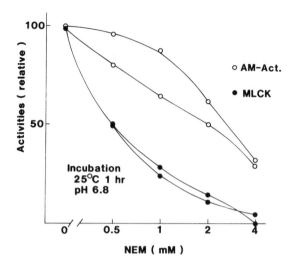

Fig. 4 Effect of N-ethylmaleimide (NEM) on LA (○) and MA (●) of crude 155K protein of bovine stomach.

pH 6.8 to 8.0 (Fig. 3); however, the results are varied from preparation to preparation, sometimes like Fig. 5,

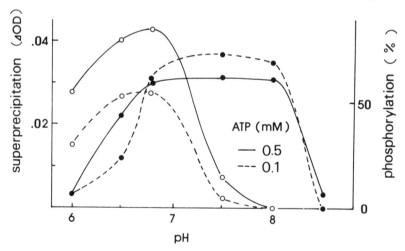

Fig. 5. Relationships between pH's and LA (●) or MA (○) of crude 155K protein of bovine stomach (quoted from Ebashi et al., submitted).

but in other cases a considerable MA, about 20% of the maximum, are retained.

c) The dependence on the ATP concentration is very much different between LA and MA, i.e., the concentrations for half maximum activation are about 2×10^{-5}M for LA and more than 0.5×10^{-3}M for the latter (cf. Ebashi, in press). At 5×10^{-5}M, LA exhibits almost full activity, whereas MA shows only a trace.

DISCUSSION

Thus the two activities, LA and MA, can be distinguished by various procedures. This clearly indicates that there is a process(es) other than MA to activate the actin-myosin-ATP interaction of smooth muscle.

A question is then raised whether or not the enzymatic activity itself of MLCK can activate the interaction. The above findings seem to make it difficult to accept the view favoring the active role of MLCK. To reconcile with this view, we must assume: a) A proteolytic fragment may repress the actin-myosin-ATP interaction, so that the activation of the interaction could not be induced by MA; b) At higher pH's and at low ATP concentrations, a faint phosphorylation is enough to activate the interaction. All these assumptions, though not impossible, are rather unlikely.

The next question may be why our results are at variance with other groups'. We have completely confirmed the results of Chacko et al.: phosphorylated myosin, separated from actin, MLCK and other components, could fully be activated by actin irrespective of Ca^{2+}. One of the secrets of our results may reside in the fact that we are using desensitized natural actomyosin as the material for assessment. Freshly prepared myosin, being soluble at low ionic strength, say 0.06., rather quickly becomes insoluble once detached from actin, whereas natural actomyosin even after desensitization, the myosin molecule has retained for a long time its solubility like freshly prepared myosin. This might be one of the keys for explaining this puzzling discrepancy.

Then, what kind of mechanism other than myosin phosphorylation could be supposed? At present we have entirely no idea. It is enigmatic that a very small amount of leiotonin, or a factor contained in MLCK, can exert its effect on actomyosin system. This might allure us to think of some enzymatic mechanism. On the other hand the initial rate of superprecipitation increases with increase in the amount of the factor and requiring a fairly large amount for saturation, though the degree of phosphorylation saturates at much lower concentration of the factor. This may indicate that the factor may influence structural arrangement of the thin filament. Studies in this direction are urgently required.

CONCLUSION

The fraction that was prepared from bovine stomach and aorta following the principles to prepare MLCK certainly activates the myosin-actin-ATP interaction in the presence of Ca^{2+}, but this activating effect can be dissociated from MLCK activity under various conditions; the activation can be exhibited only with a trace of phosphorylation of myosin light chain, whereas virtually no activation can be seen even with full phosphorylation of the light chain.

These results are not inconsistent with the early concept that the activation of smooth muscle contractile system is carried out by an actin-linked factor, leiotonin.

REFERENCES

Aksoy MO, Williams D, Sharkey EM, Hartshorne DJ (1976). Relationship between Ca^{2+} sensitivity and phosphorylation of gizzard actomysin. Biochem biophys Res Commun 69:35-41.
Chacko S, Conti MA, Adelstein RS (1977). Effect of phosphorylation of smooth muscle myosin on actin activation and Ca^{2+} regulation. Proc Natl Acad Sci USA 74:129-133.
Ebashi S (1980). Regulation of muscle contraction (Croonian Lecture, 1979). Proc R Soc Lond B 207:259-286.
Ebashi S (1986). Contractile and regulatory protein in cardiovascular system. In Dhalla NS and Beamish RE (eds) "Proc 8th Ann Meeting of the International Soci for Heart Res Am Sec,": Martius Nijhoff, in press.
Ebashi S, Mikawa T, Hirata M, Toyo-oka T, Nonomura Y (1977). Regulatory proteins of smooth muscle, In Casteels R et al. (eds): "Excitation Contraction Coupling in Smooth Muscle,"

Amsterdam: Elsevier North-Holland Biomedical Press, pp 325-334.

Ebashi S, Nakasone H (1981). Notes on preparation of Leiotonin. Proc Japan Acad 57:217-221.

Ebashi S, Nonomura Y, Nakamura S, Nakasone H, Kohama K (1982). Regulatory mechanism in smooth muscle actin-linked regulation. Federation Proc 41:2863-2867.

Ebashi S, Nonomura Y (1984). Ca Regulation in Smooth Muscle Contraction. In Ebashi S et al. (eds): "Calcium Regulation in Biological Systems Academic Press," New York: pp 59-69.

Ebashi S, Suzuki M, Koga R, Ishizaki Y (1986). Effects of proteases on actomyosin-activating activities of myosin light chain kinase preparations. J Biochem submitted.

Hartshorne DJ, Siemankowski RF (1981). Regulation of smooth muscle actomyosin. Ann Rev Physiol 43:519-530.

Kohama K, Saida K, Hirata M, Kitaura T, Ebashi S (1986). Superprecipitation a model for in vitro contraction superior to ATPase activity. Jap J Pharmacol in press.

Kuwayama H, Suzuki M, Koga R, Ebashi S (1986). Actomyosin-activating activities of myosin light chain kinase preparations derived from bovine stomach and aorta. J Biochem submitted.

Mikawa T, Toyo-oka T, Nonomura Y, Ebashi S (1977). Essential factor of gizzard 'Troponin' fraction. J Biochem 81:273-275.

Sobieszeck A and Small JV (1976). Myosin-linked calcium regulation in vertebrate smooth muscle. J Mol Biol 112:559-576.

CALDESMON, A MAJOR ACTIN- AND CALMODULIN-BINDING PROTEIN OF SMOOTH MUSCLE

Michael P. Walsh

Department of Medical Biochemistry, University of Calgary, Calgary, Alberta, Canada T2N 4N1

INTRODUCTION

Since the initial observation made by Sobieszek (1977) that phosphorylation of smooth muscle myosin is required for actin-activation of its Mg^{2+}-ATPase activity, a great deal of biochemical and physiological experimentation has been directed towards establishing the importance, or lack thereof, of myosin phosphorylation in regulating smooth muscle contraction. As a consequence, a substantial body of evidence has accumulated which supports a fundamental role for myosin phosphorylation in the regulation of smooth muscle contraction (see Adelstein & Eisenberg, 1980; Walsh & Hartshorne, 1982; Walsh, 1985; Kamm & Stull, 1985 for reviews). The resultant model (Fig. 1) is based on primary regulation of contraction-relaxation by reversible phosphorylation of myosin as follows. Nervous or hormonal stimulation of the smooth muscle cell induces an elevation of cytosolic [Ca^{2+}] originating mainly from intracellular stores (the sarcoplasmic reticulum (SR)) and possibly the extracellular milieu (Somlyo, 1985). Ca^{2+} binds to calmodulin (CaM) forming the Ca_4^{2+}-CaM complex. This binding of Ca^{2+} induces a conformational change in CaM which involves exposure of a hydrophobic site(s). In its altered conformation, CaM can bind with high affinity ($K_d \sim$ 1 nM) to the inactive apoenzyme myosin light chain kinase (MLCK) to form the active holoenzyme, Ca_4^{2+}-CaM-MLCK. The activated kinase catalyzes the transfer of the terminal phosphoryl group of ATP to serine-19 on each of the two 20 kDa light chains of myosin. In the phosphorylated state, myosin can interact with actin and hydrolyse ATP at a fast rate which provides

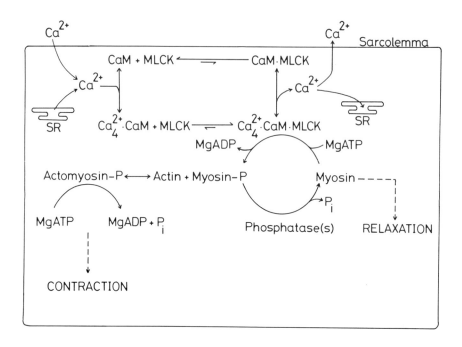

Figure 1. The role of Ca^{2+}/calmodulin-dependent myosin phosphorylation in the regulation of smooth muscle contraction.

the energy for rapid crossbridge cycling and the development of tension. Most simplistically, relaxation occurs when Ca^{2+} is removed from the sarcoplasm which leads to dissociation of Ca^{2+} from the $Ca_4^{2+} \cdot CaM \cdot MLCK$ complex and inactivation of the kinase. Myosin phosphorylation therefore ceases and myosin which had been phosphorylated during the activation phase of the contractile cycle is dephosphorylated by the action of one or more phosphatases (Pato, 1985). In this state, actin and myosin dissociate and the muscle relaxes.

Several physiological and biochemical observations, however, indicate that the scheme in Fig. 1 is an oversimplification. Nevertheless, it is widely recognized that myosin phosphorylation plays an important role in the regulation of smooth muscle contraction and the vast majority of evidence indicates that myosin phosphorylation is a prerequisite for the development of tension. For example, a correlation between the level of myosin phosphorylation and the maximum velocity of shortening (crossbridge cycling rate) has been observed (Dillon et al, 1981; Aksoy et al, 1982). However, support is growing in favor of the notion that at least one other Ca^{2+}-dependent regulatory mechanism is operative in smooth muscle. For example, numerous physiological studies have indicated that developed tension can be maintained in a variety of smooth muscle tissues during prolonged stimulation in spite of the fact that myosin becomes dephosphorylated (Aksoy et al, 1982, 1983, 1986; Butler et al, 1983; Dillon et al, 1981; Gerthoffer & Murphy, 1983; Silver & Stull, 1982). This tension maintenance (the so-called "latch state") is Ca^{2+}-dependent and is characterized by a relatively low rate of ATP hydrolysis, i.e., non-cycling or slowly-cycling crossbridges. Such a control mechanism would allow energy conservation during prolonged tonic contractions. The latch mechanism appears to have a greater sensitivity to Ca^{2+} than does myosin phosphorylation and the evidence indicates that prior phosphorylation of myosin, with consequent development of tension, is required for expression of the latch state (Chatterjee & Murphy, 1983).

At a biochemical level, several instances of thin filament-linked regulation of actomyosin Mg^{2+}-ATPase have been reported. For example, Marston et al (1980) have reported results of studies with smooth muscle actomyosin, using the myosin competition test devised by Lehman and Szent-Györgyi (1975), which support the existence of a thin filament-linked regulatory mechanism in smooth muscle, in addition to the myosin-linked phosphorylation system. It was originally proposed by Driska and Hartshorne (1975) that a 130 kDa protein found in gizzard actomyosin and thin filament preparations represents a thin filament-linked regulatory protein. Similar thin filament preparations from a variety of smooth muscles have been described which exhibit Ca^{2+}-dependent activation of the Mg^{2+}-ATPase of rabbit skeletal myosin and thiophosphorylated smooth muscle myosin

(Marston & Smith, 1984, 1985; Marston & Lehman, 1985).

We have been interested during the past three years in the biochemical basis of such a thin filament-linked regulatory system in vertebrate smooth muscle and, on the basis of work done in several laboratories, including our own, conclude that the protein caldesmon is a strong candidate for such a control mechanism.

DISCOVERY

Caldesmon was first described by Kakiuchi and his colleagues in 1981 who named the protein (which means "calmodulin binding") on the basis of its demonstrated Ca^{2+}-dependent interaction with calmodulin (Sobue et al, 1981a). They isolated caldesmon from chicken gizzard as the major calmodulin-binding protein in this tissue and demonstrated its ability also to interact with F-actin. We initially came across caldesmon during attempts to purify the native (M_r 136,000) form of gizzard MLCK (Adachi et al, 1983; Ngai et al, 1984; Walsh et al, 1983). A protein of subunit M_r 141,000 copurified with MLCK through calmodulin-Sepharose affinity chromatography but the two proteins could be separated by ion-exchange chromatography (Ngai et al, 1984). The 141 kDa protein was shown to be devoid of MLCK activity and did not cross-react with monoclonal antibodies to gizzard MLCK; it was identified as caldesmon (Ngai et al, 1984) and we determined the tissue concentrations of these two calmodulin-binding proteins to be 4.6 µM (MLCK) and 11.1 µM (caldesmon) (Ngai & Walsh, 1985a).

PHYSICOCHEMICAL PROPERTIES

Purification

Several methods have been developed for the isolation of caldesmon (Bretscher, 1984; Clark et al, 1986; Fürst et al, 1986; Ngai et al, 1984; Ngai & Walsh, 1985a; Sobue et al, 1981a), all of which are rather simple and exploit the unusual features of caldesmon, e.g., its Ca^{2+}-dependent interaction with calmodulin, association with the thin filament and heat-stability. We use two methods routinely, one which incorporates a heat-treatment step as described by Bretscher (1984) and one which avoids heat-treatment. Briefly, the former method consists of heating a 0.3 M KCl-extract of minced chicken gizzard to 85-90°C and removal of denatured proteins by centrifugation. Ammonium sulfate fractionation (30-50%) is then followed by ion-exchange chromatography which yields an electrophoretically homogeneous preparation (Ngai & Walsh, 1984). In view of the possibility that heat-treatment may affect the functional properties of caldesmon, we also use an alternative isolation procedure which avoids this step. This protocol involves the preparation of washed myofibrils from frozen chicken gizzard followed by extraction of caldesmon, MLCK, filamin and a few other proteins with 25 mM $MgCl_2$. This extract is then subjected to ion-exchange chromatography, which separates the MLCK, followed by affinity chromatography on a column of AffiGel Blue and a second ion-exchange chromatography under different conditions (Ngai & Walsh, 1985a).

Molecular Weight and Native Structure

There appear to be two principal forms of caldesmon, which are immunologically cross-reactive and exhibit similar functional properties. One class has a subunit molecular weight as determined by SDS-polyacrylamide gel electrophoresis of 130-155 kDa, which varies slightly from one tissue to another, and the other class 70-80 kDa which has been identified in some, but not all, tissues examined and which appears to be a true isoform of caldesmon and not a proteolytic fragment of the high molecular weight species. The low molecular weight form of caldesmon is particularly abundant in cultured cells (Bretscher & Lynch, 1985; Owada et al, 1984) and has been isolated from bovine adrenal medulla (Sobue et al, 1985a) and human platelets (Dingus et al, 1986). Burgoyne et al (1986) isolated a 70 kDa protein from adrenal chromaffin cells which bound to chromaffin granule membranes in the absence of Ca^{2+} but not at higher than micromolar concentrations of Ca^{2+}. Immunochemical techniques identified this protein as caldesmon.

In some tissues, e.g., chicken gizzard and hog stomach, the high molecular weight form of caldesmon appears as two closely-spaced bands on SDS-polyacrylamide gels (Bretscher, 1984; Fürst et al, 1986; Sobue et al, 1982). There is some controversy as to the subunit composition of native caldesmon which has been described as a dimer of two ~150 kDa polypeptides (Fürst et al, 1986; Sobue et al, 1981a) or a monomer of ~150 kDa (Bretscher, 1984; Clark et al, 1986). Caldesmon is a highly asymmetric molecule as shown by sucrose density gradient centrifugation (sedimentation coefficient = 2.7S), gel filtration (Stokes radius = 91 Å) (Bretscher, 1984) and low-angle rotary shadowing (Fürst et al, 1986). The hydrodynamic properties of gizzard caldesmon suggest the native structure to be a monomer (Bretscher, 1984) and indeed Fürst et al (1986) have observed elongated, highly flexible monomers of hog stomach caldesmon in the electron microscope. However, they also observed a tendency to form end-to-end dimers and higher-order aggregates. Bretscher (1986) has shown that isolated caldesmon is highly susceptible to oxidation with formation of intermolecular disulfide bonds and oligomerization. The protein must be stored in the presence of high concentrations of dithiothreitol to avoid such aggregation.

Amino Acid Composition

The amino acid compositions of chicken gizzard and bovine aorta caldesmons are shown in Table 1. Both proteins contain a very large proportion of glutamic acid + glutamine residues (25.4% of total residues in chicken gizzard caldesmon and 22.3% of total residues in the bovine aorta protein). They have comparable amounts of lysine, aspartic acid and several other residues, but have noticeable differences in the amounts of histidine, glycine, tyrosine and phenylalanine. One-dimensional peptide mapping of chicken gizzard and bovine aorta caldesmons using either α-chymotrypsin or S. aureus V8 protease revealed very few common peptides, consistent with significant differences in amino acid sequence between the two proteins (Clark et al, 1986). Amino-terminal sequence analysis revealed that both chicken gizzard and bovine aorta caldesmons are N-terminally blocked.

Calmodulin and Actin Binding

As indicated above, caldesmon was originally discovered as a calmodulin-binding protein which also interacts with F-actin (Sobue et al, 1981a). Calmodulin binding to caldesmon is Ca^{2+}-dependent and has been demonstrated by calmodulin-Sepharose affinity chromatography (Bretscher, 1984; Clark et al, 1986; Fürst et al, 1986; Ngai et al, 1984; Sobue et al, 1981a, 1982), non-denaturing polyacrylamide gel electrophor-

TABLE 1. Amino Acid Compositions of Chicken Gizzard and Bovine Aorta Caldesmons

Amino Acid	residues/mol	
	Chicken Gizzard	Bovine Aorta
Lysine	167	150
Histidine	5	25
Arginine	126	98
Aspartic acid	90	99
Threonine	51	76
Serine	58	90
Glutamic acid	312	302
Proline	37	55
Glycine	51	103
Alanine	135	112
Cysteine	7	11
Valine	56	68
Methionine	17	8
Isoleucine	22	33
Leucine	61	80
Tyrosine	7	15
Phenylalanine	15	29
Tryptophan	9	+

esis of a mixture of [^{3}H]calmodulin and caldesmon (Sobue et al, 1981a, b) and changes in tryptophan fluorescence emission spectrum (Dabrowska & Galazkiewicz, 1986). The stoichiometry of calmodulin binding to caldesmon is 1 calmodulin:1 caldesmon (150 kDa polypeptide) (Sobue et al, 1981b) with K_d = 1.4-1.7 µM (Marston et al, 1985; Sobue et al, 1981b). Interaction between caldesmon and F-actin has been demonstrated by ultracentrifugation (Bretscher, 1984; Clark et al, 1986; Ngai et al, 1984; Sobue et al, 1981b) and viscosimetry (Sobue et al, 1981b); this interaction is independent of Ca^{2+}, and is saturable, with maximal binding of 1 caldesmon:6 (Bretscher, 1984) or 9-10 (Clark et al, 1986) actin monomers. From studies with isolated thin filaments, Marston and Lehman (1985) have determined that caldesmon is associated with the thin filament in vivo at a stoichiometry of 1 caldesmon (150 kDa):26 actin monomers. Caldesmon binds to high-affinity sites on actin with $K_d \sim 0.2$-1 µM and to actin/

tropomyosin with $K_d \sim 50$ nM (Marston & Smith, 1985). Most interestingly, calmodulin and F-actin compete for binding to caldesmon in the presence of micromolar concentrations of Ca^{2+}; at lower levels of Ca^{2+}, caldesmon binds exclusively to F-actin (Bretscher, 1984; Clark et al, 1986; Sobue et al, 1981a). This led Sobue et al (1981a, 1982) to suggest that caldesmon can "flip-flop" between actin and calmodulin as a function of the sarcoplasmic Ca^{2+} concentration, a phenomenon which may be related to its physiological function(s) (see below). There is some controversy concerning the effect, if any, of tropomyosin on the caldesmon-actin interaction. We have found no effect of tropomyosin on this interaction or on the capacity of calmodulin to remove either bovine aorta or chicken gizzard caldesmon from F-actin in the presence of Ca^{2+} (Clark et al, 1986). Similarly, Bretscher (1984) observed little effect of tropomyosin on caldesmon binding to F-actin. Sobue et al (1982), however, claimed that caldesmon binding to F-actin required tropomyosin, although flow birefringence studies indicated no effect of tropomyosin on caldesmon-actin interaction (Maruyama et al, 1982). The important consensus, however, is that caldesmon interacts with the actin/tropomyosin complex and can be removed by calmodulin only in the presence of Ca^{2+}. Our in vitro binding studies (Clark et al, 1986) indicated that half-maximal removal of bovine aorta caldesmon (1.3 µM) from actin (11.9 µM) was effected at 5.3 µM calmodulin in the presence of 0.1 mM Ca^{2+}. Identical results were obtained in the presence of 2 µM tropomyosin. At physiological ratios of actin to caldesmon to calmodulin (26:1:3), however, one would expect most of the caldesmon to be bound to the actin filament, even in the presence of Ca^{2+}.

TISSUE AND SPECIES DISTRIBUTION

Several studies involving isolation of caldesmon and the use of polyclonal antibodies raised against chicken gizzard caldesmon have indicated a broad tissue and species distribution of this protein. Caldesmon has been isolated from chicken gizzard (Bretscher, 1984; Ngai et al, 1984; Sobue et al, 1981b), bovine (Clark et al, 1986) and sheep (Smith & Marston, 1985) aorta, bovine adrenal medulla (Sobue et al, 1985a), hog stomach (Fürst et al, 1986) and human platelets (Dingus et al, 1986). With the use of immunoblotting techniques, caldesmon has been identified in a variety of chicken (Ngai & Walsh, 1985b), rat (Ban et al, 1984) and bovine

(Clark et al, 1986) tissues, including various smooth muscle and non-muscle tissues and striated muscles, human platelets (Kakiuchi et al, 1983) and a variety of cultured cells (Bretscher & Lynch, 1985; Owada et al, 1984). The demonstration of immunologically cross-reactive caldesmons in a variety of tissues from diverse species suggests a degree of structural conservation consistent with an important functional role.

SUBCELLULAR LOCALIZATION

The topographical distribution of caldesmon was initially studied in rat tissues using light-microscopic immunochemistry (Ban et al, 1984). Caldesmon immunostaining was observed in all smooth muscle cells examined (oesophagus, intestine, aorta, uterus, ovary) although no clear-cut subcellular localization could be deduced. Caldesmon was also observed in uterus, oviduct, submandibular gland, pancreas and renal tubules, in the peripheral cytoplasm of adrenal medullary cells (Ban et al, 1984), in the apical part of follicle epithelial cells of the rat thyroid (Fujita et al, 1984) and in the apical cytoplasm of the absorptive epithelial cells of the intestine (Ishimura et al, 1986). The distribution of caldesmon in these tissues coincided with that of actin, consistent with the suggestion from _in vitro_ binding studies discussed above that caldesmon may be localized on the thin filament.

Bretscher and Lynch (1985) have used indirect immunofluorescence microscopy to demonstrate the location of caldesmon in the terminal web, but not in the microvilli, of the brush border of intestinal epithelial cells and in the stress fibers and ruffling membranes of cultured cells. A periodic distribution of caldesmon along stress fibers, which coincides with the distribution of tropomyosin and is complementary to that of α-actinin, was observed by light microscopy. This led Bretscher and Lynch (1985) to suggest the association of caldesmon with a "contractile unit" composed additionally of actin, myosin, tropomyosin and MLCK. Such a location would be consistent with a functional role for caldesmon in the regulation of actin-myosin interaction (see below). In support of such a notion, Small et al (1986) have distinguished two structurally distinct actin-containing domains in smooth muscle: an actomyosin domain, which is made up of continuous longitudinal arrays of actin and myosin filaments, and an actin-intermediate filament domain, which forms longitudinal fibrils containing actin, filamin, desmin and α-actinin-rich dense bodies but which are free of myosin. Caldesmon was found to be located exclusively in the actomyosin domain by immunocytochemistry of ultrathin sections of smooth muscle at the light and electron microscope levels (Fürst et al, 1986).

Marston and his colleagues succeeded recently in isolating native thin filaments from several smooth muscles: gizzard, stomach, trachea and aorta (Marston & Smith, 1984; Marston & Lehman, 1985). Such preparations, which resemble those described earlier by Driska and Hartshorne (1975), contain three major protein components: actin, tropomyosin and caldesmon in molar ratios of 1:1/7:1/26. We have been able to confirm these observations, identifying the caldesmon by Western immunoblotting and co-electrophoresis with isolated caldesmon (Ngai, P.K. & Walsh, M.P., unpublished observations). Marston's thin filament purification procedure is carried out in the presence of EGTA, conditions expected to encourage caldesmon association with actin. However, Lehman (1986) has recently shown that chicken gizzard thin filaments isolated either in the presence or absence of Ca^{2+} contain identical amounts of caldesmon. This supports the conclusion (see above) that the amount of calmodulin in smooth muscle is insufficient to cause significant dissociation of caldesmon from actin in the presence of Ca^{2+}.

EFFECTS OF CALDESMON ON ACTIN-MYOSIN INTERACTION

Many of the experimental observations described above suggest, albeit indirectly, that caldesmon may function in regulating actin-myosin interaction, i.e., the contractile state of smooth muscle. This question has been addressed by several investigators using well-characterized in vitro systems reconstituted from purified contractile and regulatory proteins. Sobue et al (1982) were the first to provide evidence of the regulation of actin-myosin interaction by caldesmon by the demonstration that isolated caldesmon inhibited superprecipitation of desensitized chicken gizzard actomyosin. This inhibitory effect could be overcome by Ca^{2+}/calmodulin. Subsequently, we showed that caldesmon inhibited the actin-activated myosin Mg^{2+}-ATPase activity of a system reconstituted from purified actin, myosin, tropomyosin, calmodulin and MLCK without affecting myosin phosphorylation

(Ngai & Walsh, 1984). These observations have since been confirmed by several investigators using chicken gizzard, bovine aorta or adrenal medullary caldesmons and smooth muscle or skeletal muscle myosins or bovine adrenal medullary actomyosin (Clark et al, 1986; Dabrowska et al, 1985; Lim & Walsh, 1986; Marston & Lehman, 1985; Sobue et al, 1985b).

In our studies with isolated bovine aorta caldesmon, we observed half-maximal inhibition of the actomyosin Mg^{2+}-ATPase at approximately 1 mol caldesmon/12 mol actin (Clark et al, 1986). This inhibition was unaffected by tropomyosin or Ca^{2+}/calmodulin since identical inhibition was seen in the presence and absence of Ca^{2+}/calmodulin using prephosphorylated gizzard myosin. The Mg^{2+}-ATPase activity of gizzard myosin in the absence of actin was not inhibited by caldesmon (Clark et al, 1986). Dabrowska et al (1985) observed maximal (80%) inhibition of rabbit skeletal muscle actomyosin Mg^{2+}-ATPase activity at a molar ratio of gizzard caldesmon to actin of 1:10-13. Caldesmon isolated from thin filaments is a more potent inhibitor of actomyosin Mg^{2+}-ATPase activity: maximal inhibition of rabbit skeletal muscle myosin Mg^{2+}-ATPase in the presence of aorta actin and tropomyosin was observed at a molar ratio of caldesmon to actin of approximately 1:30, i.e., similar to the in vivo ratios of these proteins (Marston & Lehman, 1985). Caldesmon isolated by other methods may have an altered affinity for F-actin.

While most functional studies have utilized caldesmon isolated from tissue extracts or washed myofibrils, Marston and coworkers have worked mainly with thin filament preparations and caldesmon purified from these thin filaments. Such thin filament preparations conferred a high degree of Ca^{2+}-sensitivity to the Mg^{2+}-ATPase of skeletal myosin or thiophosphorylated gizzard myosin (Marston & Smith, 1984, 1985; Marston & Lehman, 1985). For example, reconstitution of sheep aorta thiophosphorylated myosin with thin filaments of sheep aorta, rabbit stomach and chicken gizzard resulted in Mg^{2+}-ATPase activities which were 91.6-96.8% Ca^{2+}-sensitive (Marston & Lehman, 1985), compared with only 18.3% Ca^{2+}-sensitivity when the myosin was reconstituted with aorta actin/tropomyosin. Caldesmons isolated from these three thin filament preparations were equally potent inhibitors of skeletal muscle myosin Mg^{2+}-ATPase activated by aorta actin/tropomyosin. This inhibition, however, was unaffected by Ca^{2+} which led to the suggestion that native thin filaments contain, in addition to actin, tropomyosin and caldesmon, a Ca^{2+}-binding protein which may be calmodulin and which, in conjunction with caldesmon, confers Ca^{2+}-sensitivity to the system (Marston & Smith, 1985).

Caldesmon has been known for some time to enhance the viscosity of F-actin (Sobue et al, 1981a). Bretscher (1984) observed, by light and electron microscopy, the formation of massive F-actin bundles made up of hundreds of filaments in the presence of caldesmon. This raised the possibility that inhibition of the actomyosin Mg^{2+}-ATPase discussed above may be an in vitro phenomenon resulting from actin filament bundling. However, Sobue et al (1985c) showed that caldesmon aggregates upon concentration or freeze-thawing and only this aggregated form of caldesmon is capable of inducing actin bundling. Furthermore, Moody et al (1985) observed inhibition of actomyosin Mg^{2+}-ATPase activity at concentrations of caldesmon at which no actin filament bundling occurred; bundling was observed only at higher caldesmon concentrations. It is concluded, therefore, that regulation of actin-myosin interaction by caldesmon is unrelated to its actin bundling activity. Caldesmon is also capable of inducing polymerization of G-actin; this effect is prevented by Ca^{2+}/calmodulin (Galazkiewicz et al, 1985).

From a physiological standpoint, very little has been done to date to probe the function of caldesmon. Szpacenko et al (1985) observed relaxation of chicken gizzard skinned fibers by caldesmon at pCa 5.9 but not pCa 5.2 without affecting the level of myosin phosphorylation. This caldesmon-induced relaxation could be reversed by a large molar excess of calmodulin over caldesmon in the presence of Ca^{2+}.

PHOSPHORYLATION OF CALDESMON

In our initial studies of the effects of caldesmon on the actin-activated gizzard myosin Mg^{2+}-ATPase, we observed inhibition by an electrophoretically homogeneous preparation of caldesmon purified by Mg^{2+}-extraction of washed myofibrils, ion-exchange chromatography, Affi-Gel Blue affinity chromatography and a second ion-exchange chromatography (Ngai & Walsh, 1984). However, no inhibition was observed using a preparation of caldesmon obtained by a similar method which substituted calmodulin-Sepharose affinity chromatography for the Affi-Gel Blue and second ion-exchange chromatography steps

(Ngai & Walsh, 1984). This preparation was approximately 90% pure as judged by SDS-polyacrylamide gradient slab gel electrophoresis (Ngai & Walsh, 1985a). These observations suggested to us that the latter preparation of caldesmon contained a factor(s) which blocked the inhibitory action of caldesmon on the actomyosin Mg^{2+}-ATPase and we found that incubation of this caldesmon with $Mg^{2+}[\gamma-^{32}P]$ATP led to the incorporation of $^{32}P_i$ (2-4 mol/mol) into caldesmon (Ngai & Walsh, 1984, 1985a). No phosphorylation was detected when the electrophoretically homogeneous preparation of caldesmon was incubated with $Mg^{2+}[\gamma-^{32}P]$ATP. The factor appeared therefore to be a protein kinase which phosphorylates caldesmon. More detailed investigation revealed that the caldesmon kinase required Ca^{2+} and calmodulin for activity (Ngai & Walsh, 1984, 1985a). Furthermore, we demonstrated the existence in smooth muscle of a phosphatase which catalyzes the removal of phosphate incorporated into caldesmon. Caldesmon is similarly phosphorylated whether in the free state, bound to actin or actin/tropomyosin, or in native thin filaments (Ngai, P.K., Sutherland, C. & Walsh, M.P., unpublished observations) We have also observed rapid stoichiometric phosphorylation of caldesmon in a reconstituted system made from actin, myosin, tropomyosin, calmodulin and MLCK (Lim, M.S., Ngai, P.K. & Walsh, M.P., unpublished observations). Chicken gizzard smooth muscle contains, therefore, all the necessary enzymatic machinery to achieve the reversible phosphorylation of caldesmon in a Ca^{2+}-regulated manner. The available evidence suggests that caldesmon in the non-phosphorylated state inhibits the actin-activated myosin Mg^{2+}-ATPase whereas, upon phosphorylation, this inhibitory effect is abolished (Table 2).

Umekawa and Hidaka (1985) observed phosphorylation of chicken gizzard caldesmon by protein kinase C (the Ca^{2+}- and phospholipid-dependent protein kinase) to approximately 8 mol P_i/mol caldesmon. Maximal phosphate incorporation was reduced to approximately 6 mol P_i/mol caldesmon in the presence of bound calmodulin. Phosphorylated but not unphosphorylated, caldesmon inhibited MLCK activity partially. The physiological significance of these observations is unclear at present.

A POSSIBLE PHYSIOLOGICAL ROLE FOR CALDESMON IN LATCHBRIDGE FORMATION

TABLE 2. The Effect of Caldesmon Phosphorylation on Inhibition of Smooth Muscle Actin-Activated Myosin Mg^{2+}-ATPase

CONDITIONS	ATPase rate nmol P_i/mg myosin/min	% control	Superprecipitation
1. No preincubation			
+ Ca^{2+}	89.0	100	+
− Ca^{2+}	2.3	2.6	−
+ CaD/CaD kinase + Ca^{2+}	89.3	100.3	+
+ heat-treated CaD + Ca^{2+}	57.6	64.7	−
2. Preincubated			
+ Ca^{2+}	92.1	103.5	+
+ CaD/CaD kinase + Ca^{2+}	83.2	93.5	+
+ heat-treated CaD + Ca^{2+}	46.8	52.6	−

The effect of caldesmon containing endogenous caldesmon kinase activity on the actin-activated Mg^{2+}-ATPase of gizzard myosin was compared with that of heat-treated caldesmon either without preincubation or immediately following preincubation at 30°C for 60 min under the following conditions: 20 mM Tris-HCl (pH 7.5), 5 mM $MgCl_2$, 0.1 mM $CaCl_2$, 60 µg/ml calmodulin, 0.5 mM [γ-^{32}P]ATP (~2,000 cpm/nmol) in the absence and presence of 0.3 mg/ml caldesmon (heat-treated CaD or CaD containing endogenous CaD kinase). ATPase assay conditions were: 25 mM Tris-HCl (pH 7.5), 60 mM KCl, 10 mM $MgCl_2$, 0.1 mM $CaCl_2$ or 1 mM EGTA, 1.2 µM calmodulin, 73.5 nM MLCK, 0.74 µM tropomyosin, 0.91 µM myosin, 5.95 µM actin, 1 mM [γ-^{32}P]ATP (~2,000 cpm/nmol) in the presence or absence of 0.71 µM caldesmon. ATPase rates were calculated by linear regression analysis of the time course assays obtained by withdrawing aliquots of reaction mixtures at t = 1, 2, 3, 4, 5, 6 and 7 min for quantification of $^{32}P_i$ release. The occurrence of superprecipitation was determined by visual inspection.

As discussed in the Introduction, there is compelling evidence that myosin phosphorylation plays a central role in the regulation of smooth muscle contraction. Most investigators are in agreement that myosin phosphorylation is a prerequisite for tension development and that there is a correlation between the level of myosin phosphorylation and the maximum velocity of shortening (crossbridge cycling rate). There is now growing experimental evidence indicative of at least one secondary Ca^{2+}-dependent regulatory mechanism which can modulate actin-myosin interaction in smooth muscle, i.e., the contractile state of the muscle. This review has focussed on the protein caldesmon as such a potential secondary control mechanism. Before discussing the possible molecular mechanism describing the physiological role of caldesmon in the regulation of smooth muscle contraction, it is appropriate to introduce a third Ca^{2+}-dependent regulatory mechanism which may be involved in smooth muscle, i.e., the direct binding of Ca^{2+} to myosin. In scallop muscle, the direct binding of Ca^{2+} to myosin is the primary control mechanism (Lehman & Szent-Györgyi, 1975) and such direct Ca^{2+} binding to myosins of mammalian striated and smooth muscles has been demonstrated (see Ikebe & Hartshorne, 1985 for refs.). This is believed to be a consequence of structural homology of the myosin light chains to calmodulin, troponin C and related Ca^{2+}-binding proteins (Kretsinger, 1980). The possibility that Ca^{2+} binding directly to smooth muscle myosin can affect actin-myosin interactions has been shown by several groups who have found that the actin-activated Mg^{2+}-ATPase activity of phosphorylated smooth muscle myosin can be further enhanced by Ca^{2+} (Chacko & Rosenfeld, 1982; Chacko et al, 1977; Kaminski & Chacko, 1984; Nag & Seidel, 1983; Rees & Frederiksen, 1981). Ikebe and Hartshorne (1985), however, showed that the Ca^{2+}-sensitivity of this effect was relatively low and therefore of questionable physiological significance. However, it is possible that the Ca^{2+}-sensitivity of this system *in vivo* is significantly greater than that observed with isolated myosin.

In light of these studies of direct Ca^{2+} binding to myosin and the knowledge of caldesmon and its properties which has emerged in the past few years, we have developed a working model for the possible physiological role of caldesmon and direct Ca^{2+} binding to myosin in the regulation of smooth muscle contraction. We suggest that these two systems act in concert to enable the formation of latch-bridges at intermediate concentrations of Ca^{2+} and independently of myosin phosphorylation. This mechanism is depicted in Fig. 2. At stage 1, the sarcoplasmic [Ca^{2+}] is low (10^{-7} M), crossbridges are detached and the muscle is relaxed. Binding studies predict that, under these circumstances, caldesmon will be bound to the thin filament. Stimulation of the muscle leads to a transient increase in sarcoplasmic [Ca^{2+}] to approximately 3-5 µM (Morgan & Morgan, 1982; Williams & Fay, 1986). We suggest that four Ca^{2+}-dependent processes are activated at this level of Ca^{2+}: myosin phosphorylation, calmodulin binding to caldesmon, phosphorylation of caldesmon and direct binding of Ca^{2+} to myosin (stage 2). Of these four processes, myosin phosphorylation is responsible for the development of tension, rapid crossbridge cycling and a high ATPase rate. The binding of calmodulin to caldesmon and the phosphorylation of caldesmon serve to remove caldesmon from the thin filament to allow rapid crossbridge cycling which would presumably be hindered by the presence of caldesmon on the thin filament. As mentioned earlier, *in vitro* binding experiments suggest that there is insufficient calmodulin in smooth muscle to remove a significant amount of caldesmon

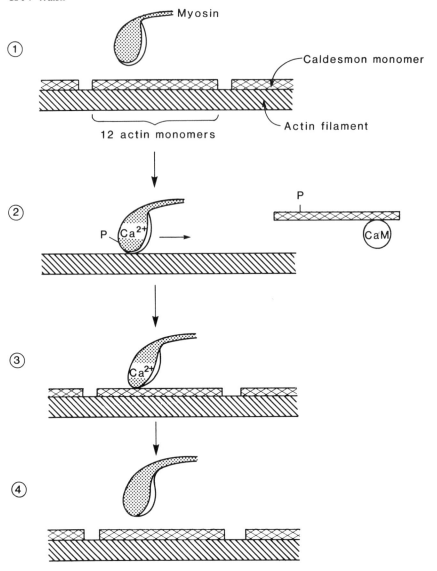

Figure 2. A working model postulating the role of caldesmon and direct Ca^{2+} binding to myosin in formation of latchbridges.

from F-actin in the presence of Ca^{2+}. However, recent binding studies we have performed comparing phosphorylated and nonphosphorylated caldesmons indicate that the phosphorylated protein has a reduced affinity for actin and an increased affinity for calmodulin (Ngai, P.K. & Walsh, M.P., unpublished observations).

Maintenance of the stimulus can lead to the latch state which is believed to occur when the sarcoplasmic [Ca^{2+}] falls to levels which are insufficient to maintain myosin in the phosphorylated state but are sufficiently elevated above resting levels to maintain at least one of the Ca^{2+}-dependent mechanisms in an activated state (stage 3). We suggest that, under these conditions, myosin is dephosphorylated, calmodulin dissociates from caldesmon, and caldesmon becomes dephosphorylated. The Ca^{2+} concentration is sufficient, however, to maintain Ca^{2+} bound to myosin heads. Caldesmon therefore returns to the thin filament and the conformation of the myosin heads is such that the crossbridge remains attached to the thin filament via caldesmon. This latch state is characterized by a low ATPase rate, slowly- or non-cycling crossbridges and the maintenance of tension. We have evidence that caldesmon can interact with myosin as well as with actin since several of the ATPase activities of skeletal muscle myosin are affected by caldesmon in the absence of actin (Lim & Walsh, 1986). Finally, as the sarcoplasmic free [Ca^{2+}] returns to resting levels, Ca^{2+} dissociates from the myosin heads which change conformation causing the crossbridges to detach and the muscle to relax (stage 4).

In considering this possible mechanism of action of caldesmon, the stoichiometry of caldesmon to actin and its arrangement on the thin filament are important considerations. Let us consider the 150 kDa caldesmon polypeptide as the basic unit. From the electron microscopic measurements of Fürst et al (1986), this would be ~70 nm long. The length of an actin monomer in the thin filament is 5.9 nm (Marston & Smith, 1985) so that a single molecule of caldesmon could span 12 actin monomers. The actual *in vivo* molar ratio of caldesmon:actin is 1:26 (Marston & Lehman, 1985). Two possible scenarios could account for this approximately 2-fold difference in stoichiometry. The proportions of the actomyosin and actin-intermediate filament domains may be more equal than estimated by Small et al (1986), in which case caldesmon could span the length of the actin filaments located in the actomyosin domain. Alternatively, a single continuous strand of caldesmon may run the length of the actin double helix, unlike the two-stranded distribution of tropomyosin, so that each caldesmon polypeptide contacts 24 actin monomers rather than only 12.

Obviously, Fig. 2 depicts a hypothetical model which serves primarily as a guide to our future investigations. Several questions remain to be answered: What is the affinity of myosin for Ca^{2+} in vivo? What are the Ca^{2+}-dependencies of the binding of phosphorylated and nonphosphorylated caldesmons to F-actin on the one hand and to calmodulin on the other? Can formation of a myosin-caldesmon-actin complex at low $[Ca^{2+}]$ be demonstrated? Does phosphorylation of caldesmon occur in vivo and does this assist in its removal from the thin filament? What is the precise location of caldesmon in the resting muscle, the actively contracting muscle and the muscle in latch? Clearly there is much still to be learnt about the molecular mechanisms involved in the regulation of smooth muscle contraction.

ACKNOWLEDGEMENTS

I would like to express my thanks to several colleagues who have made invaluable contributions to our original work referred to in this review: Dr. Kazuo Adachi, Cheryl A. Carruthers, Timothy Clark, Dr. Ute Gröschel-Stewart, Megan S. Lim, Philip K. Ngai, Gisele C. Scott-Woo and Cindy Sutherland. Our work is supported by grants from the Medical Research Council of Canada, the Alberta Heritage Foundation for Medical Research and the Canadian Heart Foundation.

REFERENCES

Adachi K, Carruthers CA, Walsh MP (1983). Identification of the native form of chicken gizzard myosin light chain kinase with the aid of monoclonal antibodies. Biochem Biophys Res Commun 115:855-863.
Adelstein RS, Eisenberg E (1980). Regulation and kinetics of the actin-myosin-ATP interaction. Annu Rev Biochem 49: 921-956.
Aksoy MO, Murphy RA, Kamm KE (1982). Role of Ca^{2+} and myosin light chain phosphorylation in regulation of smooth muscle. Am J Physiol 242:C109-C116.
Aksoy MO, Mras S, Kamm KE, Murphy RA (1983). Ca^{2+}, cAMP and changes in myosin phosphorylation during contraction of smooth muscle. Am J Physiol 245:C255-C270.
Aksoy MO, Stewart GJ, Harakal C (1986). Myosin light chain phosphorylation and evidence for latchbridge formation in norepinephrine stimulated canine veins. Biochem Biophys

Res Commun 135:735-741.
Ban T, Ishimura K, Fujita H, Sobue K, Kakiuchi S (1984). Immunocytochemical demonstration for caldesmon and actin in the striated and smooth muscle cells and non-muscular cells of various organs of rats. Acta Histochem Cytochem 17:331-338.
Bretscher A (1984). Smooth muscle caldesmon. Rapid purification and F-actin cross-linking properties. J Biol Chem 259:12873-12880.
Bretscher A (1986). J Muscle Res Cell Motility, in press (abstract).
Bretscher A, Lynch W (1985). Identification and localization of immunoreactive forms of caldesmon in smooth and non-muscle cells: A comparison with the distributions of tropomyosin and α-actinin. J Cell Biol 100:1656-1663.
Burgoyne RD, Cheek TR, Norman K-M (1986). Identification of a secretory granule-binding protein as caldesmon. Nature 319:68-70.
Butler TM, Siegman MJ, Mooers SU (1983). Chemical energy usage during shortening and work production in mammalian smooth muscle. Am J Physiol 244:C234-C242.
Chacko S, Rosenfeld A (1982). Regulation of actin-activated ATP hydrolysis by arterial myosin. Proc Natl Acad Sci USA 79:292-296.
Chacko S, Conti MA, Adelstein RS (1977). Effect of phosphorylation of smooth muscle myosin on actin activation and Ca^{2+} regulation. Proc Natl Acad Sci USA 74:129-133.
Chatterjee M, Murphy RA (1983). Calcium-dependent stress maintenance without myosin phosphorylation in skinned smooth muscle. Science 221:464-466.
Clark T, Ngai PK, Sutherland C, Gröschel-Stewart U, Walsh MP (1986). Vascular smooth muscle caldesmon. J Biol Chem 261:in press.
Dabrowska R, Galazkiewicz B (1986). Possible role of caldesmon in the regulation of smooth muscle and nonmuscle motile systems. Biomed Biochim Acta 45:S153-S158.
Dabrowska R, Goch A, Galazkiewicz B, Osinska H (1985). The influence of caldesmon on ATPase activity of the skeletal muscle actomyosin and bundling of actin filaments. Biochim Biophys Acta 842:70-75.
Dillon PF, Aksoy MO, Driska SP, Murphy RA (1981). Myosin phosphorylation and the cross-bridge cycle in arterial smooth muscle. Science 211:495-497.
Dingus J, Hwo S, Bryan J (1986). Identification by monoclonal antibodies and characterization of human platelet caldesmon. J Cell Biol 102:1748-1757.

Driska S, Hartshorne DJ (1975). The contractile proteins of smooth muscle. Properties and components of a Ca^{2+}-sensitive actomyosin from chicken gizzard. Arch Biochem Biophys 167:203-212.

Fujita H, Ishimura K, Ban T, Kurosumi M, Sobue K, Kakiuchi S (1984). Immunocytochemical demonstration of caldesmon and actin in thyroid glands of rats. Cell Tissue Res 237:375-377.

Fürst DO, Cross RA, DeMey J, Small RV (1986). Caldesmon is an elongated, flexible molecule localized in the actomyosin domains of smooth muscle. EMBO J 5:251-257.

Galazkiewicz B, Mossakowska M, Osinska H, Dabrowska R (1985). Polymerization of G-actin by caldesmon. FEBS Lett 184:144-149.

Gerthoffer WT, Murphy RA (1983). Myosin phosphorylation and regulation of cross-bridge cycle in tracheal smooth muscle. Am J Physiol 244:C182-C187.

Ikebe M, Hartshorne DJ (1985). Effects of Ca^{2+} on the conformation and enzymatic activity of smooth muscle myosin. J Biol Chem 260:13146-13153.

Ishimura K, Fujita H, Ban T, Matsuda H, Sobue K, Kakiuchi S (1986). Immunocytochemical demonstration of caldesmon (a calmodulin-binding, F-actin-interacting protein) in smooth muscle fibers and absorptive epithelial cells in the small intestine of the rat. Cell Tissue Res 235:207-209.

Kakiuchi R, Inui M, Morimoto K, Kanda K, Sobue K, Kakiuchi S (1983). Caldesmon, a calmodulin-binding, F-actin-interacting protein, is present in aorta, uterus, and platelets. FEBS Lett 154:351-356.

Kaminski EA, Chacko S (1984). Effects of Ca^{2+} and Mg^{2+} on the actin-activated ATP hydrolysis by phosphorylated heavy meromyosin from arterial smooth muscle. J Biol Chem 259:9104-9108.

Kamm KE, Stull JT (1985). The function of myosin and myosin light chain kinase phosphorylation in smooth muscle. Ann Rev Pharmacol Toxicol 25:593-620.

Kretsinger RH (1980). Structure and evolution of calcium-modulated proteins. CRC Crit Rev Biochem 8:119-174.

Lehman W (1986). Caldesmon association with smooth muscle thin filaments isolated in the presence and absence of calcium. Biochim Biophys Acta 885:88-90.

Lehman W, Szent-Györgyi AG (1975). Regulation of muscular contraction. Distribution of actin control and myosin control in the animal kingdom. J Gen Physiol 66:1-30.

Lim MS, Walsh MP (1986). The effects of caldesmon on the ATPase activities of rabbit skeletal muscle myosin. Biochem J, in press.

Marston SB, Lehman W (1985). Caldesmon is a Ca^{2+}-regulatory component of native smooth-muscle thin filaments. Biochem J 231:517-522.

Marston SB, Smith CWJ (1984). Purification and properties of Ca^{2+}-regulated thin filaments and F-actin from sheep aorta smooth muscle. J Muscle Res Cell Motility 5:559-575.

Marston SB, Smith CWJ (1985). The thin filaments of smooth muscles. J Muscle Res Cell Motility 6:669-708.

Marston S, Lehman W, Moody C, Smith CWJ (1985). Ca^{2+}-dependent regulation of smooth muscle thin filaments by caldesmon. Adv Prot Phosphatases 2:171-189.

Marston SB, Trevett RM, Walters M (1980). Calcium ion-regulated thin filaments from vascular smooth muscle. Biochem J 185:355-365.

Maruyama K, Sobue K, Kakiuchi S (1982). Effect of the calmodulin-caldesmon system on the physical state of actin filaments. In Kakiuchi S, Hidaka H, Means AR (eds): "Calmodulin and Intracellular Ca^{2+} Receptors", New York and London: Plenum Publishing Corp, pp 183-188.

Moody CJ, Marston SB, Smith CWJ (1985). Bundling of actin filaments by aorta caldesmon is not related to its regulatory function. FEBS Lett 191:107-112.

Morgan JP, Morgan KG (1982). Vascular smooth muscle: The first recorded Ca^{2+} transients. Pflügers Archiv 395:75-77.

Nag S, Seidel JC (1983). Dependence on Ca^{2+} and tropomyosin of the actin-activated ATPase activity of phosphorylated gizzard myosin in the presence of low concentrations of Mg^{2+}. J Biol Chem 258:6444-6449.

Ngai PK, Walsh MP (1984). Inhibition of smooth muscle actin-activated myosin Mg^{2+}-ATPase activity by caldesmon. J Biol Chem 259:13656-13659.

Ngai PK, Walsh MP (1985a). Properties of caldesmon isolated from chicken gizzard. Biochem J 230:695-707.

Ngai PK, Walsh MP (1985b). Detection of caldesmon in muscle and non-muscle tissues of the chicken using polyclonal antibodies. Biochem Biophys Res Commun 127:533-539.

Ngai PK, Carruthers CA, Walsh MP (1984). Isolation of the native form of chicken gizzard myosin light-chain kinase. Biochem J 218:863-870.

Owada MK, Hakura A, Iida K, Yahara I, Sobue K, Kakiuchi S (1984). Occurrence of caldesmon (a calmodulin-binding protein) in cultured cells: Comparison of normal and transformed cells. Proc Natl Acad Sci USA 81:3133-3137.

Pato MD (1985). Properties of the smooth muscle phosphatases from turkey gizzards. Adv Prot Phosphatases 1:367-382.

Rees DD, Frederiksen DW (1981). Calcium regulation of porcine aortic myosin. J Biol Chem 256:357-364.
Silver PJ, Stull JT (1982). Regulation of myosin light chain and phosphorylase phosphorylation in tracheal smooth muscle. J Biol Chem 257:6145-6150.
Small JV, Fürst DO, DeMey J (1985). Localization of filamin in smooth muscle. J Cell Biol 102:210-220.
Smith CWJ, Marston SB (1985). Disassembly and reconstitution of the Ca^{2+}-sensitive thin filaments of vascular smooth muscle. FEBS Lett 184:115-119.
Sobieszek A (1977). Vertebrate smooth muscle myosin. Enzymatic and structural properties. In Stephens NL (ed): "The Biochemistry of Smooth Muscle", Baltimore: University Park, pp 413-443.
Sobue K, Morimoto K, Inui M, Kanda K, Kakiuchi S (1982). Control of actin-myosin interaction of gizzard smooth muscle by calmodulin- and caldesmon-linked flip-flop mechanism. Biomed Res 3:188-196.
Sobue K, Muramoto Y, Fujita M, Kakiuchi S (1981a). Purification of a calmodulin-binding protein from chicken gizzard that interacts with F-actin. Proc Natl Acad Sci USA 78:5652-5655.
Sobue K, Muramoto Y, Fujita M, Kakiuchi S (1981b). Calmodulin-binding protein from chicken gizzard that interacts with F-actin. Biochem Int 2:469-476.
Sobue K, Takahashi K, Tanaka T, Kanda K, Ashino N, Kakiuchi S, Maruyama K (1985c). Crosslinking of actin filaments is caused by caldesmon aggregates, but not by its dimers. FEBS Lett 182:201-204.
Sobue K, Tanaka T, Kanda K, Ashino N, Kakiuchi S (1985a). Purification and characterization of caldesmon$_{77}$: A calmodulin-binding protein that interacts with actin filaments from bovine adrenal medulla. Proc Natl Acad Sci USA 82:5025-5029.
Sobue K, Tanaka T, Kanda K, Takahashi K, Ito K, Kakiuchi S (1985b). A dual regulation of the actin-myosin interaction in adrenal medullary actomyosin by actin-linked and myosin-linked systems. Biomed Res 6:93-102.
Somlyo AP (1985). Excitation-contraction coupling and the ultrastructure of smooth muscle. Circ Res 57:497-507.
Szpacenko A, Wagner J, Dabrowska R, Rüegg JC (1985). Caldesmon-induced inhibition of ATPase activity of actomyosin and contraction of skinned fibres of chicken gizzard smooth muscle. FEBS Lett 192:9-12.
Umekawa H, Hidaka H (1985). Phosphorylation of caldesmon by protein kinase C. Biochem Biophys Res Commun 132:56-62.

Walsh MP (1985). Calcium regulation of smooth muscle contraction. In Marmé D (ed): "Calcium and Cell Physiology", Berlin: Springer-Verlag, pp 170-203.

Walsh MP, Hartshorne DJ (1982). Actomyosin of smooth muscle. In Cheung WY (ed): "Calcium and Cell Function Vol 3", New York: Academic Press, pp 223-269.

Walsh MP, Hinkins S, Muguruma M, Hartshorne DJ (1983). Identification of two forms of myosin light chain kinase in turkey gizzard. FEBS Lett 153:156-160.

Williams DA, Fay FS (1986). Calcium transients and resting levels in isolated smooth muscle cells as monitored with quin 2. Am J Physiol 250:C779-C791.

MODULATION OF ACTOMYOSIN ATPase BY THIN FILAMENT-ASSOCIATED PROTEINS

Samuel Chacko, Hidetake Miyata and Kurumi Y. Horiuchi

Department of Pathobiology, University of Pennsylvania, 3800 Spruce Street, Philadelphia, Pennsylvania 19104

INTRODUCTION

Phosphorylation of the 20,000 dalton light chain of myosin is a prerequisite for actin-activation of the Mg-ATPase activity of smooth muscle myosin (Gorecka et al., 1976; Chacko et al., 1977; Sobieszek and Small, 1977). Maximal activation of the Mg-ATPase of phosphorylated myosin by actin requires stoichiometric binding of tropomyosin to actin (Miyata and Chacko, 1986) and the presence of Ca^{2+} at free Mg^{2+} below 3 mM (Chacko and Rosenfeld, 1982; Nag and Seidel, 1983; Heaslip and Chacko, 1985). The mechanism by which free Ca^{2+} modulates actin-activated ATP hydrolysis at low Mg^{2+} is not clear; however, conformational change of the myosin (Onishi and Wakabayashi, 1982; Trybus et al., 1982; Trybus and Lowey, 1984; Ikebe and Hartshorne, 1985) may play a role. The conformational change induced by Mg^{2+} has been shown to influence the ATPase activity of smooth muscle myosin (Ikebe et al., 1983).

In addition to tropomyosin, other actin binding proteins have been shown to influence the actin-myosin interaction and actomyosin ATPase in smooth muscle (Sobue et al., 1981; Hinssen et al., 1984; Dabrowska et al., 1985). The effects of some of these actin binding proteins on actomyosin ATPase are due to the aggregation of actin filaments induced by these proteins (Hinssen et al., 1984; Dabrowska et al., 1985). Caldesmon, an actin binding protein first reported by Kakiuchi and his colleagues (Sobue et al., 1981), also binds to calmodulin in the presence of calcium. Caldesmon has been shown to inhibit the actin-activated ATPase activity of smooth muscle myosin at concentrations not sufficient to aggregate the actin filaments (Ngai and Walsh, 1984; Marston et al., 1985; Sobue et al., 1985; Horiuchi et al., 1986).

In this paper we intend to review primarily our data on the role of tropomyosin on the ATPase activity of the reconstituted smooth muscle actomyosin and present evidence to show that the caldesmon inhibits the tropomyosin potentiated actin-activated ATP hydrolysis by phosphorylated smooth muscle myosin. This inhibition is released by calcium-calmodulin. Hence, the actin-activated ATPase of phosphorylated myosin is modulated by thin filament associated proteins.

RESULTS AND DISCUSSION

Effect of Ca^{2+} and Mg^{2+} on tropomyosin binding to actin and the effect of tropomyosin binding on actin-activation of the Mg-ATPase of myosin. Although the Mg-ATPase of phosphorylated myosin is activated by actin alone, the addition of tropomyosin to reconstituted actomyosin causes a 3-7 fold increase in actin-activated ATPase (Chacko et al., 1977; Sobieszek and Small, 1977; Hartshorne et al., 1977). The level of actin-activated ATPase in the presence of tropomyosin depends on the degree of phosphorylation; the actin-activated ATPase activity of unphosphorylated myosin is not increased by the addition of tropomyosin (Chacko, 1981). The enhancement of the actin-activated ATPase activity by tropomyosin is observed both in the presence and absence of Ca^{2+}, irrespective of the source of actin and tropomyosin (Heaslip and Chacko, 1985). The potentiation of the actin-activated ATPase activity of fully phosphorylated gizzard myosin by gizzard and skeletal muscle tropomyosin is shown in Figure 1. The ATPase activity increased on raising the actin concentration. The activity begins to plateau when the molar ratio of myosin:actin reaches 1:5. Further increase in actin concentration caused only a slight increase in the activity. When the actin used for preparing the reconstituted actomyosin contained tropomyosin, the activities, at all actin concentrations tested, are 3-4 fold higher than those of the actomyosin containing no tropomyosin. The potentiation was slightly greater with skeletal muscle tropomyosin.

This finding is not surprising since the binding of skeletal muscle tropomyosin to smooth muscle actin is also found to be higher under the conditions used in our study (Miyata and Chacko, 1986). The inhibition of skeletal muscle acto-HMM ATPase, which requires tropomyosin binding to actin, occurs at a higher free Mg^{2+} when smooth muscle tropomyosin is used instead of skeletal muscle tropomyosin (Williams et al., 1984).

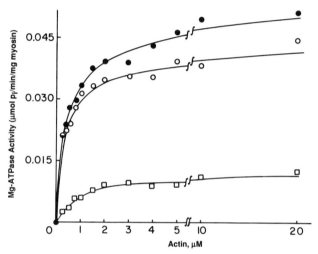

Figure 1. Effect of tropomyosin on the actin-activated ATPase activity of phosphorylated gizzard myosin. Phosphorylated myosin was reconstituted with varying concentrations of either actin (□) or actin containing smooth (○) or skeletal muscle (●) tropomyosin at molar ratio of tropomyosin to actin of 1:3. Assay conditions: myosin, 0.134 mg/ml; ionic strength, 0.05 M with KCl; imidazole-HCl, 15 mM; pH 7.0; Mg-ATP, 2 mM; free Mg^{2+}, 2 mM ($MgCl_2$); pCa 4; DTT, 2.5 mM and temperature, 25°C.

Since the ATPase activity of phosphorylated myosin reconstituted with actin and tropomyosin in low free Mg^{2+} (< 3 mM) is higher in the presence of Ca^{2+} than in its absence (Chacko and Rosenfeld, 1982; Nag and Seidel, 1983), it is possible that the increased actin-activated ATPase may be due to increased binding of tropomyosin to actin in the presence of Ca^{2+}. The effect of Ca^{2+} on the binding of smooth muscle tropomyosin to smooth muscle actin at various free Mg^{2+} concentrations is shown in Figure 2A. The molar ratios of tropomyosin bound to actin monomer at varying tropomyosin concentrations are determined and these ratios are plotted vs. free tropomyosin concentration. At 0.5 mM free Mg^{2+}, only a small amount of tropomyosin is bound to actin even at very high ratios of tropomyosin:actin. Full saturation of tropomyosin to actin (i.e., 1 mol of tropomyosin to 7 mol of actin monomer) is reached at free Mg^{2+} concentrations of 2, 4, and 8 mM. However, full saturation at 2 and 4 mM requires a molar ratio of 1:3 (tropomyosin:actin) while it was achieved at a ratio of 1:6 in 8 mM Mg^{2+}. The binding of tropomyosin to actin at all Mg^{2+} is independent of Ca^{2+} and it is highly cooperative as evidenced by the convex nature of the Scatchard plot (Figure 2B).

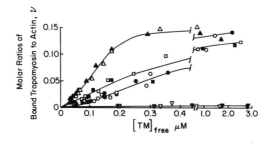

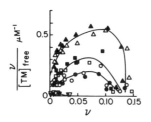

Figure 2A & B. Binding of gizzard tropomyosin to gizzard actin at various concentrations of free Mg^{2+} ions in the presence and absence of Ca^{2+}. Upper figure (A) shows the binding as a function of free tropomyosin concentration. Tropomyosin was iodinated with ^{125}I according to the procedure by Marchalonis (1969). Actin concentration was 20 μM. Free Mg^{2+} concentrations were 0.5 (▽,▼), 2 (○,●), 4 (□,■), and 8 mM (△,▲). Open symbols show the result at pCa 8, and closed symbols show the result at pCa 4. Buffer conditions were the following: ionic strength 0.05

M; imidazole hydrochloride 15 mM (pH 7.0); Mg-ATP 2 mM; DTT 2.5 mM. Lower figure (B) shows the Scatchard plot. Meaning of the symbols is the same. Taken from Miyata and Chacko (1986).

Binding of skeletal muscle tropomyosin to skeletal muscle actin is dependent not only on the Mg^{2+} concentration but also on the ionic strength (Eaton et al., 1975). In order to determine if the Mg^{2+} dependence on the binding of smooth muscle tropomyosin to smooth muscle actin is altered by varying the ionic strength (I), the binding at 0.5 and 2 mM free Mg^{2+} are determined as a function of ionic strength (Figure 3). At 2 mM free Mg^{2+} and 0.05 M I, the actin is saturated with tropomyosin at stoichiometric level. Raising the ionic strength to higher than 0.06 M causes a decrease in the binding. The binding at 0.5 mM free Mg^{2+} is very low at 0.05 M I (as shown also in Figure 2A). The binding at 0.5 mM Mg^{2+}

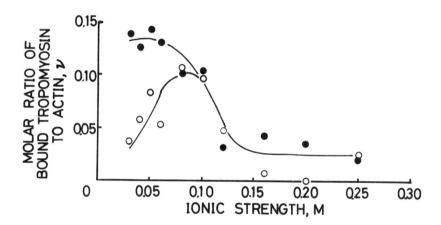

Figure 3. Binding of gizzard tropomyosin to gizzard F·actin as a function of ionic strength at two different free Mg^{2+} concentrations and in the presence of Ca^{2+} (pCa 4). Actin concentration was 10 µM. Tropomyosin concentration (iodinated with ^{125}I) was 3.3 µM. Molar ratio of tropomyosin:actin = 1:3. Free Mg^{2+} concentrations were 0.5 (O) or 2 mM (●). Buffer conditions were the same as described in the legend to Figure 1. Ionic strength was adjusted by adding appropriate amounts of KCl according to the calculation based on the procedure described by Fabiato and Fabiato (1979). Taken from Miyata and Chacko (1986).

increases on raising the ionic strength. At 0.1 M I, bindings at 0.5 mM and 2 mM Mg^{2+} are similar. Increasing the ionic strength to 0.2 M decreases the binding. Slightly higher binding of tropomyosin at ionic strengths between 0.1 and 0.2 M is obtained at both 0.5 and 2 mM free Mg^{2+} when noniodinated tropomyosin is used for determining the binding (data not shown). The reason for the decreased binding of the iodinated tropomyosin at higher ionic strengths is not clear; however the possibility that the part of the tropomyosin is oxidized during the lactoperoxidase reaction for iodination of the protein (Marchalonis, 1969) is not ruled out. The binding of skeletal muscle tropomyosin to skeletal muscle actin is also altered by modifying the Cys-190 of tropomyosin (Ishii and Lehrer, 1985; Walsh and Wegner, 1980). In any case, a decrease in the binding of smooth muscle tropomyosin to smooth muscle actin is observed on raising the ionic strength close to physiological.

The actin-activated ATPase activity is highly dependent on ionic strength (Figure 4). Although the actin is fully saturated with tropomyosin at 2 mM Mg^{2+} and ionic strength below 0.04 M, the ATPase is low. The maximum activity is observed at ionic strengths between 0.04 and 0.08 M. If the ATPase depends only on tropomyosin binding to actin, the activity between 0.08 and 0.1 M I would be expected to be the same at 0.5 and 2 mM free Mg. As shown in Figure 4, irrespective of the similar level of tropomyosin binding,

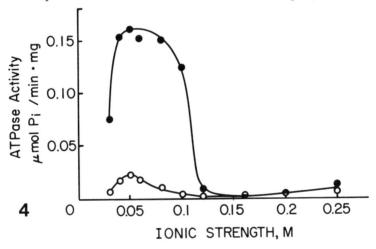

actin-activated ATPase in 2 mM free Mg^{2+} is several fold higher than in 0.5 mM free Mg^{2+}. This indicates that the free Mg^{2+} modulates the actin-activated ATPase not only by its effect on tropomyosin binding to actin but also by a direct effect on actin-activated ATP hydrolysis (Chacko and Rosenfeld, 1982; Nag and Seidel, 1983), presumably through its effect on the myosin conformation (Ikebe and Hartshorne, 1985). Since the effect of free Mg^{2+} on actin-activated ATP hydrolysis is observed also with heavy meromyosin (Kaminski and Chacko, 1984; Ikebe and Hartshorne, 1985), the conformational change involving the LMM portion of the myosin molecule (Ohnishi and Wakabayashi, 1982; Trybus et al., 1982) is not involved in the Mg^{2+} dependence for actin-activation. However, conformational change involving the myosin heads (Suzuki et al., 1985) may play an important role in the actin-activation.

The finding that there is less binding than stoichiometric ratio of smooth muscle tropomyosin bound to smooth muscle F·actin, at ionic strengths close to physiological conditions, appears to be a characteristic of the actomyosin system reconstituted with pure proteins since an ionic strength higher than 0.5 M is required to remove the tropomyosin from homogenized muscle fibers and native thin filaments (unpublished observation). This preliminary observation raises the possibility that other proteins may be important in keeping the tropomyosin bound to actin in the thin filaments in smooth muscle cells. Further studies are required to determine if any proteins associated with thin filaments of smooth muscle are important in the binding of tropomyosin to F·actin under physiological conditions.

Effect of caldesmon on actomyosin ATPase. Caldesmon, a protein which binds to calmodulin in the presence of Ca^{2+}, has been shown to be associated with thin filaments

---

Figure 4. Effect of ionic strength on the actin-activated ATPase activity of myosin under conditions which are the same as in Figure 3. Appropriate amounts of KCl were added to adjust the ionic strength. Solid circles and open circles indicate 2 and 0.5 mM free Mg^{2+}, respectively. Myosin:actin molar ratio is 1:20. Actin-activated ATPase was determined as in Figure 2. Taken from Miyata and Chacko (1986).

in smooth muscle cells from a variety of sources (Kakiuchi et al., 1983). In order to determine if the binding of caldesmon modifies the actin-activated ATP hydrolysis, ATPase activity of reconstituted actomyosin is determined in the presence of caldesmon. The effect of caldesmon, at varying molar ratios of caldesmon to actin, on the ATPase activities of thiophosphorylated myosin reconstituted with either pure actin or actin containing tropomyosin is shown in Figure 5.

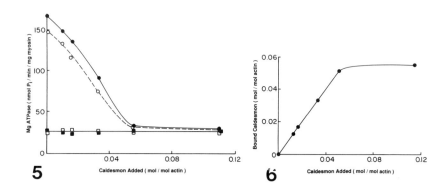

Figure 5. Inhibition of actin-activated Mg-ATPase of thiophosphorylated myosin by caldesmon. ATPase activity was measured with 0.13 mg/ml of gizzard myosin, 0.26 mg/ml of gizzard actin, and 0.14 mg/ml of gizzard tropomyosin (A:Tm=3:1) in 15 mM imidazole buffer (pH 7.0), 4 mM MgCl2, 2 mM (γ-^{32}P)-ATP, 2.5 mM DTT, 50 mM I (adjusted with KCl) at 37°C. In the presence (●,○) or absence (■,□) of tropomyosin at pCa 4 (●,■) and pCa 8 (○,□). Caldesmon concentration was calculated using the molecular weight of 300 K daltons. Taken from Horiuchi et al. (1986).

Figure 6. Binding of caldesmon to actin containing tropomyosin at pCa 4. Conditions were the same as described in the legend to Figure 1. The binding of caldesmon levelled off at the caldesmon to actin molar ratio of around 1:18, a molar ratio at which the ATPase was maximally inhibited (Figure 5). Taken from Horiuchi et al. (1986).

The addition of caldesmon to pure actin has no effect on the actomyosin ATPase. However, the activity of actomyosin containing tropomyosin decreases on addition of caldesmon and at caldesmon:actin molar ratio of 1:18, it reaches the level of activity of the actomyosin which contains no tropomyosin. The effect of caldesmon is independent of Ca^{2+}. The inhibition of the ATPase activity of reconstituted actomyosin containing tropomyosin by caldesmon is associated with the binding of caldesmon to actin as shown in Figure 6.

Measurements of tropomyosin bound to actin in the presence and absence of caldesmon showed that the binding of tropomyosin to actin is not affected by caldesmon. Stoichiometric binding of tropomyosin to actin is obtained under the conditions of the ATPase assay. Hence, the lack of enhancement of the actin-activated ATPase by tropomyosin in the presence of caldesmon is not due to an inhibition of the tropomyosin binding to actin by caldesmon.

The inhibition of ATPase by caldesmon is reversed by the addition of calmodulin. The effect of varying concentrations of calmodulin on the reversal of the actin-activated ATPase inhibited by caldesmon is shown in Figure 7.

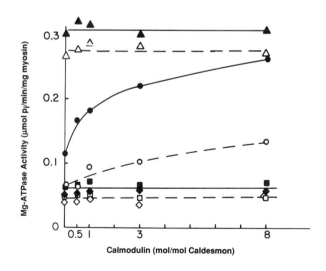

Figure 7

Figure 7. Reversal of the caldesmon inhibited ATPase activities by calmodulin. Conditions of the assay are the same as in Figure 5. Thiophosphorylated myosin reconstituted with actin-tropomyosin in the presence (●,○) and absence (▲,△) of caldesmon; thiophosphorylated myosin reconstituted with actin in the presence (■,□) and absence (◆,◇) of caldesmon. ●,▲,■,◆ in pCa 4 and ○, △, □, ◇ in pCa 8.

Caldesmon induced inhibition of tropomyosin potentiated activity begins to disappear on the addition of calmodulin in the presence of Ca^{2+}. The amount of calmodulin required to obtain maximum reversal of the activity inhibited by caldesmon is not consistent for the various preparations of the caldesmon; but the reversed level of activity plateaus around a molar concentration (caldesmon: calmodulin) of 1:10. The reversal of the inhibition of the ATPase requires Ca^{2+}. The low level of reversal in pCa 8 on the addition of calmodulin (Figure 7) can be abolished by adding EGTA to the ATPase mixture to obtain a final concentration of 1 mM EGTA (data not shown).

As shown in Figure 5, the Mg-ATPase of phosphorylated myosin, activated by actin containing no tropomyosin, is not affected by the caldesmon at molar concentrations of actin:caldesmon up to 9:1, concentrations at which the maximum inhibition of the tropomyosin potentiated actin-activated activity is observed. The data from this experiment demonstrate that caldesmon inhibits the activation of Mg-ATPase by smooth muscle actin containing stoichiometric amount of bound tropomyosin. Calmodulin reverses this inhibition in the presence of Ca^{2+}. Thus, calmodulin modulates the activation of myosin ATPase by F·actin containing tropomyosin and caldesmon.

It has been suggested that the caldesmon, in combination with calmodulin, regulates the actin-activated Mg-ATPase in a "flip-flop" manner depending on the free Ca^{2+} concentration (Kakiuchi and Sobue, 1983; Sobue et al., 1985). This model for the regulation of actin-activation requires caldesmon to be released from the actin in the presence of calcium-calmodulin in order for actin to activate the Mg-ATPase of myosin. The finding that caldesmon inhibits only the tropomyosin potentiated actin-activated ATPase raises the possibility that tropomyosin favors the binding of caldesmon to actin. The effect of tropomyosin on the binding of caldesmon to actin in the presence and absence of calcium-calmodulin is determined under the conditions of the actin-activated ATPase assays (Table 1).

TABLE 1: EFFECT OF TROPOMYOSIN ON CALDESMON BINDING TO ACTIN

		CALDESMON BOUND TO ACTIN (MOL/MOL)			
		-CALMODULIN		+CALMODULIN	
CAD PREP #	MOLAR RATIO OF TROPOMYOSIN:ACTIN	pCa 4	pCa 8	pCa 4	pCa 8
23	0	0.051	0.050	0.021	0.050
	1:3	0.050	0.057	0.036	0.050
24	0	0.048	0.051	0.028	0.051
	1:3	0.053	0.058	0.039	0.054

CONDITIONS: Thiophosphorylated gizzard myosin was reconstituted with gizzard actin or actin-tropomyosin, caldesmon and either with or without calmodulin (molar ratio of caldesmon:calmodulin, 1:8). The buffer conditions are similar to that of ATPase assays (ionic strength, 50 mM; Mg-ATP, 2 mM; free Mg^{2+}, 2 mM; pCa 4 or 8; DTT, 2 mM; imidazole-HCl, 15 mM; pH 7.0 and temperature 23°C. The samples were incubated for 10 minutes and centrifuged using an Airfuge for 20 minutes. The supernatants and pellets were collected separately, dissolved in SDS sample buffer and electrophoresed (Fairbanks et al., 1971). The amounts of caldesmon bound to sedimented actin (mol per mol actin) in each case were determined by scanning densitometry of the gels stained with Coomassie brilliant blue.

Mole of caldesmon bound per mole of actin is not altered by the binding of tropomyosin to actin at stoichiometric levels both in the presence and absence of Ca^{2+}. Caldesmon is partially (60%) released from actin containing no tropomyosin to the supernatant on addition of calmodulin in the presence of Ca^{2+} (pCa 4). However, the release of caldesmon from actin is lower (20-26%) when the actin contained bound tropomyosin indicating that the tropomyosin increases the affinity of caldesmon to actin in the presence of calcium-calmodulin. As shown in Table 1, although the same level of caldesmon binding to actin is obtained with pure actin and actin containing tropomyosin, only the tropomyosin potentiated activity is inhibited. Furthermore, caldesmon remains bound to actin when the inhibition is released in the presence of calcium-calmodulin. Hence this data are not in agreement with the "flip-flop" mechanism (Kakiuchi and Sobue, 1983). Neither the binding nor the release of caldesmon, in the presence of calcium-calmodulin, from actin devoid of tropomyosin has any effect on the actin-activation.

Reversal of the caldesmon induced inhibition of actomyosin ATPase has been shown to be associated with the phosphorylation of the caldesmon by a specific kinase which requires calcium-calmodulin (Ngai and Walsh, 1984). Caldesmon preparation is often contaminated by caldesmon kinase; hence phosphorylation of the caldesmon occurs with the reversal of the inhibition of actomyosin ATPase on the addition of calcium-calmodulin (Horiuchi et al., 1986). However, the inhibition of actomyosin ATPase by caldesmon preparations which are free of kinase activity is also reversed and the caldesmon is partially released from actin on the addition of calcium-calmodulin. Therefore, the role of caldesmon phosphorylation on the binding of caldesmon to actin and the actomyosin ATPase is not clear.

SUMMARY

Phosphorylation of the myosin light chain is a prerequisite for actin-activation of the Mg-ATPase of smooth muscle myosin. However, maximal activation of the Mg-ATPase by actin requires stoichiometric binding of tropomyosin to actin filaments and Ca^{2+} at free Mg^{2+} below 3 mM. The requirement for Ca^{2+} for actin-activation is not due to a calcium-mediated binding of tropomyosin to actin since the binding of tropomyosin to actin is not dependent on Ca^{2+}. Caldesmon, an actin and calmodulin binding protein, at caldesmon:actin molar ratio of 1:18, binds equally to pure actin and actin containing stoichiometric amounts of bound tropomyosin. The Mg-ATPase of myosin reconstituted with actin is not affected by the caldesmon; on the other hand, the activity of actomyosin containing tropomyosin is inhibited. The inhibition of activity by the caldesmon is reversed by the addition of calmodulin (caldesmon:calmodulin molar ratio, 1:8) in the presence of Ca^{2+}. The amount of caldesmon bound to actin in the presence of calcium-calmodulin is 50% more when actin filaments contain tropomyosin, indicating that the release of inhibition of the activity inhibited by caldesmon does not require complete release of caldesmon from actin.

ACKNOWLEDGEMENTS

We are grateful to Janet Flinn for expert secretarial assistance and Dr. William S. Filler for valuable discussions. This work was supported by HL 22264 and PCM 83-09139.

REFERENCES

Chacko S (1981). Effects of phosphorylation, calcium ion, and tropomyosin on actin-activated adenosine 5'-triphosphatase activity of mammalian smooth muscle myosin. Biochemistry 20:702-707.

Chacko S, Conti MA, Adelstein RS (1977). Effect of phosphorylation of smooth muscle myosin on actin activation and Ca^{2+} regulation. Proc Natl Acad Sci USA 74:129-133.

Chacko S, Rosenfeld A (1982). Regulation of actin-activated ATP hydrolysis by arterial myosin. Proc Natl Acad Sci USA 79:292-296.

Dabrowska R, Goch A, Galazkiewicz B, Osinska H (1985). The influence of caldesmon on ATPase activity of the skeletal muscle actomyosin and bundling of actin filaments. Biochemica et Biophysica Acta 842:70-75.

Eaton BL, Kominz DR, Eisenberg E (1975). Correlation between the inhibition of the acto-heavy meromyosin ATPase and the binding of tropomyosin to F-actin: Effects of Mg^{2+}, KCl, troponin I and troponin C. Biochemistry 14:2718-2725.

Fabiato A, Fabiato F (1979). Calculator programs for computing the composition of the solutions containing multiple metals and ligands used for experiments in skinned muscle cells. J Physiol (Paris) 75:479-494.

Fairbanks GT, Steck TL, Wallach DFH (1971). Electrophoretic analysis of the major polypeptides of the human erythrocyte membrane. Biochemistry 10:2606-2617.

Gorecka A, Aksoy MO, Hartshorne DJ (1976). The effect of phosphorylation of gizzard myosin on actin-activation. Biochem Biophys Res Commun 71:325-331.

Hartshorne DJ, Gorecka A, Aksoy MO (1977). Aspects of regulatory mechanism in smooth muscle. In Casteels R, Godfraind T, Ruegg JC (eds): "Excitation-Contraction Coupling in Smooth Muscle," Amsterdam: Elsevier/North Holland, p 377.

Heaslip RJ, Chacko S (1985). The effects of Ca^{2+} and Mg^{2+} on the actomyosin ATPase of stably phosphorylated gizzard myosin. Biochemistry 24:2731-2736.

Hinssen H, Small JV, Sobieszek A (1984). A Ca^{2+}-dependent actin modulator from vertebrate smooth muscle. FEBS Lett 166:90-95.

Horiuchi KY, Miyata H, Chacko S (1986). Modulation of smooth muscle actomyosin ATPase by thin filament associated proteins. Biochem Biophys Res Commun 136:962-968.

Ikebe M, Hartshorne DJ (1985). Effects of Ca^{2+} on the conformation and enzymatic activity of smooth muscle myosin. J Biol Chem 260:13146-13153.

Ikebe M, Hinkins S, Hartshorne DJ (1983). Correlation of enzymatic properties and conformation of smooth muscle myosin. Biochemistry 22:4580-4587.

Ishii Y, Lehrer SS (1985). Fluorescence studies of the conformation of pyrene-labeled tropomyosin: Effects of F-actin and Myosin subfragment-1. Biochemistry 24:6631-6639.

Kakiuchi S, Sobue K (1983). Control of the cytoskeleton by calmodulin and calmodulin-binding proteins. Trends Biochem Sci 8:59-62.

Kakiuchi R, Inui M, Morimoto K, Kanda K, Sobue K, Kakiuchi S (1983). Caldesmon, a calmodulin-binding, F·actin-interacting protein, is present in aorta, uterus and platelets. FEBS Lett 154:351-356.

Kaminski EA, Chacko S (1984). Effects of Ca^{2+} and Mg^{2+} on the actin-activated ATP hydrolysis by phosphorylated heavy meromyosin from arterial smooth muscle. J Biol Chem 259:9104-9108.

Marchalonis JJ (1969). An enzymic method for the trace iodination of immunoglobulins and other proteins. Biochem J 113:299-305.

Marston SB, Lehman W (1985). Caldesmon is a Ca^{2+}-regulatory component of native smooth-muscle thin filaments. Biochem J 231:517-522.

Marston SB, Lehman W, Moody C, Smith C (1985). Ca^{2+}-dependent regulation of smooth muscle thin filaments by caldesmon. Adv Prot Phosphatases II, 171 189.

Miyata H, Chacko S (1986). Role of tropomyosin in smooth muscle contraction: Effect of tropomyosin binding to actin on actin-activation of myosin ATPase. Biochemistry 25:2725-2729.

Nag S, Seidel JC (1983). Dependence on Ca^{2+} and tropomyosin of the actin-activated ATPase activity of phosphorylated gizzard myosin in the presence of low concentrations of Mg^{2+}. J Biol Chem 258:6444-6449.

Ngai PK, Walsh MP (1984). Inhibition of smooth muscle actin-activated myosin Mg^{2+}-ATPase activity by caldesmon. J Biol Chem 259:13656-13659.

Onishi H, Wakabayashi T (1982). Electron microscopic studies of myosin molecules from chicken gizzard muscle I: The formation of the intramolecular loop in the myosin tail. J Biochem (Japan) 92:871-879.

Sobieszek A, Small JV (1977). Regulation of the actin-myosin interaction in vertebrate smooth muscle: Activation via a myosin light chain kinase and the effect of tropomyosin. J Mol Biol 112:559-576.

Sobue K, Muramoto Y, Fujita M, Kakiuchi S (1981). Purification of a calmodulin-binding protein from chicken gizzard that interacts with F-actin. Proc Natl Acad Sci USA 78:5652-5655.

Sobue K, Takahashi K, Wakabayashi I (1985). $Caldesmon_{150}$ regulates the tropomyosin-enhanced actin-myosin interaction in gizzard smooth muscle. Biochem Biophys Res Commun 132:645-651.

Suzuki H, Stafford WF, Slayter HS, Seidel JC (1985). A conformational transition in gizzard heavy meromyosin involving the head-tail junction resulting in changes in sedimentation coefficient, ATPase activity and orientation of heads. J Biol Chem 260:14810-14817.

Trybus KM, Huiatt TW, Lowey S (1982). A bent monomeric conformation of myosin from smooth muscle. Proc Natl Acad Sci USA 79:6151-6155.

Trybus KM, Lowey S (1984). Conformational states of smooth muscle myosin. J Biol Chem 259:8564-8571.

Walsh TP, Wegner A (1980). Effect of the state of oxidation of cysteine 190 of tropomyosin on the assembly of the actin-tropomyosin complex. Biochemica et Biophysica Acta 626:79-87.

Williams DL, Greene LE, Eisenberg E (1984). Comparison of effects of smooth and skeletal muscle tropomyosin on interactions of actin and myosin subfragment 1. Biochemistry 23:4150-4155.

Yamaguchi M, Ver A, Carlos A, Seidel JC (1984). Modulation of the actin activated adenosine triphosphatase activity of myosin by tropomyosin from vascular and gizzard smooth muscles. Biochemistry 23:774-779.

MgATPase ACTIVITY OF VERTEBRATE SMOOTH MUSCLE ACTOMYOSIN:
STIMULATION BY TROPOMYOSIN IS MODIFIED BY MYOSIN PHOS-
PHORYLATION AND ITS CONFORMATIONAL STATE

Apolinary Sobieszek

Institute of Molecular Biology, Austrian
Academy of Sciences
Billrothstr. 11, A-5020 Salzburg, Austria

INTRODUCTION

In all actomyosin systems tropomyosin is a constant component of the thin filaments, independent of their mode of Ca^{2+}-dependent regulation. Its role in skeletal muscle has been well established (Ebashi, 1980; Gergely, 1980; Perry, 1983) where it propagates certain conformational changes along the actin filament in response to the binding of Ca^{2+} to troponin (Potter and Gergely, 1974; Leavis and Gergely, 1984). From studies of the influence of various muscle and non-muscle tropomyosins on the actin-activation of myosin ATPase using mainly skeletal muscle actomyosin as a model system, three tropomyosin effects were established. Firstly, at micromolar ATP concentrations tropomyosin potentiates F-actin cofactor activity. This potentiation is absent at physiological ATP concentrations and arises from the presence of a population of rigor type myosin head attachments to actin, together with normal myosin-ATP-actin complexes (Bremel et al., 1972). Secondly, under certain conditions tropomyosin can inhibit the ATPase activity, this inhibition being more a characteristic of striated muscle tropomyosin (Eaton et al., 1975; Williams et al., 1974). Thirdly, at high myosin to actin ratios no inhibition is generally observed (Sobieszek, 1982; Lehrer and Morris, 1982): instead the presence of tropomyosin results in a stimulation of the actin-activated ATPase activity. Such stimulation is a characteristic feature of smooth and non-muscle tropomyosin (Sobieszek and Small, 1981; Yamaguchi et al., 1984; Mayata and Chacko, 1986) and seems to be unaffected by the occupancy of myosin heads on F-actin (Sobieszek, 1982).

For the present discussion the role of tropomyosin in vertebrate smooth muscle is of particular interest and is not presently fully understood (Marston and Smith, 1985). In this type of muscle, phosphorylation of myosin plays a primary role in the regulation of actin-myosin interaction (Adelstein and Eisenberg, 1980; Small and Sobieszek, 1980; Marston, 1982; Hartshorne and Mrwa, 1982; Kamm and Stull, 1985). The controversial question, at present, is whether or not there are other regulatory and/or modulatory components associated with actin (Marston and Smith, 1985). If so, these components must functionally be incorporated into the tropomyosin modulatory system.

The aim of the present study was to establish more precisely the role of tropomyosin in the interaction of myosin and actin in smooth muscle to provide a basis for on-going investigations of possible modulators associated with actin. In particular, it has been established that the stimulation of the ATPase activity by tropomyosin depends on the level of myosin phosphorylation, the number of myosin heads attached to F-actin and is greatly modified by the conformational state of myosin.

MATERIALS AND METHODS

Actomyosin and Myosin

Similarly to numerous current studies (for ref. see Marston, 1982), turkey and chicken gizzard or pig stomach myofibrils were used as a starting preparation. These are well homogenized muscle fragments, extensively washed prior to extraction of actomyosin (Sobieszek and Bremel, 1975). The freshly extracted actomyosin was either directly fractionated for purification of myosin (Small and Sobieszek, 1983; Sobieszek, 1985) or precipitated overnight after addition of divalent cations (Sobieszek and Small, 1976).

The removal of tropomyosin and myosin light chain kinase (MLCKase) from actomyosin was carried out as before (Sobieszek and Small, 1977) by at least two selective precipitations at 25% ammonium sulfate saturation of complexed actin and myosin. The actomyosin (pellet) and myosin (suspension) aliquots were stored at -70°C after their initial freezing in liquid nitrogen.

Actin

As an actin source we used an acetone powder made directly from "myofibrils" or from actomyosin extracted "myofibrils". Extraction and purification was essentially carried out as described previously (Strzelecka-Colaszewska and Sobieszek, 1981). This procedure was, however, significantly improved by a single purification step. This involved a precipitation at 25% ammonium sulfate saturation of semi-polymerized actin from normally very diluted extract. Relatively pure actin from this pellet was subsequently depolymerized by extensive dialysis (G-actin conditions) and most of the remaining impurities removed by 2 h high speed centrifugation (45,000 rpm). The supernatants which had a G-actin concentration of about 8 to 12 mg/ml, could then be directly polymerized by addition of KCl and/or $MgCl_2$ or after further purification (Sephadex G-150 column) and concentration (Sobieszek, in preparation). After freezing in liquid nitrogen F-actin was stored at -70°C as a 8 to 16 mg/ml solution.

Tropomyosin

Some "by-product" supernatants obtained during actomyosin, myosin or actin ammonium sulfate fractionation contained considerable amounts of tropomyosin and were therefore used as a convenient source of tropomyosin. The tropomyosin was collected by precipitation at pH 4.7. When a considerable amount of this material was accumulated, it was dissolved in 1 M KCl solution and tropomyosin purified from it by at least three cycles of purification; 40-55% ammonium sulfate followed by an isoelectric point precipitation (Bailey, 1948). Tropomyosin was stored at -25°C as a 10 mg/ml stock solution.

Enzymes

Myosin light chain kinase (MLCKase) and myosin light chain phosphatase (MLCPase) were purified as described in detail previously (Sobieszek and Barylko, 1984). Small aliquots of concentrated solutions of these enzymes were stored at -70°C after their initial freezing in liquid nitrogen. Some of their activity was lost even after the first freezing and therefore the aliquots were never frozen again.

Concentrated preparations of the MLCKase were obtained by including a calmodulin affinity column as a final purification step. A sufficiently pure and concentrated preparation of the MLCPase was obtained by terminating the purification after a small volume, AcA 34 gel filtration step (Sobieszek and Barylko, 1984).

Calmodulin

Gizzard calmodulin was obtained as a by-product of the MLCKase purification (Sobieszek and Barylko, 1984). Pig stomach calmodulin was purified from a 55% ammonium sulfate supernatant of a whole muscle, EGTA-containing extract. After addition of $CaCl_2$, the supernatant was applied into a Phenyl-Sepharose column and eluted with EGTA containing buffer (Gopalakrishna and Anderson, 1982). The eluted calmodulin was then subjected to three further purification steps. This included isoelectric point precipitation, gel filtration and DEAE-ionic exchange chromatography.

ATPase and Phosphorylation Assays

ATPase assays were carried out essentially as before (Sobieszek and Small, 1976) but with the final assay volume reduced to 1550 µl. The ATPase reaction was initiated by pipetting 100 to 150 µl of ATP stock solution (17.5 mM), and terminated by the addition of 150 µl to 50% (w/v) TCA solution. Incubation times were varied according to the myosin concentration and ranges between 2 to 10 min. The assay temperature was 25°C.

The assay medium contained 60 mM KCl, 2 mM Mg Cl_2, 0.5 mM DTT, 0.1 mM Ca Cl_2 and 10 mM imidazole with the pH adjusted to 7.0, 7.3 or 6.6. For the assay at pH 6.6, the medium contained additionally 10 mM BIS TRIS and only 40 mM KCl. Unless otherwise stated myosin concentration was 2 to 4 mg/ml.

Protein concentrations were determined by the Biuret-method using a protein standard from Sigma (St. Louis, USA) for calibration.

Myosin phosphorylation was carried out in the same medium as used for the ATPase measurements with a final assay volume of 120 µl. The phosphorylation reaction was initiated by addition of 10 to 20 µl of γ^{32}P ATP stock solution which had been made from a 100 to 200 fold dilution of γ^{32}P ATP (3000 Ci/ m mol; NEM Boston, USA) with the 17.5 mM cold ATP solution as used in the ATPase assays. The reaction was stopped by addition of 102 mg of solid urea. After solubilization of the urea, 150 µl of the urea-mixture was applied onto 2 cm x 4 cm pieces of 3 MM (Whatman, Maidstone, U.K.) paper. The 3 MM pieces were washed in several changes of 5% TCA and after a final rinse in ethanol they were counted in water (Sobieszek and Barylko, 1984).

The remaining 30 to 50 µl of the urea-mixture was stored in a freezer until its analysis by urea-glycerol gel electrophoresis (Sobieszek and Jertschin, 1986).

Tropomyosin Binding Assays

Tropomyosin binding experiments were done by adding increasing amounts of iodinated tropomyosin to F-actin or actomyosin, and subsequent pelleting of the formed complexes. To obtain tight pellets, a 30 min centrifugation was necessary using a Beckman Airfuge (Fullerton, USA) for F-actin, and Biofuge A (Heraeus-Crist, Osterode, F.R.G.) for actomyosin. A 2 mg/ml stock solution of ^{125}I labelled tropomyosin was prepared according to the method of Eaton et al., (1975). The specific activity was high enough so that it could be diluted a further 3 to 9 fold with cold tropomyosin.

RESULTS

Tropomyosin-, and MLCKase-Free Actomyosin

Gizzard actomyosin isolated by our published procedure (via Mg or Ca precipitation: Sobieszek and Small, 1976; 1977) contains relatively high levels of myosin light chain kinase (MLCK) activity. These levels are sufficient to obtain phosphorylation up to 2 mol P_i per mol myosin within 5 to 10 s after addition of Ca^{2+} and ATP. Tropomyosin and MLCKase was removed from such actomyosin by at least two ammonium sulfate precipitations at 25% saturation (Sobieszek and Small, 1977; Small and Sobieszek, 1983). Since the second 25% actomyosin pellet is often very loose, its sedimentation is best achieved at higher speeds (19,000 for 1 to 2 h) than previously indicated.

As shown previously (Sobieszek and Small, 1977), no traces of tropomyosin can be detected in this actomyosin by

SDS-gel electrophoresis. Using $\gamma^{32}P$ ATP in phosphorylation assays we established that there was about 200 to 400 times less MLCKase activity present as compared to the parent actomyosin. Only with prolonged incubation times (30 min or more) did this residual activity produce significant levels of myosin phosphorylation. Moreover, these levels always approached 50%, independent of the initial, residual phosphorylation rate, suggesting that in this actomyosin and for such low levels of endogenous kinase, only one myosin head may be phosphorylated.

Addition of calmodulin did not produce any change in phosphorylation rate. Since the MLCKase activity is calmodulin dependent we concluded that the tropomyosin-free actomyosin contains stoichiometrical amounts of calmodulin and MLCKase. In contrast, the addition of tropomyosin to this actomyosin resulted in a 30 to 50% reduction of the phosphorylation rate. The optimal inhibition was obtained at approximately physiological molar ratios of tropomyosin to actin (1 to 7).

Myosin Phosphorylation and Tropomyosin Stimulation

The actomyosin depleted of both tropomyosin and MLCKase, described above, represents a very convenient model system for investigation of correlations between phosphorylation and actin-activation of myosin. By varying the amount of MLCKase and calmodulin added and/or the incubation time, a stepwise increase of myosin phosphorylation could be obtained (Fig. 1 A and B). Taking advantage of this possibility to control the phosphorylation level we have been measuring the actin-activated ATPase activity as a function of phosphorylation (Fig. 1 A and B). In the absence of tropomyosin, the relationship was clearly parabolical, indicating a random phosphorylation of the two myosin heads with a requirement that both heads have to be phosphorylated before the molecule could be activated by actin.

With added tropomyosin the phosphorylation rates were approximately the same but, as expected, the ATPase activity was generally about 2-fold higher. Notably, however, a differential stimulation of ATPase was observed with tropomyosin such that the relationship between phosphorylation and ATPase activity became linear (Fig. 1 A and B). Such a linear relationship implies a strong positive cooperativity in phosphorylation of the two myosin heads.

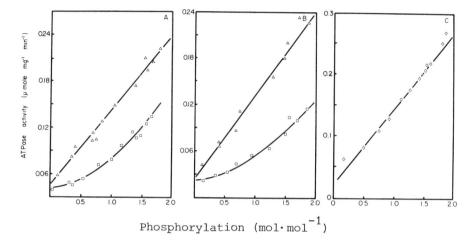

Fig. 1 Relationship between myosin phosphorylation and Mg ATPase activity. (A) Mg- and (B) Ca-precipitated chicken gizzard actomyosins free of both tropomyosin and MLCKase. (C) - Ca-precipitated native actomyosin. For (A) and (B) the experiments were performed in the absence (squares) and presence (triangles) of turkey gizzard tropomyosin. Phosphorylation levels were measured by the addition of γ-^{32}P ATP into a separate 100 µl aliquot of the ATPase assay mixture. Abscisse point-values corresponded to the final phosphorylation levels while their ordinate values were the average initial rates. Actomyosin concentration was 5.0 mg/ml and pH 7.0. For more details see Materials and Methods and the text.

A similar linear relationship was obtained for a native actomyosin containing tropomyosin (Fig. 1 C). In this latter case the phosphorylation rate was varied by controlling the free Ca^{2+} concentration using a Ca/EGTA buffer system.

Actin-Activation; pH Dependence

In addition to other medium factors, pH greatly influences the actin-activated ATPase activity of smooth muscle myosin. That this aspect is not generally taken into account is indicated by the wide pH range used in published studies (see for example: Lebowitz and Cooke, 1979; Ebashi et al., 1979; Persechini and Hartshorne, 1981; Sellers et al., 1983;

Yamaguchi et al., 1984). We have investigated this pH dependence more systematically.

In agreement with earlier reports (Wachsberger and Kaldor, 1971; Driska and Hartshorne, 1975) the ATPase activity of myosin alone was also pH sensitive (Fig. 2) but over a relatively wide pH range. Our measurements were carried out under exactly the same conditions as for the actin-activation measurements for which myosin was phosphorylated and tropomyosin was present. In this situation it was found that tropomyosin did not result in any change of the ATPase activity of myosin alone, like it does for actomyosin.

As can be seen in Fig. 2 myosin ATPase had a broad maximum at a pH around 5.0. This maximum decreased relatively slowly towards neutral and alkaline conditions. Thus, the activity of myosin alone represented a significant portion of the actin-activated ATPase activity especially at pH

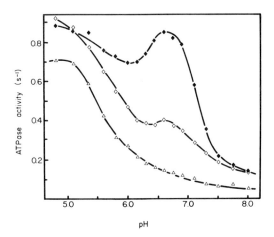

Fig. 2 pH dependence of Mg ATPase activity of reconstituted turkey gizzard actomyosin in the presence (closed diamonds) and absence (open diamonds) of tropomyosin. The activities of myosin alone are indicated by triangles. The proteins were first resuspended or dissolved in a solution containing 60 mM KCl and 2 mM Mg Cl_2. They were then pipetted into assay tubes containing 150 μl of buffer mixture (150 mM histidine and 100 mM imidazole) with the pH adjusted to the required values. The starting ATP-containing solution was not buffered. The final concentrations were 1.2, 1.5 and 0.5 mg/ml respectively for myosin, actin and tropomyosin.

values below 7.0. Under the assay conditions at pH above 7.3 (see figure legend) myosin was in monomeric form whereas at a pH around 6.6 it was assembled into filaments (data not shown, see also Kendrick-Jones et al., 1983).

In contrast to the data with myosin alone, the activation of ATPase by F-actin was relatively insensitive of pH (Fig. 2, middle curve). There was, however, a small maximum at pH values around 6.6. As pointed out before, these were the most favourable conditions for the formation of myosin filaments. Significantly, tropomyosin stimulation of the ATPase activity was again strongly pH dependent and occurred only in the pH range between 6.0 to 7.0 with a clear maximum at pH 6.6 (Fig. 2). Thus, it can be concluded that tropomyosin stimulation was optimal for myosin assembled into filaments, stable in the presence of ATP.

Actin-Activation, KCl Dependence

In contrast to the relatively simple pH dependence of the ATPase activity, effects of ionic strength were more complex and often depended on the myosin to actin ratio (see also next section). We investigated a range of salt concentrations from 20 to 200 mM KCl since within this range myosin undergoes major transformational changes (Onishi and Wakabayashi, 1982; Trybus et al., 1982; Craig et al., 1983).

Our earlier measurements of the ATPase activity as a function of KCl concentration showed a narrow maximum between 40 to 80 mM KCl (Sobieszek and Small, 1976; Fig. 3 A). In these experiments, myosin phosphorylation and its activation by actin was not carried out separately. At low KCl concentration this does not significantly affect the final result since under the assay conditions maximal phosphorylation was obtained after about 10% of the incubation time of the assays. However, at high KCl concentrations, the phosphorylation rate was reduced and decayed in parallel with the actin-activation.

Fig. 3 B shows a related assay but for which myosin was first phosphorylated and then added to F-actin or to F-actin-tropomyosin complexes for different KCl concentrations. At pH 7.3 and as follows from the data in Fig. 3 A activation by actin alone was almost non-existent around 120 mM KCl. With tropomyosin present there was an overall stimulation of ATPase activity even at the 120 mM KCl. As seen in the figure tropomyosin widened and shifted the KCl dependence profile towards physiologically relevant KCl concentrations.

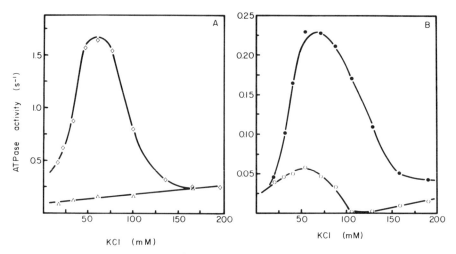

Fig. 3 KCl dependence of Mg ATPase activity of reconstituted actomyosin from pig stomach (A) and turkey gizzard (B) muscles. In (A), the pH was 7.0, and phosphorylation and ATPase activity were initiated at the same time. The activities in the presence and absence of Ca^{2+} are represented by "diamonds" and "triangles", respectively. In (B), the pH was 7.3 and myosin was first phosphorylated and then added to F-actin (o - o) or F-actin tropomyosin complexes (● - ●).

With myosin in filamentous form (pH 6.6) the tropomyosin stimulation occurred at very narrow and lower ranges of KCl concentrations (Fig. 4 A). At higher KCl concentration no TM stimulation was seen (Fig. 4 B) or tropomyosin produced an inhibition rather than a stimulation (Fig. 4 A).

In Fig. 4 B Ca^{2+} and ATP was added to unphosphorylated myosin, and phosphorylation was taking place during the ATPase assays. At this pH, the activities of myosin alone (Fig. 4 B, lower curve) were relatively high. As can be seen they can not simply be substracted from the total activities (actin with or without tropomyosin). At this pH and high KCL concentrations it is not clear at all how to evaluate the myosin"alone"controls. Perhaps, the activities of myosin in the absence of Ca^{2+} represent a better control, despite of the fact that this activity increased about 2 fold after addition of Ca^{2+}. Following phosphorylation the Mg ATPase activity of myosin was in contrast again about 2 fold lower.

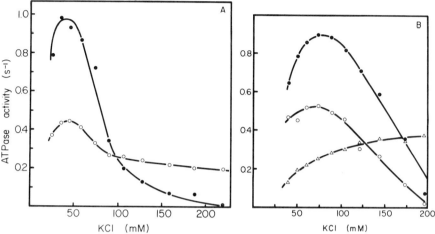

Fig. 4 KCl dependence of Mg ATPase activity of reconstituted turkey gizzard actomyosin containing stable filamentous myosin (pH 6.6). In (A), myosin was first phosphorylated and then added to F-actin in the presence (o - o and absence (● - ●) of tropomyosin. In (B), phosphorylation and ATPase activity were initiated at the same time. The activity of myosin alone, under exactly the same conditions, is indicated by triangles.

V_{max} and K_m Changes

The stimulatory effect of tropomyosin in the activation of myosin ATPase by actin has been investigated in a number of studies (for ref. see Marston and Smith, 1985). Very few of them, however, attempted to establish whether tropomyosin modifies the apparent affinity of the thin filament for myosin (K_m effect) or simply accelerates, in some way the ATP hydrolysis rate (V_{max} effect).

From the reevaluation of already published data (Sobieszek and Small, 1977) it appears that tropomyosin modifies the apparent affinity as well as the hydrolysis rate. As shown in Fig. 5 A lines corresponding to the ATPase activity in the presence and absence of tropomyosin cross the ordinate at the same point. Thus, the V_{max} and K_m increase in parallel so that their ratio remains constant. This enzyme kinetic situation can be termed as "uncompetitive activation" by analogy with the formerly used term "uncompetitive inhibition".

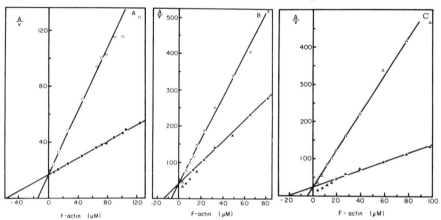

Fig. 5 Hanes-Woolf plots of actin-activated Mg ATPase activity of gizzard myosin in the presence (closed symbols) and absence (open symbols) of tropomyosin. (A) represents a replot of data from Sobieszek and Small (1977) in which the tropomyosin-MLCKase fraction was added to F-actin (● - ●). Chicken gizzard proteins were used and the myosin concentration was 2.0 mg/ml. In (B), MLCKase was present and the activities were measured in the presence (▲-▲) and absence (△-△) of tropomyosin. Turkey gizzard proteins were used and the myosin concentration was 3.0 mg/ml. In (C), rabbit skeletal muscle F-actin was used and turkey gizzard myosin concentration was 1.0 mg/ml.

The figure (5 A) shows a replot of some early data in which very fresh preparations of myosin and actin were used but for which the tropomyosin contained some MLCKase activity (a tropomyosin-MLCKase fraction; Sobieszek and Small, 1977). This data may be compared with that from a similar experiment using tropomyosin purified according to Bailey (MLCKase-free) (Fig. 5 B and C). The changes in V_{max} and K_m seen in this case were essentially the same; as before, the K_m to V_{max} ratio remained constant. Interestingly, the tropomyosin stimulation (V_{max} increase) was notably higher for skeletal muscle actin (Fig. 5 C) than for its smooth muscle counterpart (Fig. 5 B).

It is also clear from the latter figures that the points at low actin concentrations fall below the average lines.

A more detailed analysis into this problem showed that tropomyosin stimulation also depended on the level of occupancy of the myosin heads on actin, that is on myosin to actin ratio. At the molar ratios below 1 to 6 the stimulation was practically absent while it was optimal at a ratio of one myosin head per actin monomer (Fig. 6).

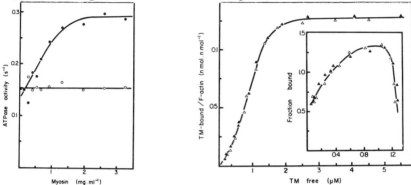

Fig. 6 (left) Tropomyosin stimulation of the Mg ATPase activity as a function of myosin (turkey gizzard) concentration. F-actin (turkey gizzard) concentration was 15 µM, and the activities in the presence and in the absence of tropomyosin are shown by closed and open circles respectively.
Fig. 7 (right) Binding of turkey gizzard tropomyosin to chicken gizzard F-actin in the presence (Δ - Δ) and absence (▲ - ▲) of (Mg) ATP (1.6 mM). The insert represents a Scatchart plot of the same data. Unlabelled abscisse corresponds to "tropomyosin bound (n mol)".

Binding of Smooth Muscle Tropomyosin to Smooth Muscle Actin

Binding of smooth muscle tropomyosin to skeletal muscle actin has been characterized by Sanders and Smillie (1984). As with skeletal muscle actin the binding of smooth muscle actin was positively cooperative (sigmoidal; Fig. 7 and 8 A) with an apparent binding constant of about 1.0 µM. This binding was completely unaffected by the presence of 1 to 2 mM ATP (Fig. 7). Thus, any tropomyosin dependent effects of ATP on actin-myosin interaction must result from the binding of ATP to myosin which, in turn, modifies the binding of myosin to the actin-tropomyosin complex. As shown in Fig. 8 A the addition of unphosphorylated smooth muscle heavy meromyosin (HMM) to F-actin, resulted in a dramatic increase in tropomyosin binding especially at very low concentrations of tropomyosin. In addition, the nature of the binding was

modified by the attachment of the myosin heads to actin. As
the figure shows, at low tropomyosin concentrations, the
binding to F-actin alone was sigmoidal. Under exactly the
same conditions, the binding to actin-HMM complexes was
clearly hyperbolic (Fig. 8 A).

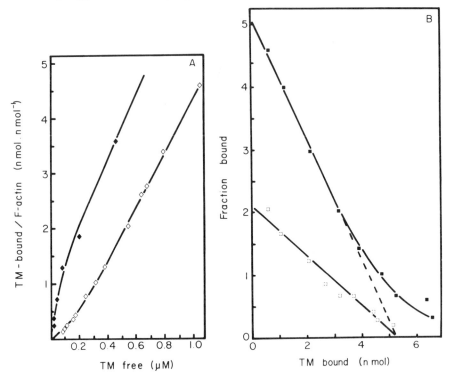

Fig. 8A Effect of unphosphorylated HMM on binding of tropomyosin to F-actin in the absence of ATP. Note that the binding in the presence of HMM was parabolic (closed diamonds) while in its absence was sigmoidal (open diamonds). The concentration of F-actin (chicken gizzard) and HMM was 28μM and 4 μM respectively.

Fig. 8B Effect of myosin phosphorylation on binding of tropomyosin to F-actin. The figure shows a Scatchard plot of the binding to turkey gizzard tropomyosin-free actomyosin with its myosin component either in phosphorylated (closed squares) or dephosphorylated (open squares) state. Actomyosin concentration was 2.6 mg/ml.

The effect of myosin phosphorylation on the binding of tropomyosin to actomyosin is shown in Fig. 8 B. Dephosphorylation of the myosin component by purified myosin light chain phosphatase resulted in an at least 2.5 fold increase in the apparent affinity of tropomyosin for actin. There was also some indication of either a negative cooperativity of this binding or for the phosphorylated actomyosin having an increased capacity to bind higher amounts of tropomyosin. The latter possibility is considered to be more likely, since free tropomyosin appears to be in a fast exchangeable equilibrium with actomyosin. This was indicated by experiments in which radioactively labelled tropomyosin was fully exchangeable with the tropomyosin already bound to the F-actin component.

DISCUSSION

Fresh and Frozen Preparations

In vitro studies on the regulation of the actin-myosin interaction in vertebrate smooth muscle necessarily involve recombination experiments that, in turn, require the parallel preparation and purification of several contractile proteins and enzymes. In our earlier studies (Sobieszek and Small, 1976; 1977) only fresh preparations were used. However, in order to make possible more complex experiments involving the simultaneous combination of several proteins, the use of one or more frozen preparations is virtually unavoidable. Our measurements showed that such preparations had generally up to 50% lower phosphorylation and Mg ATPase rates as compared to equivalent fresh preparations. This problem, faced in numerous previous and current studies in the field adds further to the already significant variability in the reported values of phosphorylation and ATPase rates and should not be overlooked.

Filamentous and Monomeric Myosin; Actin-Activation

More relevant to the present discussion is the activation of ATPase activity of myosin by actin and the influence of myosin conformational changes. Since under physiological conditions smooth muscle myosin is assembled into filaments (Lowey et al., 1973; Shoenberg and Haselgrove, 1974) the relevant conformational changes are those permitted within the filament. These changes are, however, difficult to investigate and in consequence, the first reported changes likely relate to the configuration of monomeric myosin. Thus, at moderate and high ionic strength, smooth muscle myosin can exist in an open (straight: 6S) configuration. In contrast, at lower ionic strength (100 - 150 mM KCl, pH 7.3) the myosin adopts a looped (folded: 10S) configuration in which the myosin tail bends back on itself to contact the neck region of the myosin head (Onishi and Wakabayashi, 1982; Trybus and Lowey, 1982; Craig et al., 1983).

At present it is not clear how these two monomeric configurations are related to the myosin configuration on the filaments. On the one hand and for symmetry reasons only 6S straight molecules can be considered as a filament building unit. On the other, the head and neck portion of the molecule have a relatively high degree of freedom even with the tail immobilized within the filament backbone. Less excessive folding could then possibly also take place within molecules in a filament, but with the neck contacting a neighbouring molecule rather than its own tail (Craig et al., 1983). As was shown in recent studies (Cross et al., 1986; see also Suzuki et al., 1982) the folded molecules have an extremely low ATP hydrolysis rate and practically trap $ADP \cdot P_i$; the hydrolysis products. The trapping was not detected for myosin filaments under similar conditions (Cross et al., 1986) and as shown here, the ATPase hydrolysis rate for myosin in this state was higher and often comparable to that in the additional presence of F-actin. Thus, gross confirmational changes similar to folding seem not to take place for the molecules assembled into filaments.

For conditions yielding folded myosin a low ATPase activity was seen also in the presence of actin (Figs 2 and 3). This could be a result of a relatively low affinity of folded myosin for F-actin. When tropomyosin was added, the affinity became higher and a significant binding of myosin to actin took place: whether or not this was accompanied by unfolding of the molecules has yet to be established.

It is suggested that a similar interpretation would explain the inhibition of the Mg ATPase at neutral and slightly alkaline pH values (Fig. 2). Under these pH conditions there is no reduction in the amount of tropomyosin bound to actin. A similar conclusion was drawn by Yamaguchi et al. (1984) but for data covering a narrower pH range. Yamaguchi et al., did not attempt, however, to give an explanation for the enhanced tropomyosin stimulation under conditions (around pH 6.6) for which ATP-stable myosin filaments are formed.

Relationship between Phosphorylation and Mg ATPase

The relationship between phosphorylation of myosin and its activation by actin in the presence and absence of tropomyosin was not investigated previously. This relationship has variously been reported to be linear (Sobieszek, 1977; Chacko, 1981; Wagner et al., 1985), semi-parabolical (Persechini and Hartshorne, 1981; Cole et al., 1983) or even sigmoidal (Ikebe et al., 1982) in form. As discussed in detail elsewhere (Sellers et al., 1983; see also Results), it should be possible to establish, from this type of data, the mode of phosphorylation of the two myosin heads. The only general agreement here, however, is that only myosin with both heads phosphorylated, can be activated by actin. Our observation that a change from a parabolical to a linear type of relationship occurs after addition of tropomyosin might serve to explain the present controversy. Thus, the data obtained for actin and myosin alone were indeed classified as "semi-parabolical" (Sellers et al., 1983) while those obtained with tropomyosin were "linear" (Chacko, 1981).

The change from a parabolical to a linear relationship implies, in addition, an introduction of a positive cooperativity in the phosphorylation of the two myosin heads. The mechanism of influence of the actin-tropomyosin complex on the phosphorylation process is unclear. It could include a binding of the MLCKase to the complex after phosphorylation of the first myosin head with a subsequent reorientation of the MLCKase enabling phosphorylation of the second head. These considerations are consistent with the MLCKase having comparable affinities for myosin and for actin (Sellers and Pato, 1984; Sobieszek, 1985, and unpublished observations).

The very recent discovery of the second site phosphorylation on smooth muscle myosin (Ikebe and Hartshorne, 1985; Cole et al., 1985) and, related to this, a further increase in the Mg ATPase activity (Tanaka et al., 1985; Ikebe and Hartshorne, 1985) provides an additional parameter to be considered. Double phosphorylation is particularly prevalent at relatively high concentrations of the MLCKase and calmodulin. Under our experimental conditions no phosphorylation at the second site could be seen and this was routinely checked by a modified urea glycerol electrophoresis method (Sobieszek and Jertschin, 1986; see also Materials and Methods).

Tropomyosin Stimulation; Enzyme Kinetic Analysis

The effect of tropomyosin on actin-myosin interaction has been investigated in numerous studies (Sobieszek and Small, 1977; Chacko, 1981; Yamaguchi et al., 1984; Merkel at al., 1985; Heaslip and Chacko, 1985; Cavadore et al., 1985). These were, however, restricted to the measurements of the Mg ATPase in the absence and presence of tropomyosin in model systems for which no enzyme kinetic conclusions could be drawn. The common result was that smooth muscle tropomyosin effects a 2 - 3 fold stimulation of the actin-activated ATPase activity of myosin. From the present data it appears that this is not a simple stimulation but a complex modulation depending on a number of factors. These include the configuration of the myosin molecule within a filament, the degree of myosin phosphorylation, the number of myosin heads attached to F-actin and the affinity of tropomyosin for actin.

Steady state enzyme kinetic analysis (Fig. 5) demonstrated that tropomyosin induced a parallel increase of the hydrolysis rate (V_{max}) and the apparent affinity (K_m) such that the V_{max}/K_m ratio remained constant. A similar situation was formerly described for the inhibitory effect of skeletal and invertebrate muscle tropomyosins (Sobieszek, 1982). In the light of the present work we conclude that the stimulation of smooth muscle tropomyosin represents a reduction of the average number of undissociated actin plus myosin complexes. This could arise from the acceleration of product release at the rate limiting step of the ATPase hydrolysis cycle. Consistently with the present work it was recently concluded from studies on smooth muscle acto-HMM that phosphorylation dependent regulation likewise operates at the same kinetic step (Sellers, 1985). However, similar experiments carried out on thymus myosin and skeletal muscle actin indicate that in non-muscle systems this regulation in the presence of thymus tropomyosin could be realised via modification of the affinity of myosin for actin (Wagner and George, 1986).

Tropomyosin Binding to Actin

We show here that the affinity of smooth muscle tropomyosin for smooth muscle F-actin is 2 fold higher than for skeletal muscle F-actin (Sanders and Smillie, 1984) although otherwise the binding was very similar. The weaker binding would require a lower head saturation on actin in order to "push" tropomyosin into a stimulatory position. Consistently the stimulation with skeletal muscle actin (Fig. 5C) was higher and required lower myosin concentrations. More significant, however, was the effect of the attachment of myosin heads to actin on the binding of tropomyosin to actin. We observed that such attachment results not only in a great increase in the binding affinity but also in a change from a sigmoidal to a hyperbolic type of binding. A simple increase in the binding of skeletal muscle tropomyosin under similar conditions was already noted (Eaton, 1976).

Our binding experiments, including these with phosphorylated and dephosphorylated actomyosin (Fig. 8B) were performed in the absence of ATP. Therefore, they relate to a physiologically less relevant situation in which myosin heads are complexed with actin in a very high affinity, rigor-type binding. Nevertheless, the significant effects of tropomyosin, reported here, indicate the level at which tropomyosin modulation may be operating. Clearly, with ATP present, the affinity of myosin for actin will be dramatically reduced and tropomyosin action should be more effective. It is our intention to test this hypothesis.

ACKNOWLEDGEMENTS

I thank Dr. J.V. Small for improving the manuscript and Dr. M. Matzke for additional comments. The excellent technical assistance of Ms. U. Müller and Mrs. A. Jankela as well as the general assistance of Mrs. G. McCoy are also acknowledged. These studies were supported by a grant from the Muscular Dystrophy Association, Inc.

REFERENCES

Adelstein RS, Eisenberg E, (1980). Regulation and kinetics of the actin-myosin-ATP interaction. Ann Rev.Biochem 49: 921-56.
Bailey K, (1948). Tropomyosin: a new asymmetric protein component of the muscle fibril. Biochem J 43:271-281.
Bremel RD, Murray JM, Weber A, (1972). Manifestations of cooperative behavior in the regulated actin filament during actin-activated ATP hydrolysis in the presence of calcium. Cold Spring Harbor Symp Quant Biol 37:267-275.

Cavadore JC, Berta P, Axelrud-Cavadore C, Haiech J. (1985). Calcium binding of arterial tropomyosin: involvement in the thin filament regulation of smooth muscle. Biochem 24: 5216-5221.

Chacko S, (1981). Effects of phosphorylation, calcium ion, and tropomyosin on actin-activated adenosine 5'-triphosphatase activity of mammalian smooth muscle myosin. Biochem 20:702-707.

Cole HA, Griffiths HS, Patchell VB, Perry SV, (1985). Two-site phosphorylation of the phosphorylatable light chain (20-kDa light chain) of chicken gizzard myosin. FEBS Lett 180:165-169.

Cole HA, Patchell VB, Perry SV, (1983) Phosphorylation of chicken gizzard myosin and the Ca^{2+}-sensitivity of the actin-activated Mg^{2+}-ATPase. FEBS Lett 158:17-20.

Craig R, Smith R, Kendrick-Jones J, (1983). Light-chain phoshporylation controls the conformation of vertebrate non-muscle and smooth muscle myosin molecules. Nature 302: 436-439.

Cross RA, Cross KE, Sobieszek A, (1986). ATP-linked monomer-polymer equilibrium of smooth muscle myosin: the free folded monomer traps $ADP \cdot P_i$. EMBO J in press.

Driska S, Hartshorne DJ, (1975). The contractile proteins of smooth muscle. Properties and components of a Ca^{2+}-sensitive actomyosin from chicken gizzard. Arch Biochem Biophys 167:203-212.

Eaton BL, (1976). Tropomyosin binding to F-actin induced by myosin heads. Science 192:1337-1339.

Eaton BL, Kominz, DR, Eisenberg, E, (1975). Correlation between the inhibition of the acto-heavy meromyosin ATPase and the binding of tropomyosin to F-actin: effects of Mg^{2+}, KCl, troponin I, and troponin C. Biochemistry 14:2718-2724.

Ebashi S, (1979). Regulation of muscle contraction. Proc R Soc Lond 207:259-286.

Ebashi S, Mikawa T, Hirata M, Toyo-Oka T, Nonomura Y, (1977). Regulatory proteins of smooth muscle. In Casteels R, Godfraind T, Ruegg JC (eds): "Excitation-Contraction Coupling in Smooth Muscle". Amsterdam: Elsevier/North Holland, pp 325-334.

Gergely J, (1980). Ca^{2+} control of actin-myosin interaction. Basic Res Cardiol 75:18-25.

Gopalakrishna R, Anderson WB, (1982). Ca^{2+}-induced hydrophobic site on calmodulin: application for purification of calmodulin by phenyl-sepharose affinity chromatography. Biochem Biophys Res Comm 104:830-836.

Hartshorne DJ, Mrwa U, (1982). Regulation of smooth muscle actomyosin. Blood Vessels 19:1-18.

Heaslip RJ, Chacko S, (1985). Effects of Ca^{2+} and Mg^{2+} on the actomyosin adenosine-5'-triphosphatase of stably phosphorylated gizzard myosin. Biochemistry 24:2731-2736.

Ikebe M, Hartshorne DJ, (1985). Phosphorylation of smooth muscle myosin at two distinct sites by myosin light chain kinase. J Biol Chem 260:10027-10031.

Ikebe M, Ogihara S, Tonomura Y, (1982). Nonlinear dependence of actin-activated Mg^{2+}-ATPase activity on the extent of phosphorylation of gizzard myosin and H-meromyosin. J Biochem 91:1809-1812.

Kamm KE, Stull JT, (1985). The function of myosin and myosin light chain kinase phosphorylation in smooth muscle. Ann Rev Pharmacol Toxicol 25:593-620.

Kendrick-Jones J, Cande WZ, Tooth PJ, Smith RC, Scholey JM, (1983). Studies on the effect of phosphorylation of the 20,000 M_r light chain of vertebrate smooth muscle myosin. J Mol Biol 165:139-162.

Leavis PC, Gergely J, (1984). Thin filament proteins and thin filament-linked regulation of vertebrate muscle contraction. Critical Reviews in Biochemistry 16:235-305.

Lebowitz EA, Cooke R, (1979). Phosphorylation of uterine smooth muscle permits actin activation. J Biochem (Tokyo) 85:1489-1494.

Lehrer SS, Morris EP, (1982). Dual effects of tropomyosin and troponin-tropomyosin on actomyosin subfragment 1 ATPase. J Biol Chem 257:8073-8080.

Lowy J, Vibert PJ, Haselgrove JC, Poulsen FR, (1973). The structure of the myosin elements in vertebrate smooth muscles. Phil Trans R Soc Lond 265:191-196.

Marston SB, (1982). The regulation of smooth muscle contractile proteins. Prog Biophys molec Biol 41:1-41.

Marston SB, Smith CWJ, (1985). The thin filaments of smooth muscles. J Muscle Res Cell Motil 6:669-708.

Merkel L, Meisheri KD, Pfitzer G, (1984). The variable relation between myosin light-chain phosphorylation and actin-activated ATPase activity in chicken gizzard smooth muscle. Eur J Biochem 138:429-434.

Miyata H, Chacko S, (1986). Role of tropomyosin in smooth muscle contraction: effect of tropomyosin binding to actin on actin activation of myosin ATPase. Biochemistry 25:2725-2729.

Onishi H, Wakabayashi T, (1982). Electron microscopic studies of myosin molecules from chicken gizzard muscle I: the formation of the intramolecular loop in the myosin tail. J Biochem 92:871-879.

Perry SV, (1983). Phosphorylation of the myofibrillar proteins and the regulation of contractile activity in muscle. Phil Trans R Soc Lond 302:59-71.

Persechini A, Hartshorne DJ, (1981). Phosphorylation of smooth muscle myosin: evidence for cooperativity between the myosin heads. Science 213:1383-1385.

Potter JD, Gergely J, (1974). Troponin, tropomyosin and actin interactions in the Ca^{2+} regulation of muscle contraction. Biochemistry 13:2697-2703.

Sanders C, Smillie LB, (1984). Chicken gizzard tropomyosin: head-to-tail assembly and interaction with F-actin and troponin. Can J Biochem Cell Biol 62:443-448.

Sellers JR, (1985). Mechanism of the phosphorylation-dependent regulation of smooth muscle heavy meromyosin. J Biol Chem 260:15815-15819.

Sellers JR, Chock PB, Adelstein RS, (1983). The apparently negatively cooperative phosphorylation of smooth muscle myosin at low ionic strength is related to its filamentous state. J Biol Chem 258:14181-14188.

Sellers JR, Pato MD, (1984). The binding of smooth muscle myosin light chain kinase and phosphatase to actin and myosin. J Biol Chem 259:7740-7746.

Shoenberg CF, Haselgrove JC, (1974). Filaments and ribbons in vertebrate smooth muscle. Nature 249:152-154.

Small JV, Sobieszek A, (1980). The contractile apparatus of smooth muscle. Int Rev Cytol 64:241-306.

Small JV, Sobieszek A, (1983). Contractile and structural protein of smooth muscle. In Stephens NL (ed): "Biochemistry of Smooth Muscle". Boca Raton: CRC Press, pp 84-140.

Sobieszek A, (1977). Ca-linked phosphorylation of a light chain of vertebrate smooth muscle myosin. Eur J Biochem 73: 477-483.

Sobieszek A, (1982). Steady-state kinetic studies on the actin activation of skeletal muscle heavy meromyosin subfragments: effects of skeletal, smooth and non-muscle tropomyosins. J Mol Biol 157:275-286.

Sobieszek A, (1985). Phosphorylation reaction of vertebrate smooth muscle myosin: an enzyme kinetic analysis. Biochemistry 24:1266-1274.

Sobieszek A, Barylko B, (1984). Enzymes regulating myosin phosphorylation in vertebrate smooth muscle. In Stephens NL (ed): "Smooth Muscle Contractions". New York: Marcel Dekker, pp 283-316.

Sobieszek A, Bremel RD, (1975). Preparation and properties of vertebrate smooth-muscle myofibrils and actomyosin. Eur J Biochem 55:49-60.

Sobieszek A, Jertschin P, (1986). Urea-glycerol-acrylamide gel electrophoresis of low molecular weight muscle, acidic proteins. Rapid determination of myosin light chain phosphorylation in myosin, acto-myosin and whole muscle samples. Electrophoresis, in press.

Sobieszek A, Small JV, (1976). Myosin-linked calcium regulation in vertebrate smooth muscle. J Mol Biol 102:75-92.
Sobieszek A, Small JV (1977). Regulation of the actin-myosin interaction in vertebrate smooth muscle: activation via a myosin light-chain kinase and the effect of tropomyosin. J Mol Biol 112:559-576.
Sobieszek A, Small JV, (1981). Effect of muscle and non-muscle tropomyosins in reconstituted skeletal muscle actomyosin. Eur J Biochem 118:533-539.
Strzelecka-Golaszewska H, Sobieszek A, (1981). Activation of smooth muscle myosin by smooth and skeletal muscle actins. FEBS Lett 134:197-202.
Suzuki H, Takahashi K, Onishi H, Watanabe S, (1982). Reversible changes in the state of phosphorylation of gizzard myosin, in that of gizzard myosin assembly, in the ATPase activity of gizzard myosin, in that of actomyosin and in the superprecipitation activity. J Biochem 91:1687-1698.
Tanaka T, Sobue K, Owada MK, Hakura A, (1985). Linear relationship between diphosphorylation of 20 kDa light chain of gizzard myosin and the actin-activated myosin ATPase activity. Biochem Biophys Res Comm 131:987-993.
Trybus KM, Huiatt TW, Lowey S, (1982). A bent monomeric conformation of myosin from smooth muscle. Proc Natl Acad Sci USA 79:6151-6155.
Wachsberger P, Kaldor G, (1971). Studies on uterine Myosin A and actomyosin. Arch Biochem Biophys 143:127-137.
Wagner PD, George JN, (1986). Phosphorylation of thymus myosin increases its apparent affinity for actin but not its maximum adenosinetriphosphatase rate. Biochemistry 25:913-918.
Wagner PD, Vu ND, George JN, (1985). Random phosphorylation of the two heads of thymus myosin and the independent stimulation of their actin-activated ATPase. J Biol Chem 260: 8084-8089.
Williams DL, Greene LE, Eisenberg E, (1984). Comparison of effects of smooth and skeletal muscle tropomyosin on interactions of actin and myosin subfragment 1. Biochemistry 23: 4150-4155.
Yamaguchi M, Ver A, Carlos A, Seidel JC, (1984). Modulation of the actin-activated adenosinetriphosphatase activity of myosin by tropomyosin from vascular and gizzard smooth muscles. Biochemistry 23:774-779.

MYOSIN LIGHT CHAIN KINASES AND KINETICS OF MYOSIN PHOSPHORYLATION IN SMOOTH MUSCLE CELLS

Kristine E. Kamm, Sancy A. Leachman, Carolyn H. Michnoff, Mary H. Nunnally, Anthony Persechini, Andrea L. Richardson and James T. Stull
Department of Pharmacology and Moss Heart Center, The Unisity of Texas Health Science Center at Dallas, Dallas, Texas 75235

INTRODUCTION

Myosin, the primary protein found in thick filaments, plays a key role in smooth muscle contractility. Thick filaments are composed of myosin monomers, and each myosin monomer consists of two heavy chain subunits (200 kDa) and two pairs of light chain subunits (17 kDa and 20 kDa, respectively). The 20 kDa light chains are also referred to as regulatory or phosphorylatable light chains. A key event in activation of smooth muscle contraction involves phosphorylation of this light chain by the calcium- and calmodulin-dependent myosin light chain kinase (Kamm and Stull, 1985a; Stull et al., 1986). Myosin light chain kinases are unique protein kinases that catalyze the phosphorylation of myosin light chains with no other identified physiologically significant protein substrate. Myosin light chain kinases have been purified from a number of vertebrate muscles and their biochemical properties have been described. We have identified two classes of myosin light chain kinases that differ in immunological and catalytic properties that are represented by enzymes from avian and mammalian skeletal muscles versus avian gizzard smooth muscle, respectively (Nunnally and Stull, 1984; Nunnally et al., 1985; Michnoff et al., 1986; Nunnally et al., 1987). The biochemical properties of myosin light chain kinases from a variety of smooth muscles have recently been examined. In addition, results on the kinetic mechanism by which myosin phosphorylation occurs in smooth muscle cells has been determined.

MYOSIN LIGHT CHAIN KINASES

The relative masses of myosin light chain kinases from skeletal muscles have been examined by Western blotting procedures (Nunnally and Stull, 1984; Nunnally et al., 1985). These myosin light chain kinases vary considerably in relative masses determined by SDS-PAGE and range from 68 kDa (human) to 150 kDa (chicken). The relative mass of myosin light chain kinase is constant within an animal species, regardless of the skeletal muscle fiber type (slow-twitch, red fibers versus fast-twitch, white). Western blot analyses have demonstrated that the masses of kinases in rabbit (87 kDa) and chicken (150 kDa) skeletal muscles in situ were identical to the purified enzymes. Thus, the marked differences in sizes of myosin light chain kinases in skeletal muscle do not appear to be related to limited proteolysis of purified enzymes. Considering the marked differences in the physiological properties of different smooth muscles, it was of interest to determine if there was significant mass heterogeneity in smooth muscle myosin light chain kinases from different animal species or from different types of smooth muscle within a particular animal.

The relative masses of three purified smooth muscle myosin light chain kinases from bovine trachealis, porcine carotid artery and avian (chicken) gizzard smooth muscle was determined by SDS-PAGE (Fig. 1). The masses of the three kinases were chicken gizzard, 130 kDa; bovine trachealis, 150 kDa; and porcine carotid artery, 140 kDa. Thus, although there appear to be differences in masses, there are not marked differences as found in skeletal muscle myosin light chain kinases.

1 SMOOTH MUSCLE MYOSIN LIGHT CHAIN KINASES
SDS-POLYACRYLAMIDE GEL ELECTROPHORESIS

Protein Stain   Immunoblot
  A B C   A B C   Tissue
                                 A. Chicken Gizzard
                                 B. Steer Trachea
                                 C. Hog Carotid Artery

The three purified myosin light chain kinases were subjected to Western blot analysis to determine the cross-reactivity of kinases to polyclonal antibodies raised against bovine trachealis myosin light chain kinase (Fig. 1). The immunoblot of the three purified kinases showed that the polyclonal antiserum crossreacted with all three purified enzymes. However, the antibodies did not crossreact to purified rabbit or chicken skeletal muscle myosin light chain kinases. Thus, the polyclonal antibodies appeared to bind specifically to smooth muscle myosin light chain kinases, but not skeletal muscle kinases.

Monoclonal antibodies have also been raised against myosin light chain kinase from bovine trachealis. Western blot analyses indicated that 2 of 4 monoclonal antibodies crossreacted with all three purified smooth muscle myosin light chain kinases, whereas two of the antibodies only reacted with mammalian myosin light chain kinases (data not shown). The polyclonal and monoclonal antibodies were used for analyses of myosin light chain kinases in extracts of smooth muscle tissues by the Western blotting procedure. The results from these analyses show that the relative mass of smooth muscle myosin light chain kinase was the same in different tissues within a given animal species. For example, the mass of myosin light chain kinase was 150 kDa in bladder, aorta, trachea and carotid artery from rabbit. Similar observations were made for other animal species. However, as shown in Table 1, there were small but significant differences in the relative masses among different animal species that ranged from 130 kDa to 150 kDa. The reason for these differences in relative masses is not known. They are considerably smaller than the marked differences observed with skeletal muscle myosin light chain kinases.

Table 1. Relative masses of smooth muscle mysoin light chain kinases in different animal species as determined by SDS-PAGE.

Chicken	130 kDa
Porcine	140 kDa
Canine	140 kDa
Rabbit	150 kDa
Bovine	150 kDa
Sheep	150 kDa

The monoclonal antibodies raised to rabbit skeletal muscle myosin light chain kinase did not crossreact with smooth muscle myosin light chain kinases, including enzyme in rabbit smooth muscles (Nunnally et al., 1987). Likewise, antibodies raised to the smooth muscle myosin light chain kinases do not crossreact with enzyme from skeletal muscle tissues. Nunnnally et al. (1986) raised a monoclonal antibody to the calmodulin binding domain of the rabbit skeletal muscle myosin light chain kinase that did not bind to smooth muscle myosin light chain kinases. Thus, there appear to be major structural differences between myosin light chain kinases from smooth and skeletal muscles, respectively.

A unique feature of the phosphotransferase reaction catalyzed by myosin light chain kinases is the specificity for myosin light chains (Stull et al., 1986). Myosin light chain kinases from a variety of tissues phosphorylate other protein substrates only at extremely low rates compared to the rate of phosphorylation of myosin light chains. This narrow protein substrate specificity distinguishes these kinases from other types of protein kinases. It should be noted, however, that the catalytic properties of smooth versus skeletal muscle myosin light chain kinases are significantly different. The primary determinants for catalysis are unique; the importance of primary structure around the phosphorylatable serine has been examined with synthetic peptide substrates homologous with the primary structure of chicken smooth and skeletal muscle myosin light chains:

Skeletal Muscle

P K K A K R R A A E G S S(P) N V F

Smooth Muscle

S S K R A K A K T T K K R P Q R A T S(P) N V F

Synthetic peptide substrates with specific deletions or replacement of residues have been used with myosin light chain kinases from rabbit (87 kDa) and chicken (150 kDa) skeletal muscles (Michnoff et al., 1986). The conclusion has been reached that the 6-8 basic residues toward the amino terminus from the phosphorylatable serine are primary determinants for catalysis by skeletal muscle myosin light chain kinases. In addition, basic amino acids located 10-11 residues toward the amino terminus from the phosphorylatable serine are important determinants. The basic residues that are in a homologous position with the smooth muscle myosin light chain (6-8 residues from the phosphorylatable serine) are also important determinants for gizzard smooth muscle myosin light chain kinase (Pearson et al., 1984). However, the arginine residue at position 16 in smooth muscle myosin light chain is an important determinant where this residue is a glutamic acid in the skeletal muscle light chain. These results raise the interesting possibility that there could be different determinants for myosin light chain kinases from different types of smooth muscle or from different animal species. Therefore, we used three synthetic peptide substrates homologous to the chicken gizzard smooth muscle light chain to examine the kinetic properties of myosin light chain kinases purified from chicken gizzard, porcine carotid artery and bovine trachealis. These structures included:

Peptide	Structure
1	K K R P Q R A T S(P) N V F S
2	A K R P Q R A T S(P) N V F S
3	K K R A A E A T S(P) N V F S

K_m and V_{max} values were determined for peptides 1 and 2. There were no significant differences in the V_{max} values (1000 nmol ^{32}P incorporated/min/mg kinase). The K_m values (uM) for the two different peptides were significantly different (Table 2). Thus, the basic residue that is 8 residues removed from the phosphorylatable serine appears to be an important determinant not only for the myosin light chain kinase from chicken gizzard smooth muscle, but also from porcine carotid artery and bovine trachealis.

Table 2. K_m values for synthetic peptides for myosin light chain kinases purified from different smooth muscles.

	Chicken Gizzard	Porcine Carotid Artery	Bovine Trachealis
Peptide 1	18.0 ± 4.5	18.6 ± 0.6	13.3 ± 0.38
Peptide 2	129 ± 1	173 ± 21	180 ± 6

The third peptide in which a glutamic acid residue had been replaced for the arginine residue that was three amino acids toward the amino terminus from the phosphorylatable serine was such a poor substrate that it was not possible to determine precise kinetic values. It had previously been shown that the substitution of AA for PQ had no effect on the kinetic properties (Kemp and Pearson, 1985). At a concentration of 1 mM, the rate of phosphorylation was 5–14 nmol/min/mg enzyme for the three different myosin light chain kinases, which is considerably slower than the average Vmax value of 1,000 nmol/min/mg enzyme for the other two peptides. Therefore, the two regions that are primary determinants for smooth muscle light chain substrates are similar for the different smooth muscle myosin light chain kinases. This conclusion is similar, in general, to the conclusion regarding catalytic properties of myosin light chain kinases from different skeletal muscles (Michnoff et al., 1985).

MYOSIN PHOSPHORYLATION IN TRACHEAL SMOOTH MUSCLE CELLS

The catalytic properties of smooth muscle myosin light chain kinase have been examined with myosin as well as free myosin light chain. Various investigators have shown that when gizzard smooth muscle myosin is in a filamentous form, it is phosphorylated by an ordered or negatively cooperative process (Persechini and Hartshorne, 1981; Sellers et al., 1983). First order rate constants were greater by 4–10 fold for phosphorylation of light chain in the first head of myosin as compared to the second head. However, other investigators have reported a single population of smooth muscle myosin heads with respect to the kinetic properties of phosphorylation by myosin light chain kinase (Trybus and Lowey, 1985; Sobieszek, 1985). Random phosphorylation of myosin was observed independent of the state of aggregation.

Previous studies on myosin light chain phosphorylation in intact smooth muscles have not completely resolved whether myosin phosphorylation is negatively cooperative nor whether both heads need to be phosphorylated for contraction. Silver and Stull (1984) found that the rapid and marked increase in light chain phosphorylation did not support the idea of a sequential or negatively cooperative phosphorylation mechanism. In addition, maintained isometric force in intact tissues was proportional to the maximal extent of light chain phosphorylation. However, in these types of investigations, agonist diffusion into a tissue results in asynchronous activation of smooth muscle cells. Therefore, time courses of myosin phosphorylation are dominated by these diffusion rates so that tissue averages do not provide an adequate measurement of the phosphorylation kinetic properties.

Kamm and Stull (1985b) recently demonstrated that electrical stimulation of bovine tracheal smooth muscle that results in a neurally-mediated contraction minimizes this problem with nearly synchronous activation of smooth muscle cells. Stimulation resulted in a latentcy of 500 msec before there was any significant change in myosin light chain phosphorylation, force or muscle stiffness. The delay in activation of myosin light chain phosphorylation and mechanical events was probably due to the time required from the release of neural transmitter (acetylcholine) to the time required for increasing cytoplasmic calcium concentrations and activation of myosin light chain kinase. Myosin light chain was phosphorylated from 0.04 to 0.8 mol phosphate/mol light chain with a pseudo-first order rate of 1.1 per sec. There was no evidence of an ordered or negatively cooperative process. The pseudo-first order rate could be predicted from product inhibition of the phosphorylation reaction at high substrate concentrations (Sobieszek, 1985) and is consistent with a random phosphorylation process in tracheal smooth muscle cells.

We recently developed a method for measuring nonphosphorylated, monophosphorylated and diphosphorylated forms of myosin in tissue extracts to determine the distribution of myosin phosphate in relation to the extent of light chain phosphorylation (Persechini et al., 1986). This analysis allowed us to determine directly whether the phosphorylation reaction in tracheal smooth muscle cells was random or negatively cooperative. Nondenaturing polyacrylamide gel electrophoresis in the presence of pyrophosphate of tracheal tissue extracts separates 4 protein bands (Fig. 2). When the extent of light chain phosphorylation was very low, two prominent bands, labeled F and M, were demonstratable. Upon stimulation with a cholinergic, muscarinic agonist, carbachol (Fig. 2) or with neural stimulation (Persechini et al., 1986), a decrease in the relative amount of band M and an increase in other bands labeled MP and MP2 was observed. The extent of light chain phosphorylation also increased to high values (Fig. 2). SDS-PAGE of pyrophosphate gel slices showed a protein with an Mr value of 240 kDa in the slowest migrating band, F, which is probably filamin. Immunoblots with antibodies raised to myosin light chain demonstrated reactivity with the three higher mobility bands M, MP and MP2. Measurements of the extent of light chain phosphorylation in bands M, MP and MP2 showed 0.2, 0.6 and 1.0 mol phosphate/mol light chain, respectively. Assuming that these three forms represented nonphosphorylated, monophosphorylated and diphosphorylated myosin, respectively, the extent of light chain phosphorylation was calculated and compared to data obtained by direct measurements of myosin light chain phosphorylation by urea/glycerol-PAGE of denatured tissue extracts. The results from this comparison showed that there was a linear relationship between the calculated and measured extent of light chain phosphorylation in unstimulated and electrically stimulated bovine tracheal tissues. Thus, upon pyrophosphate gel electrophoresis, the 4 protein bands represent in order of increasing mobility: filamin (F), nonphosphorylated myosin (M), monophosphorylated myosin (MP) and diphosphorylated myosin (MP2).

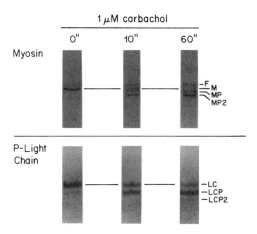

MYOSIN AND MYOSIN P-LIGHT CHAIN PHOSPHORYLATION IN BOVINE TRACHEALIS

These results are not consistent with the report on myosin isozymes identified by separation following pyrophosphate gel electrophoresis (Lema et al., 1986). These investigators found different bands upon pyrophosphate electrophoresis, but claimed that extraction of myosin from either relaxed or contracting aortic smooth muscle did not influence the mobilities of these bands. However, direct measurements of the extent of light chain phosphorylation were not made. Their freezing tissues directly in liquid nitrogen is an inadequate procedure for stopping rapid phosphorylation and dephosphorylation reactions.

In unstimulated bovine tracheal smooth muscle, the extent of light chain phosphorylation was 0.02 mol phosphate/mol light chain with no measureable amount of phosphorylated myosin (Persechini et al., 1986). With stimulation, isometric force, stiffness and light chain phosphorylation increased following a latency of about 500 msec. Light chain phosphorylation rapidly increased to a maximal value of 0.78 mol phosphate/mol light chain by 3.5 sec. During the continuous neural stimulation, there was a proportionate increase in monophosphorylated and diphosphorylated forms of myosin. The relationship between the extent of light chain phosphorylation and the relative amount of diphosphorylated myosin was consistent with a random phosphorylation mechanism in bovine tracheal smooth muscle cells (Fig. 3).

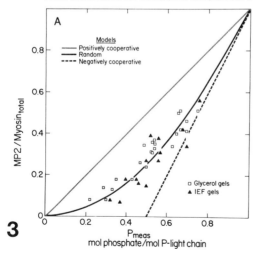

3

Thus, we have been able to quantitate the relative

amounts of nonphosphorylated, monophosphorylated and diphosphorylated myosin and have demonstrated directly that myosin phosphorylation occurs by a random process in living cells. The ability to determine directly the pattern of myosin phosphorylation in tissue samples should allow verification of a random phosphorylation process without necessarily performing analyses in neurally stimulated smooth muscle cells. Thus, activation of smooth muscle cells may be performed with agonists provided that there is a sufficient extent of light chain phosphorylation. If the relationship between the mechanical events (stress, stiffness or maximum velocity of shortening) and biochemical events (myosin phosphorylation) is to be examined mechanically, synchronous activation of smooth muscle cells is still required to evaluate the contributions of different phosphorylated forms of myosin to mechanical activation and stress maintenance. The techniques that we have recently developed will allow such an evaluation.

REFERENCES

Kamm KE, Stull JT (1985a). Function of myosin and myosin light chain phosphorylation in smooth muscle. Annu Rev Pharmacol Toxicol 25:593-620.

Kamm KE, Stull JT (1985b). Myosin phosphorylation, force and maximal shortening velocity in neurally stimulated tracheal smooth muscle. Amer J Physiol 249:C238-C247.

Kemp BE, Pearson, RB (1985). Spatial requirements for location of basic residues in peptide substrates for smooth muscle myosin light chain kinase. J Biol Chem 260:3355-3359.

Lema MJ, Pagani ED, Shemin R, Julian FJ (1986). Myosin isozymes in rabbit and human smooth muscles. Circ Res 59:115-123.

Michnoff CH, Kemp BE, Stull JT (1986). Phosphorylation of synthetic peptides by skeletal muscle myosin light chain kinases. J Biol Chem 261:8320-8326.

Nunnally MH, Stull JT (1984). Mammalian skeletal muscle myosin light chain kinases: a comparison by antiserum crossreactivity. J Biol Chem 259:1776-1780.

Nunnally MH, Rybicki SB, Stull JT (1985). Characterization of chicken skeletal muscle myosin light chain kinase: evidence for muscle specific isozyme forms. J Biol Chem 260:1020-1026.

Nunnally MH, Stull JT, Blumenthal DK, Krebs EG (1986). Properties of a monoclonal antibody raised to the calmodulin-binding domain of myosin light chain kinase. Fed Proc 45:1551.

Nunnally MH, Hsu LC, Mumby MC, Stull JT (1987). Structural studies of rabbit skeletal muscle myosin light chain kinase with monclonal antibodies. J Biol Chem (in press).

Pearson RB, Jakes R, John M, Kendrick-Jones J, Kemp BE (1984). Phosphorylation site sequence of smooth muscle myosin light chain (Mr = 20000). FEBS Lett 168:108-112.

Persechini A and Hartshorne DJ (1981). Evidence for cooperativity between the myosin heads. Science 213:1383-1385.

Persechini A, Kamm KE, Stull JT (1986). Different phosphorylated forms of myosin in contracting tracheal smooth muscle. J Biol Chem 261:6293-6299.

Sellers JR, Chock PB, Adelstein RS (1983). The apparently negatively cooperative phosphorylation of smooth muscle myosin at low ionic strength is related to its filamentous state. J Biol Chem 258:14181-14188.

Silver PJ, Stull JT (1984). Phosphorylation of myosin light chain and phosphorylase in tracheal smooth muscle in response to KCl and carbachol. Mol Pharmacol 25:267-274.

Sobieszek A (1985). Phosphorylation reaction of vertebrate smooth muscle myosin: an enzyme kinetic analysis. Biochemistry 24:1266-1274.

Stull JT, Nunnally MH, Michnoff CH (1986). Calmodulin-dependent protein kinases. In Krebs EG, Boyer PD (eds): "The Enzymes", Orlando: Academic Press, pp 113-166.

Trybus KM, Lowey S (1985) Mechanism of smooth muscle myosin phosphorylation. J Biol Chem 260:15988-15995.

AORTIC POLYCATION-MODULABLE PROTEIN PHOSPHATASE(S): STRUCTURE AND FUNCTION

Joseph Di Salvo

Department of Physiology and Biophysics
University of Cincinnati College of Medicine
Cincinnati, Ohio 45267-0576

INTRODUCTION

Diverse cellular mechanisms, including contractility and metabolism in smooth muscle, are regulated by phosphorylation and dephosphorylation of specific proteins (Krebs and Beavo, 1979; Lee et al., 1980; Cohen, 1985; Kamm and Stull, 1985). Recent studies in our laboratory have been guided by the underlying hypothesis that control of such mechanisms probably involves modulation of protein phosphatase activity in addition to modulation of kinase activity. Within this framework, our studies have focused on a class of aortic enzymes which we refer to as polycation-modulable (PCM-) phosphatases (Di Salvo et al., 1983, 1984, 1985a, 1985b). This operational term was coined to reflect the observation that in $vitro$ expression of phosphatase activity by enzymes of this kind is subject to modulation by polycationic effectors such as histone-H_1 or polylysine. Whether or not polycationic modulation of phosphatase activity occurs in $vivo$ is unknown. Nevertheless, an understanding of molecular mechanisms responsible for polycationic modulation of phosphatase activity expressed in $vitro$ offers insight into how PCM-phosphatases might be regulated in $vivo$, and it also provides a rational basis for assessing cellular changes in phosphatase activity.

In this paper a brief review is presented on the nature, polypeptide composition and regulatory properties of aortic PCM-phosphatase. Based on this information, a tentative working model of the enzyme is proposed: the native PCM-phosphatase is visualized as consisting of a single polypeptide containing functional domains rather than consisting of several noncovalently linked subunits. Effects of the enzyme on actin-myosin interactions are also discussed.

PCM-PHOSPHATASES: AN OVERVIEW

In 1982, Wilson et al. purified a protein from porcine renal extracts which stimulated phosphorylase phosphatase activity expressed by a renal cortical phosphatase: the protein was identified later as histone-H_1 (Wilson et al., 1983). In the same year, we reported that lysine-rich histone-H_1 and polycationic polylysine could be utilized for identifying so-called latent phosphorylase phosphatase in extracts from aortic smooth muscle (Di Salvo et al., 1983). We also reported that stimulation of phosphorylase phosphatase activity was limited to enzymes that were largely insensitive to protein phosphatase inhibitors 1 and 2 (see below). Further studies in our laboratory showed the influence of polycations on phosphorylase phosphatase activity was biphasic (Di Salvo et al., 1984). For example, low concentrations of polylysine (13 kDa, 5-20 nM) stimulated phosphorylase phosphatase activity 6-12 fold. However, stimulation became less pronounced at higher concentrations (40-100 nM) and activity was reduced to below control value at still higher concentrations. In contrast, only inhibition occurred in response to polylysine when purified myocardial myosin light chains were used as substrate (Fig. 1). Since expressed phosphatase activity could be either enhanced or suppressed, depending on the concentration of polycation tested and the substrate studied, we suggested that such enzymes be called polycation-modulable (PCM) phosphatases (Di Salvo et al., 1984; 1985a,b). Based on migration during sucrose density centrifugation and differences in substrate specificity, at least 2 PCM-phosphatases could be distinguished from each other in bovine aortic extracts (Di Salvo et al., 1985a). These 2 activities could represent different enzymes (i.e., products of different genes), post-translational modifications of a single enzyme, or different forms of the same enzyme which are generated during the purification procedure.

POLYPEPTIDE COMPOSITION OF PCM-PHOSPHATASES

It is now clear that protein phosphatases which respond to polycations in a substrate-dependent and/or biphasic manner (i.e., PCM-phosphatases) are present in a variety of tissues. Several of these enzymes have been purified to apparent homogeneity and shown to contain 2-4 polypeptide components.

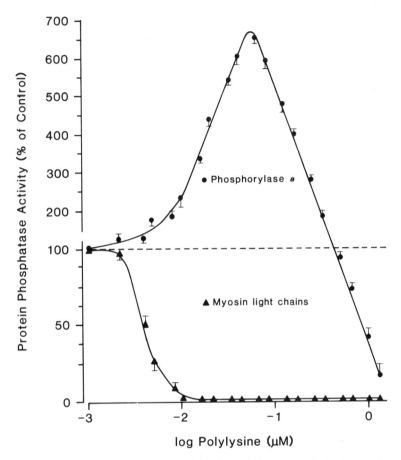

FIGURE 1. Influence of polylysine (13 kDa, 1-8000 nM) on dephosphorylation of phosphorylase a (●, 10 µM) and phosphorylated cardiac myosin light chains (▲, 4 µM) by aortic PCM-phosphatase. Each point is the mean ± SE of 3 experiments. Basal (control) phosphorylase phosphatase activity was 80 ± 9.7 mU/ml assay, whereas basal light chain phosphatase activity was 603 ± 11.4 mU/ml assay. As in previous studies, assays in this and subsequent figures were performed at 30°C in a 30 µl reaction mixture for either 10 min (phosphorylase a) or 3 min (light chain): the same amount of enzyme was present in all assays. One unit (U) of activity = 1 nmol ^{32}P released from the substrate per min under the conditions stated. The enzyme used in these studies contained 3 polypeptide components of 72-76, 53 and 35 kDa.

In 1984, Yang et al. purified a PCM-phosphatase which contained 4 polypeptides of 95, 75, 65 and 38 kDa, and a minor component of 140 kDa, from rabbit skeletal muscle. Subsequently, Tung et al. (1985), working with the same tissue, purified 3 multisubunit PCM-phosphatases which they called $2A_0$ (60, 54, 36 kDa), $2A_1$ (60, 53, 36 kDa) and $2A_2$ (60, 36 kDa). They suggested each form contained the same catalytic subunit (36 kDa) and that $2A_0$ and $2A_1$ were different enzymes. $2A_2$ was thought to be derived from $2A_0$ and/or $2A_1$ by loss of the 54 or 53 kDa component during the purification procedure. Their designation of these enzymes as 2A comes from a system of classification for protein phosphatases suggested by Ingebritsen and Cohen (1983). In this scheme, *type 1* phosphatases preferentially dephosphorylate the β-subunit of phosphorylase kinase and are inhibited by 2 proteins called inhibitor 1 and 2. Phosphatases which are insensitive to inhibitors 1 and 2, and which preferentially dephosphorylate the α-subunit of phosphorylase kinase are referred to as *type 2* enzymes. Although this scheme has certainly proven useful, it has not been universially accepted (Lee et al., 1980; Li, 1982; Merlevede et al., 1984). Nevertheless, Tung and Cohen confirmed our ealier observation showing that polycationic stimulation of phosphorylase phosphatase activity was limited to enzyme species which are largely insensitive to inhibitors 1 and 2.

Tung et al. also suggested that the enzyme described by Yang et al. (1984) might be the same as $2A_0$ or $2A_1$ if Yang's 95 kDa component was considered a contaminant. Alternatively, they suggested the Yang preparation might be the same as their $2A_2$ if both the 95 and 75 kDa components were considered contaminants. However, they apparently did not consider the possibility that the 95 kDa component described by Yang et al. might be part of the enzyme and that its absence from their 2A preparations could be ascribable to preparative losses.

In this context, we purified 2 PCM-phosphatases from bovine aortic smooth muscle. One of these contained 3 polypeptide components of 72-76, 53 and 35 kDa (Di Salvo et al., 1985b). Preparations of this kind have been used to advantage for studying effects of PCM-phosphatase on actin-myosin interaction (Bialojan et al., 1985a,b). The second aortic phosphatase contained 4 polypeptide doublets which were called A (85-94 kDa), B (70-75 kDa), C (49-54 kDa) and D (37-41 kDa). We suspect loss of the A components during purification may give rise to a form of the enzyme which contain 3 polypeptides.

Recently, a histone-stimulated phosphorylase phosphatase (i.e., a PCM-phosphatase) containing 4 polypeptides of 65, 60, 54 and 36 kDa was purified from rabbit liver (Gergely et al., 1986). However, Khandelwal and Enno purified a PCM-phosphatase from the same source which contained 2 polypeptides of 58 and 35 kDa: 2 other polypeptides in the preparation of 76 and 27 kDa were viewed as contaminants.

RELATIONSHIP BETWEEN ENZYMIC ACTIVITY AND 30-40 kDa POLYPEPTIDES

In spite of the differences noted in the number and masses of polypeptide components which have been identified in different PCM-phosphatases, all of the purified preparations contained a polypeptide of about 35 kDa. This is particularly significant because a widely held view is that catalytic activity in several native phosphatases resides in a polypeptide component of 30-40 kDa. That is, such enzymes are believed to contain several noncovalently linked subunits: catalytic activity is attributed to a 30-40 kDa subunit, whereas the other subunits present are thought to function as regulatory subunits (Lee et al., 1980; Li, 1982; Ingebritsen and Cohen, 1983; Merlevede, 1984; Cohen, 1985; Tung et al., 1985). However, recent data which we obtained with an aortic PCM-phosphatase are difficult to reconcile with this view.

Gel-filtration of an aortic PCM-phosphatase containing 4 polypeptide components (A, B, C, and D) yielded a peak of polylysine-stimulated phosphorylase phosphatase activity with a slight but obvious shoulder on the descending limb (Fig. 2). Surprisingly, fractions on the ascending limb of the peak did not contain 35-41 kDa D polypeptides (e.g., fraction 36, Fig. 4). Nevertheless, such fractions displayed basal and polylysine-stimulated phosphorylase phosphatase activities which were comparable to activities expressed by fractions on the descending limb of the peak that did contain 35-41 kDa D polypeptides (Fig. 4, fraction 42). These results show that expression of basal or polycation-modulable activity does not require the presence of a polypeptide component in the neighborhood of 35 kDa. The results obtained with fraction 36 also show that a polypeptide other than the 35-41 kDa component must be catalytically active. However, the results do not mean catalytic activity cannot be expressed by such components. Further insight into this problem was obtained from studies of the influence of limited proteolysis on enzymic activity, polypeptide composition of the preparations and responsiveness to polylysine.

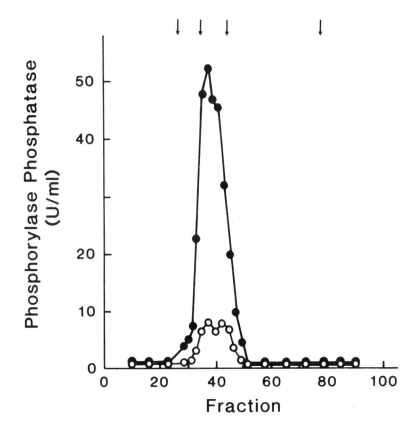

FIGURE 2. Gel filtration of an aortic PCM-phosphatase containing 4 polypeptide doublets (see text) on a TSK G 3000 SW column. The amount of enzyme applied was 80 U (200 µl) with respect to phosphorylase a and fractions of 250 µl were collected. Assays for phosphorylase phosphatase activity were performed in the presence (●) and absence (○) of polylysine: the concentration of polylysine was 160 nM. Arrows point to elution volumes of standard markers which included (from left to right) blue dextran (2×10^6 Da), catalase (240 kDa), aldolase (158 kDa) and chymotrypsinogen (25 kDa).

As shown in Fig. 3, tryptic digestion of the enzyme prior to gel-filtration resulted in (a) an increase in basal phosphorylase phosphatase activity, (b) elimination of the stimulatory response to polylysine and (c) disappearance of the B and C components: the A and D components were relatively resistant to tryptic digestion. A similar pattern of enzymic and structural changes occurred following tryptic digestion of a gel-filtration fraction which, like the starting enzyme, contained 35-41 kDa D polypeptides (Fig. 4, fraction 42).

Digestion of a gel-filtration fraction which did not contain the 35-41 kDa D components again resulted in increased basal activity and a loss of responsiveness to polylysine (Fig. 4, fraction 36). Additionally, however, digestion of this fraction also resulted in the appearance of a prominant 35 kDa polypeptide. Moreover, this new polypeptide must have been catalytically active because phosphorylase phosphatase activity was clearly evident. Thus, although catalytic activity does not require polypeptides of 35-41 kDa, proteolytic digestion can give rise to a polypeptide of about this mass which can express phosphorylase phosphatase activity.

A SUGGESTED MODEL OF THE NATIVE PCM-PHOSPHATASE

Based on these data a tentative working model of the enzyme can be proposed. The model is useful for interpreting some of the currently available data and it provides a guide for future studies. In the model, the native enzyme is visualized as consisting of a single polypeptide containing catalytic and protease-sensitive regulatory domains rather than several noncovalently linked subunits. Under basal conditions the regulatory domain probably suppresses the expression of phosphorylase phosphatase activity. In the presence of an appropriate signal, such as a polycationic effector (or perhaps a reversible covalent modification), suppression of phosphorylase phosphatase activity is relieved resulting in an apparent increase in activity.

The model suggests that proteolysis of the native enzyme results in progressive digestion of the regulatory domain and release of a 30-40 kDa segment of the catalytic domain which retains enzymic activity. In this setting, however, expressed phosphorylase phosphatase activity is enhanced because suppression by the regulatory domain is no longer possible. Similarly, polycationic stimulation of phosphorylase phosphatase activity is eliminated. The model does not exclude the possibility that some agents which influence enzymic activity act directly on the

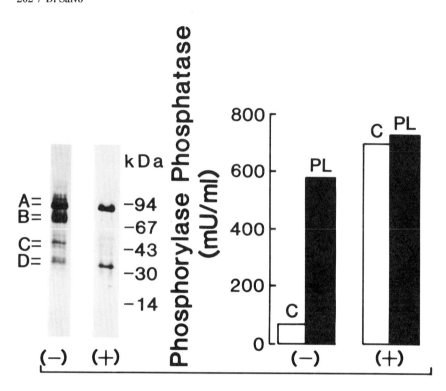

FIGURE 3. Effects of tryptic digestion on polypeptide components and phosphorylase phosphatase activities expressed by an aortic PCM-phosphatase containing 4 polypeptides. A sample of the same enzyme preparation subjected to gel-filtration (see Fig. 2) was used in this experiment. The enzyme (12.5 U/ml, basal phosphorylase phosphatase activity) was incubated at 30°C for 10 min in a reaction mixture containing 2.8 µg/ml of TPCK-trypsin. SDS-electrophoresis (left panel, Pharmacia Phastgel System) was performed before (-) and after (+) digestion: the 4 polypeptide doublets (A, B, C, D) are identified at the left of initial sample gel and the position of marker proteins is given at the right of gel showing the electrophoretic profile of the enzyme after tryptic digestion. The right panel shows phosphorylase phosphatase activities measured in the presence (dark bars, PL) and absence (clear bars, C) of 160 nM polylysine before (-) and after (+) tryptic digestion.

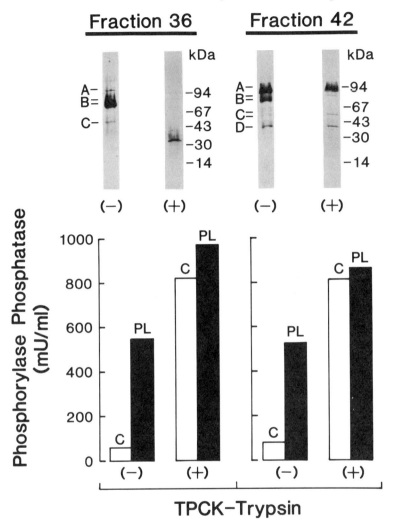

FIGURE 4. Effects of tryptic digestion on fractions 36 and 42 which were obtained during gel filtration of the 4 polypeptide-doublet aortic PCM-phosphatase (see Fig. 2). Details are the same as given in the legend to Fig. 3. Note that fraction 36 (top left) did not contain 35-41 kDa D polypeptides (-) but that a 35 kDa peptide appeared following tryptic digestion (+). Phosphorylase phosphatase activities for fraction 36 are shown at bottom left whereas enzymic activities of fraction 42 are given at bottom right.

catalytic domain or its catalytically active fragment. Within this framework, limited proteolysis of the native enzyme during purification may give rise to multiple forms of the PCM-phosphatase. Such a mechanism could account for the different number and masses of component polypeptides contained in different PCM-phosphatase preparations: it could also account for the usual presence of relatively protease-resistant 30-40 kDa components which express catalytic activity.

Accordingly, a given PCM-phosphatase preparation may represent a mixture of polypeptides including native enzyme, fragments of the regulatory domain, fragments of the catalytic domain and fragments containing portions of the regulatory and catalytic domains. With respect to the aortic PCM-phosphatase containing 4 polypeptide components (Fig. 3), the 35-41 kDa D polypeptides may represent enzymically active fragments of the catalytic domain. Conceivably, the 70-75 kDa B polypeptides may represent a fragment with functional portions of the regulatory and catalytic domains. This idea is supported by the observation that this is the dominant component present in gel-filtered fractions which display basal and stimulated phosphorylase phosphatase activities but are lacking the 35-41 kDa D components (Fig. 4, fraction 36). The idea is also supported by the fact that tryptic digestion of the B component is associated with the appearance of a 35 kDa polypeptide (Fig. 4, fraction 36), enhanced basal activity and elimination of the stimulatory response to polylysine (Figs. 3 and 4). The possible origin and functional significance of the A and C components is more difficult to interpret in terms of the suggested model. Clearly, further studies are required to assess the extent to which the model accurately reflects structural and regulatory features of the native PCM-phosphatase.

EFFECTS OF PCM-PHOSPHATASE ON ACTIN-MYOSIN INTERACTIONS

In collaborative studies with Dr. C. Bialojan and Professor J.C. Rüegg we observed that a 3 component PCM-phosphatase (72-76, 53 and 35 kDa) purified from bovine aortic muscularis exerted marked effects on actin-myosin interactions in 2 Ca^{2+}-dependent contractile models from smooth muscle. The systems studied were bovine aortic native actomyosin (actin-dependent ATPase activity) and Ca^{2+}-dependent contraction of detergent-skinned fibers from chicken gizzard and porcine carotid artery. Since the results of these studies were reported in detail (Bialojan et al., 1985a,b), only the key points will be reviewed at this time.

In both contractile models, the sensitivity to Ca^{2+} for actin-myosin interaction, as reflected by ATPase activity (actomyosin) or isometric force (skinned fibers), was markedly increased in the presence of PCM-phosphatase. Each of these effects was associated with decreased phosphorylation of the myosin light chains. Moreover, addition of the phosphatase to the medium bathing fibers contracted in low Ca^{2+} resulted in prompt relaxation. Thus, PCM-phosphatase effectively (a) inhibited initiation of actin-myosin interaction, (b) reduced the extent to which actin-myosin interaction could occur and (c) reversed interaction which had already taken place. In no instance was any evidence obtained for maintained actin-myosin interaction in the absence of phosphorylated light chains. The results also showed that PCM-phosphatase could reduce Ca^{2+}-dependent phosphorylation of the regulatory light chains in structurally and functionally intact myosin, and it could also promote dephosphorylation of myosin which had been phosphorylated previously.

Collectively, these data suggest that PCM-phosphatase is likely to be the major "myosin phosphatase" in smooth muscle. If the activity of this enzyme is subject to regulation *in vivo*, changes in expressed enzymic activity may be a mechanism for modulating the state of myosin phosphorylation and subsequent actin-myosin interaction.

These studies were supported by National Institutes of Health Grants HLB 20196 and 22619. I am grateful to Ms. A. Tolle for typing the manuscript.

REFERENCES

Bialojan C, Rüegg JC, Di Salvo J (1985a). Phosphatase-mediated modulation of actin-myosin interaction in bovine aortic actomyosin and skinned porcine carotid artery. Proc Soc Exp Biol Med 178:36-45.

Bialojan C, Rüegg JC, Di Salvo J (1985b). Influence of a polycation modulated phosphatase on actin-myosin interactions in smooth muscle preparations. Adv Prot Phosphatases 2:105-121.

Cohen P (1985). The role of protein phosphorylation in the hormonal control of enzyme activity. Eur J Biochem 151:439-448.

Di Salvo J, Waelkens E, Gifford D, Goris J, Merlevede W (1983). Modulation of latent phosphatase activity from vascular smooth muscle by histone-H_1 and polylysine. Biochem Biophys Res Commun 117:493-500.
Di Salvo J, Gifford D, Kokkinakis A (1984). Properties and function of protein phosphatases in vascular smooth muscle. Proc Soc Exp Biol Med 177:24-32.
Di Salvo J, Gifford D, Kokkinakis A (1985a). Heat-stable regulatory factors are associated with polycation-modulable phosphatases. Adv Enzyme Regul 23:103-122.
Di Salvo J, Gifford D, Kokkinakis A (1985b). Properties and function of a bovine aortic polycation-modulated protein phosphatase. Adv Prot Phosphatases 1:327-346.
Gergely P, Erdödi F, Bot G (1986). Purification and regulation of protein phosphatases in latent and active forms. Adv Prot Phosphatases 3: In Press.
Ingebritsen TS, Cohen P (1983). Protein phosphatases: Properties and role in cellular regulation. Science 221:331-338.
Kamm KE, Stull JT (1985). The function of myosin and myosin light chain kinase phosphorylation in smooth muscle. Ann Rev Pharmacol Toxicol 25:593-620.
Khandelwal RL, Enno TL (1985). Purification and characterization of a high molecular weight phosphoprotein phosphatase from rabbit liver. J Biol Chem 260:14335-14343.
Krebs EG, Beavo JA (1979). Phosphorylation-dephosphorylation of enzymes. Ann Rev Biochem 48:923-959.
Lee EYC, Silberman SR, Ganapathi MK, PEtrovic S, Paris H (1980). The phosphoprotein phosphatases: Properties of the enzymes in the regulation of glycogen metabolism. Adv Cyclic Nucleo Res 13:95-131.
Li HC (1982). Phosphoprotein phosphatases. Curr Top Cell Regul 21:129-174.
Merlevede W, Vandenheede JR, Goris J, Yang SD (1984). Regulation of the ATP, Mg-dependent protein phosphatase. Curr Top Cell Regul 23:177-215.
Tung HYL, Alemany S, Cohen P (1985). The protein phosphatases involved in cellular regulation. 2. Purification, subunit structure and properties of protein phosphatases $2A_0$, $2A_1$ and $2A_2$ from rabbit skeletal muscle. Eur J Biochem 148:253-263.
Wilson SE, Mellgren RL, Schlender KK (1983). Evidence that the heat-stable protein activator of phosphorylase phosphatase is histone H_1. Biochem Biophys Res Commun 116:581-586.
Yang SD, Vandenheede JR, Merlevede W (1984). Purification of a latent protein phosphatase from rabbit skeletal muscle. Biochem Biophys Res Commun 118:923-928.

CHARACTERIZATION OF THE SMOOTH MUSCLE PHOSPHATASES AND STUDY OF THEIR FUNCTION

Mary D. Pato and Ewa Kerc

Department of Biochemistry, University of Saskatchewan, Saskatoon, Saskatchewan, Canada S7N 0W0

INTRODUCTION

Phosphorylation-dephosphorylation of proteins is a major mechanism for the regulation of many biological processes including contractile activity in smooth muscle cells (for reviews see Kamm and Stull, 1985; Adelstein et al., 1985). In vitro studies have demonstrated a positive correlation between phosphorylation of serine -19 of the 20,000-Da light chains of myosin and its actin-activated MgATPase activity. An increase in the phosphate content of myosin is also observed as tension develops in stimulated-intact muscle strips and skinned muscle fibers. Dephosphorylation of the myosin by protein phosphatases reverses these effects.

The enzyme which catalyzes the phosphorylation of myosin is myosin light chain kinase (MLCK). It is inactive in the absence of Ca^{2+} and calmodulin. The observation from an in vitro study (Conti and Adelstein, 1981) that phosphorylation of MLCK by cAMP-dependent protein kinase inhibits its activity due to a lowered affinity for calmodulin, has provided a mechanism for the modulation of contractile activity by hormones which could alter the intracellular cAMP levels. The high activity of MLCK is restored by dephosphorylation with smooth muscle phosphatase-I.

Recently, other phosphorylation reactions which may be involved in the regulation of smooth muscle contraction have been reported. Under certain conditions, threonine-18 of the 20,000-Da light chains of myosin is also phosphorylated by MLCK (Cole et al., 1985; Ikebe et al., 1986). The second site phosphorylation of myosin causes an additional increase of its actin-activated MgATPase activity (Ikebe and Hartshorne,

1985). Myosin is also a substrate for protein kinase C
(Adelstein et al., 1985). This modification inhibits rather
than stimulates the activity of the monophosphorylated myosin.
Like cAMP-dependent protein kinase, protein kinase C
phosphorylates MLCK and inhibits its activity. The activity
of caldesmon, a protein which has been shown to affect the
rate of hydrolysis of ATP, appears to be regulated by
reversible phosphorylation (Ngai and Walsh, 1984).

Full understanding of the regulation of contractile
activity in smooth muscle requires knowledge of the enzymes
which catalyze these phosphorylation- dephosphorylation
reactions. This paper describes the properties of the smooth
muscle phosphatases we have purified from turkey gizzard and
rabbit uterine muscles and elucidation of their function.

The protein phosphatases were purified by successive
column chromatography on Sephacryl S-300, DEAE-Sephacel,
ω -aminooctyl Agarose and thiophosphorylated myosin light
chain-Sepharose. The major substrates used for the detection
of phosphatase activities were turkey gizzard intact myosin,
heavy meromyosin and isolated myosin light chains
phosphorylated with MLCK under conditions where only serine-19
is modified. Since heavy meromyosin is more soluble than
myosin, it has been used preferentially over intact myosin in
our study. We have identified 4 protein phosphatases in
turkey gizzard extract, termed Smooth Muscle Phosphatase
(SMP)-I, II, III and IV. SMP-I, II and IV have been purified
to apparent homogeneity as shown in Fig. 1 (Pato, 1985).

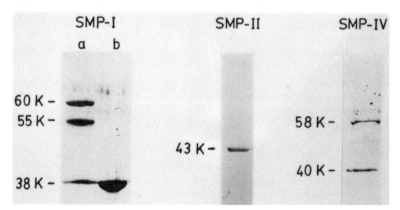

Figure 1. SDS-polyacrylamide gel of the SMP-I (a) and its
catalytic subunit (b) SMP-II and SMP-IV. Reprinted with
permission from Pato MD (1985): In "Advances in Protein Phosphatases," Vol. 1, pp
367–382. Leuven, Belgium: Leuven University Press.

CHARACTERIZATION

Physical Properties

SMP-I is composed of 3 subunits (M_r=60,000, 55,000 and 38,000) present in equimolar ratios (Pato and Adelstein, 1980). The asymmetric shape of the enzyme is suggested by the different molecular weights obtained by gel filtration on Sephadex G-200 (MW=230,000) and sedimentation equilibrium (MW=165,000) (Pato and Adelstein, 1983a). The subunits of SMP-I can be dissociated by treatment with ethanol (Pato and Kerc, 1986a) and by freezing and thawing in 0.2M mercaptoethanol (Pato and Adelstein, 1983a). Isolation of the active species following these procedures identified the 38,000-Da component as the catalytic subunit.

SMP-I is classified as protein phosphatase $2A_1$ (Pato et al., 1983) according to the criteria proposed by Ingebritsen and Cohen (Ingebritsen and Cohen, 1983). Like other type 2 phosphatases, SMP-I preferentially dephosphorylates the α-subunit of phosphorylase kinase and is insensitive to the heat stable inhibitors 1 and 2. Multisubunit protein phosphatases which are structurally similar to SMP-I have been purified from chicken gizzards (Onishi et al., 1982), pig heart (Imaoka et al., 1983), rabbit skeletal muscle (Tung et al., 1985) and rat liver (Tamura and Tsuiki, 1980). In all cases, their smallest subunits (M_r=34,000-38,000) possess the catalytic activity.

SMP-II is a single-subunit (M_r=43,000) globular protein which requires Mg^{2+} for activity (Pato and Adelstein, 1983b). Protein phosphatases exhibiting similar molecular weight and metal ion dependency as SMP-II have been purified from bovine heart (Li, 1981), canine heart (Binstock and Li, 1979) and rat liver (Hiraga et al., 1981). These enzymes are classified as protein phosphatases 2C.

SMP-IV is composed of 2 subunits (M_r=58,000 and 40,000) (Pato and Kerc, 1985). Unlike SMP-I and II, it does not conform to the classification scheme proposed by Ingebritsen and Cohen because it preferentially dephosphorylates the β-subunit of phosphorylase kinase, a property of type 1 phosphatases, and it is not inhibited by the heat stable inhibitor-2, a property of type 2 phosphatases.

Recently, we have purified to apparent homogeneity 2 (U-I and U-IV) of the 4 protein phosphatases detected in rabbit uterine muscle (Pato and Kerc, 1986b). SDS-polyacrylamide gel electrophoresis of U-I showed 3 bands (M_r=62,000, 55,000 and 38,000). Partial characterization of the uterine enzymes revealed that U-I resembles SMP-I while U-IV resembles SMP-IV. Furthermore, like SMP-II, one of the partially purified enzymes (U-II) requires Mg^{2+} for activity. These observations suggest that the avian and mammalian smooth muscle phosphatases are very similar.

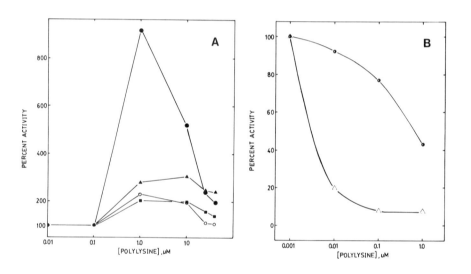

Figure 2. Effect of polylysine on the activities of the smooth muscle phosphatases. A. The activities of SMP-I toward myosin light chains (o) and the free catalytic subunit toward myosin light chains (●), heavy meromyosin (▲) and phosphorylase a (■) were determined in the presence of polylysine (MW=55,000). B. The activities of SMP-IV toward myosin light chains (●) and heavy meromyosin (▲) were determined in the presence of polylysine (MW = 14,000). The activities of the enzymes in the absence of polylysine were taken as 100%.

Effect of Polycations

The possibility that type 2 phosphatases are regulated by polycations have been proposed by several investigators based on the observation that histone and synthetic polylysine stimulate the activity of protein phosphatases toward phosphorylase $\underline{a}$ (DiSalvo et al., 1983; Mellgren and Schlender, 1983). Fig. 2A shows that low concentrations of polylysine stimulate the activities of SMP-I and its free catalytic subunit (SMP-I_c) toward phosphorylase $\underline{a}$ and the isolated myosin light chains. The dependence of the extent of stimulation on the substrate is indicated by the activity of SMP-I_c toward myosin light chains, heavy meromyosin and phosphorylase $\underline{a}$. Whereas the dephosphorylation of the myosin light chains is stimulated 9-fold, only 2-3-fold stimulation of activity is observed toward heavy meromyosin and phosphorylase $\underline{a}$. Consistent with the observation of DiSalvo and coworkers (DiSalvo et al., 1984), we found that the concentration of polylysine required to achieve maximal stimulation is inversely proportional to the chain length. Furthermore, the L-isomer of polylysine was observed to be a better effector than the D-form.

The activities of SMP-II toward myosin light chains and phosphorylase $\underline{a}$ were also slightly stimulated by polylysine. In contrast, Fig. 2B shows that polylysine inhibits rather than stimulates the activity of SMP-IV. Inhibition of 50% of the activity of SMP-IV toward heavy meromyosin and myosin light chains is caused by 0.003 and 0.6 M polylysine, respectively. Histone IIA has similar effect on the activity of SMP-IV.

Substrate Specificity

Most, if not all, of the protein phosphatases are non-specific. The smooth muscle phosphatases dephosphorylate contractile proteins as well as non-contractile proteins (Table 1). However, their activities toward contractile proteins such as myosin light chains, are much greater than the rates at which they dephosphorylate non-contractile proteins such as phosphorylase $\underline{a}$, histone IIA, phosphorylase kinase and casein.

All of the smooth muscle phosphatases have high activity toward the isolated myosin light chains. But, as isolated, only SMP-III and IV are active against intact myosin and heavy meromyosin. The ability of these enzymes to discriminate between the phosphorylated serine-19 of the isolated myosin light chains and the light chains associated with the heavy chains indicates a difference in the microenvironment of the phosphate groups in these 2 substrates. MLCK is dephosphorylated rapidly by SMP-I.

TABLE 1. Specific Activities of the Smooth Muscle Phosphatases Toward Various Substrates

	SMP-I [a]	SMP-I [a]	SMP-I [a]	SMP-II [b]	SMP-IV [c]
			μ mol/min/mg		
MLC	2.8	19.3	0.4	1.0	1.6
HMM	0	0.13	0.2	0	1.8
MLCK	0.7 [d]	ND	ND	0.01	ND
Phosphorylase a	0.2	0.5	0.1	0.01	0.02
Histone IIA	0.07 [d]	ND	ND	0.02	0.1

MLC, Myosin light chains; HMM, Heavy meromyosin;
MLCK, Myosin light chain kinase
a, data taken from Pato and Kerc, 1986a
b, data taken from Pato and Adelstein, 1983b
c, data taken from Pato and Kerc, 1985
d, data taken from Pato and Adelstein, 1983a
ND, not determined

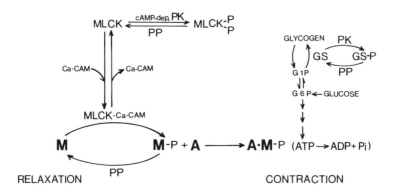

Figure 3. A general scheme for the regulation of smooth muscle contraction by Ca^{2+}-calmodulin and cAMP.

Function

Mainly on the basis of the substrate specificities of the smooth muscle phosphatases, we speculated on their possible mode of action. Fig. 3 illustrates the regulation of contractile activity in smooth muscle. This process is regulated primarily by the reversible phosphorylation of myosin. The high activity of SMP-III and IV toward myosin suggests that these enzymes catalyze the dephosphorylation of myosin in vivo to elicit relaxation.

The inability of SMP-I to dephosphorylate myosin suggests that it is involved in processes other than relaxation. The most obvious role for SMP-I is to dephosphorylate MLCK following an increase in cAMP level to restore the high activity of the kinase.

However, the possibility still exists that SMP-I is involved in relaxation because it could be activated toward myosin under certain conditions. Dissociation of the regulatory subunits from the 38,000-Da polypeptide by freezing and thawing in 0.2M mercaptoethanol or by ethanol treatment results in a form of SMP-I ($SMP-I_c$) which is active against myosin (Table 1). This observation indicates that one or both of the regulatory subunits inhibit the activity of the catalytic subunit toward myosin. To identify the inhibitory component, we subjected SMP-I to trypsin and chymotrypsin (Pato and Kerc, 1986a). Fig. 4 shows a time course of tryptic digestion of SMP-I. The protease degraded the 55,000-Da polypeptide rapidly. Slow degradation of the catalytic subunit occurred upon prolonged exposure to trypsin while the 60,000-Da polypeptide appeared to be resistant to digestion. Incubation of SMP-I with chymotrypsin gave similar digestion profile. A form of SMP-I devoid of the 55,000-Da subunit, termed $SMP-I_2$, was prepared by limited tryptic digestion and characterized. Unlike SMP-I, $SMP-I_2$ is active toward myosin (Table 1) proving that the 55,000-Da polypeptide suppresses the activity of the catalytic subunit against myosin. Thus, if there is a physiological mechanism which dissociates the catalytic subunit from the regulatory subunits and/or if there is a protease in muscle which specifically digests the 55,000-Da polypeptide, then SMP-I could play a role in the relaxation process. It is interesting to note that phosphatases which are structurally similar to $SMP-I_2$ have been purified from rabbit skeletal muscle (Paris et al., 1984; Tung et al., 1985), rat liver (Tamura et al., 1980), bovine

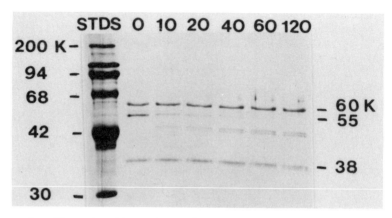

Figure 4. Tryptic digestion of SMP-I. SDS-polyacrylamide gel of SMP-I incubated with trypsin for 0, 10, 20, 40, 60 and 120 minutes.

heart (Li, 1980), bovine aorta (Werth et al., 1982) and rabbit reticulocyte lysate (Crouch and Safer, 1980).

The function of SMP-II is not obvious because it does not dephosphorylate myosin and it has low activity against MLCK. One of the proteins that serves as a substrate for SMP-II and is dephosphorylated rapidly by the other type 2C phosphatases is glycogen synthase. SMP-II may be involved in the regulation of glycogen metabolism which is essential for the production of ATP required for contraction.

To substantiate our hypothesis, we studied the binding properties of the smooth muscle phosphatases to actin and myosin to determine their intracellular localization (Sellers and Pato, 1984). Consistent with their substrate specificities, only SMP-III and IV bound to myosin with binding constants of 3.8×10^5 and $3.6 \times 10^5 M^{-1}$, respectively. When binding to thiophosphorylated myosin was determined, a dramatic increase in the binding constants was observed. None of the phosphatases bound to actin. These results suggest that none of the phosphatases are localized on the thin filaments. SMP-III and IV may be associated with the thick filaments, a concept consistent with the proposed function for SMP-III and IV.

To provide direct evidence for the role of SMP-IV, we studied the effect of SMP-IV on the contractile activity of skinned muscle fiber preparations (Hoar et al., 1985). Skinned chicken gizzard cell bundles were contracted in the absence of Ca^{2+} by treatment with Ca^{2+} insensitive MLCK prepared by limited proteolytic digestion of MLCK. Incubation of the contracted cell bundles with SMP-IV caused the muscle fibers to relax. Analysis of the phosphate content of myosin before and after incubation with SMP-IV revealed that relaxation was accompanied with dephosphorylation of myosin. This study clearly demonstrated the ability of SMP-IV to dephosphorylate myosin in vivo.

SUMMARY

The activities of some proteins involved in the process of contraction-relaxation in smooth muscle cells are regulated by reversible phosphorylation. Phosphorylation of myosin by MLCK has been shown to be a pre-requisite for muscle contraction. MLCK, itself, is a substrate for cAMP-dependent protein kinase. Relaxation is favored in the event that MLCK is phosphorylated by cAMP-dependent protein kinase because this modification inhibits the activity of MLCK.

In our attempt to understand the mechanism and regulation of contractile activity in smooth muscle cells, we purified and characterized the enzymes which catalyze the dephosphorylation of myosin and MLCK. We have purified 3 smooth muscle phosphatases termed SMP-I, II and IV to apparent homogeneity and partially purified SMP-III from turkey gizzards. Characterization of these enzymes revealed that they are distinct. They have different physical, enzymatic and immunological properties. As isolated, all 4 enzymes dephosphorylate myosin light chains rapidly but only SMP-III and IV are active toward myosin or heavy meromyosin. However, SMP-I could be activated toward myosin when its catalytic subunit is dissociated from the regulatory subunits and when the 55,000-Da regulatory subunit is digested or released from the holoenzyme. Recently we have purified to apparent homogeneity 2 protein phosphatases from rabbit uterine muscle. Partial characterization of these enzymes revealed their close similarity to the avian smooth muscle phosphatases.

Analysis of the properties of the smooth muscle phosphatases led us to speculate on their function in vivo. SMP-III and IV are most likely to dephosphorylate myosin to cause relaxation because they exhibit the highest activity toward intact myosin. SMP-I may play a role in this process if there is a physiological mechanism which dissociates the catalytic subunit from the 55,000-Da regulatory subunit or from both regulatory subunits. A more obvious role for SMP-I is to dephosphorylate MLCK following phosphorylation by cAMP-dependent protein kinase to restore the high activity of MLCK. SMP-II does not dephosphorylate myosin and has low activity toward MLCK. It is active toward glycogen synthase suggesting a role in glycogen metabolism for the production ATP required to supply the energy for contraction. We are currently undertaking experiments to verify these proposals.

ACKNOWLEDGMENT

This work is supported by the Medical Research Council of Canada.

REFERENCES

Adelstein RS, deLanerolle P, Nishikawa M, Sellers JR (1985). The regulation of smooth muscle contraction by phosphorylation. In Bevan JS et al. (eds): "Vascular Neuroeffector Mechanisms," New York: Elsevier Science Publishers, pp. 63-67.

Binstock JF, Li HC (1979). A novel glycogen synthase phosphatase from canine heart. Biochem Biophys Res Commun 87:1226-1234.

Cole HA, Griffiths HS, Patchell VB, Perry SV (1985). Two-site phosphorylation of the phosphorylatable light chain (20-kDa light chain) of ckicken gizzard myosin. FEBS Lett 180:165-169.

Conti MA, Adelstein RS (1981). The relationship between calmodulin binding and phosphorylation of smooth muscle myosin kinase by the catalytic subunit of 3 :5 cAMP-dependent protein kinase. J Biol Chem 256:3178-3181.

Crouch D, Safer B (1980). Purification and properties of eIF-2 phosphatase. J Biol Chem 255:7918-7924.

DiSalvo J, Gifford D, Kokkinakis A (1984). Modulation of aortic protein phosphatase activity by polylysine. Proc Soc Exp Biol Med 177:24-32.

DiSalvo J, Waelkens E, Gifford D, Goris J, Merlevede W (1983). Modulation of latent protein phosphatase activity from vascular smooth muscle by histone H1 and polylysine. Biochem Biophys Res Commun 117:493-500.

Hiraga A, Kikuchi K, Tamura S, Tsuiki S (1981). Purification and characterization of Mg^{2+}-dependent glycogen synthase phosphatase (Phosphoprotein phosphatase IA) from rat liver. Eur J Biochem 119:503-510.

Hoar PE, Pato MD, Kerrick WGL (1985). Myosin light chain phosphatase. Effect on the activation and relaxation of gizzard smooth muscle skinned fibers. J Biol Chem 260:8760-8764.

Ikebe M, Hartshorne DJ (1985). Phosphorylation of smooth muscle myosin at two distinct sites by myosin light chain kinase. J Biol Chem 260:10027-10031.

Ikebe M, Hartshorne DJ, Elzinga M (1986). Identification, phosphorylation and dephosphorylation of a second site for myosin light chain kinase on the 20,000-dalton light chain of smooth muscle myosin. J Biol Chem 261:36-39.

Imaoka T, Imazu M, Usui H, Kinohara N, Takeda M (1983). Resolution and reassociation of three distinct components from pig heart phosphoprotein phosphatase. J Biol Chem 258:1526-1535.

Ingebritsen TS, Cohen P (1983). Protein phosphatases: Properties and role in cellular regulation. Science 221:331-338.

Kamm KE, Stull JT (1985). The function of myosin and myosin light chain kinase phosphorylation in smooth muscle. Ann Rev Pharmacol Toxicol 25:593-620.

Li HC (1981). Purification and properties of cardiac muscle phosphoprotein phosphatase and alkaline phosphatase isozyme. In Rosen OM, Krebs EG (eds): "Protein Phosphorylation in Cold Spring Harbor Conference on Cell Proliferation," V.8 New York: Cold Spring Harbor Laboratory, pp. 441-457.

Mellgren RL, Schlender KK (1983). Histone H1-stimulated phosphorylase phosphatase from rabbit skeletal muscle. Biochem Biophys Res Commun 117:501-508.

Ngai PK, Walsh MP (1984). Inhibition of smooth muscle actin-activated myosin Mg^{2+}-ATPase activity by caldesmon. J Biol chem 259:13656-13659.

Onishi H, Umeda J, Ulchiwa H, Watanabe S (1982). Purification of gizzard myosin light chain phosphatase and reversible changes in the ATPase and superprecipitation activities of actomyosin in the presence of purified preparation of myosin light chain phosphatase and kinase. J Biochem 91:265-271.

Paris H, Ganapathi MK, Silberman SR, Aylward JH, Lee EYC (1984). Isolation and characterization of a high molecular weight protein phosphatase from rabbit skeletal muscle. J Biol Chem 259:7510-7518.

Pato MD (1985). Properties of the smooth muscle phosphatases from turkey gizzards. In Merlevede W, DiSalvo J (eds): "Adv Prot Phosphatases", V.1 Leuven, Belgium: Leuven University Press, pp. 367-382.

Pato MD, Adelstein RS (1980). Dephosphorylation of the 20,000-dalton light chain of myosin by two different phosphatases from smooth muscle. J Biol Chem 255:6535-6538.

Pato MD, Adelstein RS (1983a). Purification and characterization of a multisubunit phosphatase from turkey gizzard smooth muscle. J Biol Chem 258:7047-7054.

Pato MD, Adelstein RS (1983b). Characterization of a Mg^{2+}-dependent phosphatase from turkey gizzard smooth muscle. J Biol Chem 258:7055-7058.

Pato MD, Adelstein RS, Crouch D, Safer B, Ingebritsen TS, Cohen P (1983). The protein phosphatases involved in cellular regulation. Classification of two homogeneous myosin light chain phosphatases from smooth muscle as protein phosphatase-$2A_1$ and 2C, and a homogeneous protein phosphatase from reticulocytes active on protein synthesis initiation factor eIF-2 as protein phosphatase-$2A_2$. Eur J Biochem 132:283-287.

Pato MD, Kerc E (1985). Purification and characterization of a smooth muscle myosin phosphatase from turkey gizzards. J Biol Chem 260:12359-12366.

Pato MD, Kerc E (1986a). Limited proteolytic digestion and dissociation of smooth muscle phosphatase-I modifies its substrate specificity. J Biol Chem 261:3770-3774.

Pato MD, Kerc E (1986b). Properties of protein phosphatases from rabbit uterine smooth muscle. Fed Proc 45:1805.

Sellers JR, Pato MD (1984). The binding of smooth muscle myosin light chain kinase and phosphatases to actin and myosin. J Biol Chem 259:7740-7746.

Tamura S, Kikuchi H, Kikuchi K, Hiraga M, Tsuiki S (1980). Purification and subunit structure of a high molecular weight phosphoprotein phosphatase (Phosphatase II) from rat liver. Eur J Biochem 104:347-355.

Tamura S, Tsuiki S (1980). Purification and subunit structure of rat liver phosphoprotein phosphatase, whose molecular wight is 260,000 by gel filtration (Phosphatase 1B). Eur J Biochem 111:217-224.

Tung HYL, Alemany S, Cohen P (1985). The protein phosphatases involved in cellular regulation. Purification, subunit structure and properties of protein phosphatase $2A_o$, 2A and $2A_2$ from rabbit skeletal muscle. Eur J Biochem 148:253-263.

Werth DK, Haeberle JR, Hathaway DR (1982). Purification of a myosin phosphatase from bovine aortic smooth muscle. J Biol Chem 257:7306-7309.

ACTIVATION OF PROTEIN KINASE C AND CONTRACTION IN SKINNED VASCULAR SMOOTH MUSCLE

Meeta Chatterjee and Carolyn Foster
Schering Corporation
60 Orange St.,
Bloomfield, N.J. 07003

INTRODUCTION

Myosin light chain phosphorylation has been demonstrated to be important in triggering smooth muscle contraction. However, other mechanisms may be important for stress maintenance (Dillon et al, 1981; Chatterjee and Murphy 1983; Moreland and Murphy, in press). Reports that phorbol esters caused contractions in intact muscles (Rasmussen et al, 1984; Danthuluri and Deth, 1984) led us to consider the possibility that protein kinase C was involved in modulation of contraction. Protein kinase C is a Ca^{++} and phospholipid-dependent kinase that can be activated by phorbol esters (Castagna et al. 1982) or endogenous diacylglycerols (Nishizuka, 1983).

Since interpretation of studies of phorbol ester induced contractions in intact smooth muscle is difficult because of changes in calcium fluxes (Forder et al, 1985), we undertook the following studies to investigate the mechanism of phorbol ester induced contraction in Triton X-100 skinned porcine carotid media under conditions in which intracellular Ca^{++} could be controlled.

METHODS

Medial strips of porcine carotid arteries were dissected (Driska et al, 1981) and attached to either a Grass FT 03 force-displacement transducer or a Cambridge Technology (Model 300H) dual mode servo. After equilibration in physiological salt solution for 1 hr, muscles were stretched to their optimal length for force development (L_o) and a control response to 110 mM K^+ was obtained. Temperature was maintained at 22° throughout the experiment. Tissues were then treated with 0.5% Triton X-100 for 1 hr (Chatterjee and Murphy, 1983) using a protocol modified from Gordon (1978). Stress and phosphorylation of the 20K Da myosin light chain in response to either Ca^{++} or phorbol 12,13-dibutyrate (PDBu) were measured at times indicated (Chatterjee and Tejada, 1986). To determine the extent of myosin light chain phosphorylation, tissues were frozen in an acetone-dry ice slurry and analyzed by two-dimensional gel electrophoresis (Driska et al, 1981). Fractional phosphorylations were determined by optical densitometry of Coomassie stained gels.

Phosphorylation of proteins other than myosin light chain was studied by preincubating skinned artery strips in relaxing solution containing $\gamma^{32}P$-ATP (100uCi/300ul), contracting the strips under conditions identical to those used to measure stress, and separating the ^{32}P-labelled proteins on 7.5-20% acrylamide SDS gels. Labelled proteins were identified by autoradiography.

Peak active resistance to stretch (Singer et al, in press) was determined during imposed stretches from L_o to 1.06 L_o at a rate of 5.88 mm/sec (0.80 + 0.01 L_o/sec, n=5) in the relaxing solution (passive) and in solutions containing either varying concentrations of Ca^{++} or PDBu (total). The passive response was subtracted from the total response. The peak active resistance to stretch was the maximum active stress obtained before the tissue yielded to stretch. This parameter is proportional to the number of attached cross bridges.

Data are reported as mean $\pm$ SE. Statistical significance was determined using Students t test (P<0.05).

RESULTS

In skinned muscle strips, a trace concentration of Ca^{++} (10^{-7}M) was associated with little or no developed stress over periods up to 50 min (Fig. 1). At this Ca^{++} concentration, the peak active resistance to stretch did

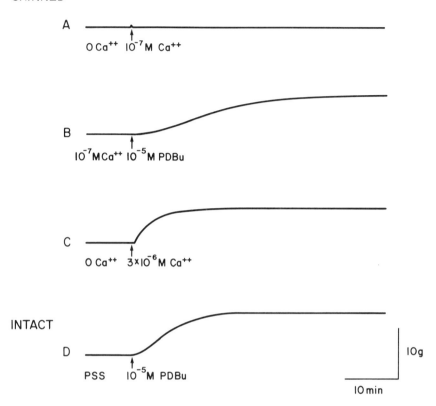

Figure 1. Stress responses obtained with Ca^{++} and PDBu (A) At 10^{-7} M Ca^{++}, muscles generated 0–6% of the responses obtained in 10^{-5} M Ca^{++}. These low values were maintained for 50 mins. (B) Addition of 10^{-5} M PDBu to 10^{-7} M Ca^{++} resulted in slow development of force. Muscles responded within 2–3 min, and maximum force was attained at ~40 min. (C) In contrast, stress development in high Ca^{++} was much more rapid. Maximal forces were achieved within 10 min when solutions containing zero added Ca^{++}, 0.1 mM EGTA were replaced with 3uM free Ca^{++} in 5 mM EGTA. Maximum values of stresses obtained with high PDBu and high Ca^{++} were comparable (see text). (D) The PDBu-dependent contractions were much faster in the intact muscles than in skinned. However, these responses were much slower than those obtained with other agonists such as norepinephrine, histamine and high K^+ (data not shown).

TABLE 1. Peak Active Resistance to Stretch

	PARS ($\times 10^4 \text{N/m}^2$)
10^{-7} M Ca^{++} (10 min)	0.82 ± 0.29 (3)
10^{-7} M Ca^{++} (40 min)	1.53 ± 0.41 (3)
10^{-5} M PDBu (45 min)	6.91 ± 0.73 (4)*
7×10^{-6} M Ca^{++} (15-20 min)	6.55 ± 0.46 (3)*

The peak active resistance to stretch (PARS) was determined in 10^{-7} M Ca^{++}, 10^{-5} M PDBu (+ 10^{-7} M Ca^{++}), and 7×10^{-6} M Ca^{++} at times indicated. Values are mean $\pm$ SE(n). * Denotes values significantly different from value at 40 min in 10^{-7} M Ca^{++} (P<0.05).

not increase significantly at 40 min (P<0.05), indicating that there was no time-dependent increase in the number of attached cross-bridges (Table 1). Addition of PDBu resulted in a slow development of stress, requiring up to 45 min to reach maximum levels (Fig. 1). This was accompanied by an increase in peak active resistance to stretch (Table 1). While the rates of stress development in PDBu were low compared to those in high Ca^{++}, maximum stresses were comparable (Fig. 1). In other experiments, the skinned muscles generated $5.72 \pm 0.59 \times 10^4$ N/m^2 (n=10) in 10^{-5} M PDBu and $5.43 \pm 0.34 \times 10^4$ N/m^2 (n=21) in 7×10^{-6} M Ca^{++} (Chatterjee and Tejada, 1986). PDBu-dependent contractions in the intact preparations were also slow (Fig. 1). Higher stresses were obtained with a combination of 3×10^{-6} M Ca and 10^{-5} M PDBu (concentrations for maximum response) than with either stimulus alone (Fig. 2). PDBu-dependent stress activation required a low level of Ca^{++} (10^{-7} M), since there was a considerable increase in developed stress at this concentration that was not observed in the absence of Ca^{++} (Fig. 3). However, only a very small increase in myosin light chain phosphorylation was observed under these conditions.

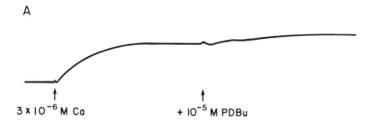

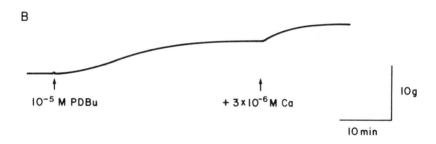

Figure 2. The response obtained in the presence of maximal Ca^{++} and maximal PDBu together was greater than that obtained with either stimulus alone. Stresses in 3×10^{-6} M Ca^{++} (A) and 10^{-5} M PDBu (B) alone were $86.1 \pm 4.7\%$ (n=5) and $73.7 \pm 1.3\%$ (n=3), respectively, of the total muscle response.

A study of the time course of stress development and myosin light chain phosphorylation further indicated that high stresses could be generated in response to PDBu without a corresponding increase in myosin light chain phosphorylation and without any indication of a transiently high phosphorylation (Fig. 4). In contrast, myosin light chain phosphorylation in 3×10^{-6} M Ca^{++} (a concentration in which tissues generate approximately 95% of maximum Ca^{++}-dependent stress) was 0.48 ± 0.01 mol P_i/mol light chain (n=7) at 20 min and 0.40 ± 0.02 mol P_i/mol light chain (n=7) at 40 min in other experiments (Chatterjee and Tejada 1986). Maximum

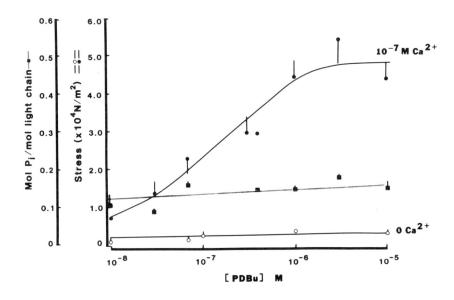

Figure 3. The dependence of steady state stress on PDBu concentration was determined in the presence of trace (●, 10^{-7} M Ca^{++}) and zero Ca^{++} (o, no added Ca^{++}, 5mM EGTA). Muscles contracting in 10^{-7} M Ca^{++} were frozen at steady state stress (■, 40-45 mins) and processed for analysis of myosin light chain phosphorylation. From Chatterjee and Tejada, 1986.

stress was obtained within 10-12 mins at these high Ca^{++} concentrations.

In order to determine if proteins other than myosin light chain were phosphorylated in response to PDBu stimulation, muscles were incubated with PDBu and γ-^{32}P-ATP under conditions identical to those used to generate contraction. A representative autoradiogram shows PDBu-dependent phosphorylation of a 25K Da protein (Fig. 5). This protein was phosphorylated slightly at concentrations of Ca^{++} that produced contraction; phosphorylation in the presence of PDBu, however, was much greater. Phosphorylation

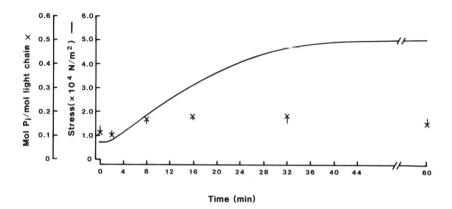

Figure 4. The time course of stress development and myosin light chain phosphorylation in response to 10^{-5}M PDBu is depicted. While stress increased from basal to values approximating 5×10^4 N/m^2 within 60 min, myosin light chain phosphorylation increased from 0.11 ± 0.02 mol P_i/mol light chain (n=4) in 10^{-7}M Ca^{++} alone to 0.15 ± 0.02 mol P_i/mol light chain (n=6) at 60 min in PDBu. Highest levels of phosphorylation observed were 0.19 ± 0.04 mol P_i/mol light chain (n=4) at 32 min. From Chatterjee and Tejada, 1986.

of the 25K Da protein was also stereospecific (Fig. 6) since 4 α-phorbol 12,13-didecanoate (α-PDD), a stereoisomer of phorbol 12,13-didecanoate that does not stimulate protein kinase C or contraction (Castagna et al 1982; Chatterjee and Tejada, 1986), did not promote phosphorylation of the 25K Da protein. The inference from these results is that PDBu activates protein kinase C, which phosphorylates a 25K Da protein.

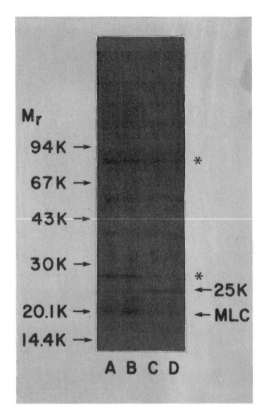

Figure 5. This autoradiogram of a 7.5 to 20% gradient gel shows increased phosphate incorporation into the 25K Da protein in the presence of two concentrations of PDBu. Note that the protein is slightly phosphorylated in the presence of contracting levels of calcium without PDBu. In the presence of PDBu, other phosphorylated species have the same or slightly less phosphate incorporation.

(A) 5×10^{-7} M free Ca^{++}; (B) 10^{-6} free Ca^{++}; (C) 5×10^{-7} M free Ca^{++} plus 10^{-5} M PDBu; (D) 5×10^{-7} M free Ca^{++} plus 3×10^{-6} M PDBu

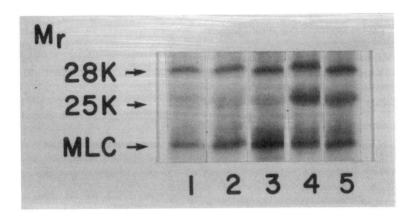

Figure 6. The autoradiographic pattern obtained for the 20-30K Da region of an SDS-polyacrylamide gel is shown. All samples were treated with 5×10^{-5} M free calcium.

(1) No additions; (2) 160 ug/ml phosphatidylserine (PS); (3) PS + 10^{-5}M 4 α-phorbol 12,13-didecanoate; (4) PS + 10^{-5}M PDBu; (5) 10^{-5}M PDBu.

DISCUSSION

The data show that high stresses can be generated with low myosin light chain phosphorylation in skinned vascular smooth muscle stimulated with PDBu. The observed Ca^{++}-dependence and stereospecificity (Chatterjee and Tejada, 1986) of the contractile response suggests that the stress response was mediated via protein kinase C by mechanisms independent of alterations in Ca^{++} flux.

In skinned porcine carotid artery stress has previously been shown to be proportional to Ca^{++} and myosin light chain phosphorylation when Ca^{++} is increased from basal values (Chatterjee and Murphy, 1983; Moreland and Murphy, in press). In the experiments described here, which are consistent with earlier findings, low Ca^{++} alone is neither associated with stress development, nor with slow development of attached dephosphorylated cross-bridges within 40-50 minutes. The addition of PDBu results in development of tension that, albeit slow, reaches levels comparable to those obtained in high Ca^{++}.

The peak active resistance to stretch (reflecting the number of attached cross-bridges) obtained in high PDBu is similar, as is stress, to that obtained in high Ca^{++} (Table 1). Since concomitant levels of myosin phosphorylation are low, (but significantly higher than values obtained in 10^{-7} M Ca^{++} alone, P<0.05, Fig. 4) it appears that protein kinase C activation leads to the formation of a significant population of attached, dephosphorylated cross-bridges. The latch state in vascular smooth muscle, described originally by Dillon et al (1981) was characterized by high stress with low myosin light chain phosphorylation and low cross-bridge cycling rates. We describe here a mechanism which, when activated, can be characterized by high stress and low myosin light chain phosphorylation. Cycling rates of these cross-bridges have yet to be determined in order to demonstrate whether this truly represents a latch-like phenomenon or is yet another modulator of vascular smooth muscle contraction.

It is not clear whether low myosin light chain phosphorylation is absolutely required for this PDBu-dependent contraction. We have shown that low myosin light chain phosphorylation is not sufficient to generate stress. In this regard recent studies in intact tissues indicate that a small pool of phosphorylated cross-bridges (15-25%) is both necessary and sufficient to generate a larger pool of dephosphorylated cross-bridges which supports significant but slowly developing stress (Ratz and Murphy, in press). Skinned preparations generate similar forces with low myosin light chain phosphorylation, within comparable periods of time (40-50 min), after treatment with activators of protein kinase C. This suggests that the enzyme plays a permissive role in the attachment of dephosphorylated cross-bridges. The observation that higher stresses are obtained with a combination of high Ca^{++} and PDBu than with either stimulus alone is consistent with this hypothesis.

The data describing the 25K Da protein introduces the notion that phosphorylation of proteins other than myosin light chain may be important for the regulation of contraction in vascular smooth muscle. Protein kinase C can phosphorylate a large number of cellular proteins including myosin light chains (Endo et al, 1982; Nishikawa et al, 1984; Nishizuka, 1984). Phosphorylation of myosin light chain by protein kinase C is at a different site from that by myosin light chain kinase and is associated with a decrease in actin-activated MgATPase activity (Nishikawa et al, 1983 and 1984). The method of two-dimensional gel electrophoresis employed in our analysis of phosphorylation cannot distinguish between sites of phosphorylation. However, PDBu-dependent changes in myosin phosphorylation are small. On the other hand, changes in phosphorylation of the 25K Da protein are relatively larger. It must be pointed out that we have not demonstrated any causal relation between phosphorylation of the 25K Da protein and stress development; the evidence to date is purely correlative. A comparable mechanism has been described in platelets, which lends credence to our speculation (Kaibuchi et al, 1983). In these cells, secretion of serotonin is mediated by a calmodulin-stimulated kinase (with phosphorylation of a 20K Da protein) as well as by protein kinase C (phosphorylation of a 40K Da protein). The two pathways are synergistic. Further experiments are required, however, to demonstrate the involvement of phosphorylation of proteins other than myosin light chain in stress maintenance of vascular smooth muscle.

ACKNOWLEDGEMENTS

The authors acknowledge the technical assistance of Manola Tejada and the secretarial assistance of Lily Shaffer.

REFERENCES

Castagna M, Takai Y, Kaibuchi K, Sano K, Kikkawa U, Nishizuka Y (1982). Direct activation of calcium-activated, phospholipid-dependent protein kinase by tumor-promoting phorbol esters. J Biol Chem 257:7847-7851.

Chatterjee M, Murphy RA (1983). Calcium-dependent stress maintenance without myosin phosphorylation in skinned smooth muscle. Science 221:464-466.

Chatterjee M, Tejada M (1986). Phorbol ester-induced contraction in chemically skinned vascular smooth muscle. Am J Physiol 251:C356-361.

Danthuluri NR, Deth RC (1984). Phorbol ester-induced contraction of arterial smooth muscle and inhibition of α-adrenergic response. Biochem Biophys Res Commun 125:1103-1109.

Dillon PF, Aksoy MO, Driska SP, Murphy RA (1981). Myosin phosphorylation and the cross-bridge cycle in arterial smooth muscle. Science 211:495-497.

Driska SR, Aksoy MO, Murphy RA (1981). Myosin light chain phosphorylation associated with contraction in arterial smooth muscle. Am J Physiol 240:C222-233.

Endo T, Naka M, Hidaka H (1982). Ca^{2+}-phospholipid dependent phosphorylation of smooth muscle myosin. Biochem Biophys Res Commun 105:942-948, 1982.

Forder J, Scriabine A, Rasmussen H (1985). Plasma membrane calcium flux, protein kinase C activation and smooth muscle contraction. J Pharmacol Exp Ther 235:267-273.

Gordon AR (1978). Contraction of detergent-treated smooth muscle. Proc Natl Acad Sci USA 75:3527-3530.

Kaibuchi K, Takai Y, Sawamura M, Hoshijima M, Fujikura T, Nishizuka Y (1983). Synergistic functions of protein phosphorylation and calcium mobilization in platelet activation. J Biol Chem 258:6701-6704.

Kojima I, Lippes H, Kojima K, Rasmussen H (1983). Aldosterone secretion: Effect of phorbol ester and A23187. Biochem Biophys Res Commun 116:555-562.

Moreland R, Murphy RA (in press). Determinants of Ca^{2+}-dependent stress maintenance in skinned swine carotid media. Am J Physiol 251 (Cell Physiol 20).

Nishikawa M, Hidaka H, Adelstein RS (1983). Phosphorylation of smooth muscle heavy meromysin by calcium-activated, phospholipid-dependent protein kinase. J Biol Chem 258:14069-14072.

Nishikawa M, Sellers JR, Adelstein RS, Hidaka H (1984). Protein kinase C modulates in vitro phosphorylation of the smooth muscle heavy meromyosin by myosin light chain kinase. J Biol Chem 259:8808-8814.

Nishizuka Y (1983). Phospholipid degradation and signal translation for protein phosphorylation. Trends Biochem Sci 8:13-6.

Rasmussen H, Forder J, Kojima I, Scriabine A (1984). TPA-induced contraction of isolated rabbit vascular smooth muscle. Biochem Biophys Res Commun 122:776-784.

Ratz P, Murphy RA (in press). Contributions of intracellular (sarcoplasmic retriculum) and extracellular (plasma membrane) Ca^{2+} pools to myoplasmic Ca^{2+} and activation of vascular smooth muscle. J Muscle Res Cell Motility (abstract).

CYCLIC AMP DEPENDENT AND MYOSIN LIGHT CHAIN KINASE: RELATIONSHIP TO ALTERED VASCULAR REACTIVITY IN HYPERTENSION AND DEVELOPMENT OF DIRECT PHARMACOLOGICAL MODULATORS

Paul J. Silver

Div. Experimental Therapeutics, Wyeth Labs, Inc., Philadelphia, PA 19101

Present address: Cardiopulmonary Section, Dept. of Pharmacology, Sterling-Winthrop Research Institute, Rensselaer, NY 12144

INTRODUCTION

Regulation of contractile activity in smooth muscle by vasoactive or bronchoactive mediators occurs via modulation of intracellular levels of Ca^{2+} or cyclic nucleotides (cAMP or cGMP). For the most part, these messengers operate via activation of specific protein kinases, which phosphorylate distinct substrates which either modulate contractile protein function or Ca^{2+} flux. The four major protein kinases in smooth muscle are two Ca^{2+}-regulated kinases [myosin light chain kinase (MLCK) and protein kinase c(PKC)] and two cyclic nucleotide-regulated kinases [cAMP dependent protein kinase (cAPK) and cGMP dependent protein kinase (cGPK)]. MLCK is regulated by Ca^{2+} and calmodulin, while PKC is regulated by Ca^{2+}, phosphatidylserine and diacylglycerol.

Because of the importance of these kinases in ultimate regulation of smooth muscle contractile activity, we have focused our initial studies on two of these kinases, MLCK and cAPK, in vascular smooth muscle. This manuscript will discuss two fundamental areas of research. These include comparisons of these enzymes in vascular smooth muscle from normotensive and hypertensive animals, and development of direct pharmacological modulators of MLCK as an approach for developing vasodilator/antihypertensive agents.

MATERIALS AND METHODS

I. cAPK and MLCK in Vascular Smooth Muscle

Twenty- to 24-week old SHR and age/sex matched normotensive WKY rats were obtained from Taconic Farms, Inc. (Germantown, NY). Systolic blood pressures (indirect tail cuff) averaged 200 mm Hg for SHR and 130 mm Hg for WKY animals. Renal hypertensive rats (RHR) were prepared from WKY animals (1 clip-Grollman); average systolic blood pressure was 181 mm Hg for this group of animals. Animals were anesthetized with ether and the desired arteries (usually aorta) were rapidly removed and immediately cleaned of visible connective tissue in a chilled Krebs-Ringer-bicarbonate buffer solution. After cleaning, arterial tissues were frozen by immersion in liquid nitrogen and stored (for up to 4 weeks) at -70°C in airtight containers before biochemical analysis.

For determination of total protein kinase activity, frozen arterial strips were homogenized in appropriate extraction buffers (see Silver et al., 1985b and Miller et al., 1983) and centrifuged at 30,000 Xg for 20 min. to obtain supernatant fractions. Over 97% of the extractable cAPK activity and 90% of MLCK activity were present in the initial extraction. Protein kinase activity was determined by the filter paper method of Corbin and Reimann (1974). Assay mixtures for cAPK included 20 µl of the supernatant fraction, 20 mM KH_2PO_4 (pH 6.8), 6 mM $MgCl_2$, 10 mM NaF, 100 µg of histone II-A, 0.1 mM 1-methyl-3-isobutylxanthine, 2 µM cAMP (a concentration which produced maximal activation of the protein kinase) and 0.4 mM ATP supplemented with [γ-^{32}P] ATP (New England Nuclear, Boston, MA) to yield a specific activity of 55 cpm/pmol of ATP. Assays were terminated after 10 min, which was determined previously to be on the linear portion of the assay curve (Silver et al., 1985b). MLCK activity was determined as described by Miller et al. (1983); the concentration of calmodulin was 2 µM, $CaCl_2$ was 10 µM and cardiac myosin light chains (300 µM) were used as the substrate.

The relative distribution of cAPK isozymes was determined by the native flatbed isoelectric focusing/autoradiographic method previously described (Silver et al., 1982, 1985b).

Intact muscle strip studies were performed with thoracic aortic strips from WKY, SHR, or RHR. The extent of activation of cAPK was assessed in paired muscle strips by calculating activity ratios, which represent the ratio of enzymatic activity measured in the absence and presence of maximal (2 µM) cAMP (Corbin et al., 1973, 1975). In these studies, the standard extraction buffer was supplemented with 100 mM NaCl and supernatant fractions were rapidly obtained by centrifugation at 10,000 Xg for 30 sec (Palmer et al., 1980; Silver et al., 1985b).

Experimental paradigms in additional intact muscle studies included concentration-response relationships to $CaCl_2$ (0.05 to 7.5 mM) in K^+-depolarized (100 mM) or norepinephrine-contracted (10 µM) aortic strips, sensitivity to nifedipine, diltiazem and W-7 (all pretreated) in K^+-depolarized aortic strips from WKY and SHR, and concentration-dependent relaxation of 100 mM KCl-contracted aortic strips by forskolin.

II. Direct MLC Phosphorylation Inhibitors

Preparation of native actomyosin, superprecipitation assays, isoelectric focusing, and quantitation of myosin light chain phosphorylation were as previously described (Silver et al., 1984a,b; 1985a). Calmodulin was purchased from Ocean Biologics, Inc. (Edmonds, WA).

Intact vascular smooth muscle strip studies were performed as previously described with rabbit aortic, guinea pig aortic, or porcine left anterior descending (LAD) coronary arterial strips (Silver et al., 1984a). Experimental paradigms included assessment of potency vs. K^+-depolarized contractions, efficacy of inhibitors vs. vascular agonists (serotonin, histamine, $PGF_{2\alpha}$, carbocyclic thromboxane A_2, leukotriene C_4), and the ability of the Ca^{2+} entry stimulator Bay K 8644 to attenuate inhibition by various antagonists.

RESULTS AND DISCUSSION

cAPK and MLCK in Hypertension

A decreased sensitivity and maximum responsiveness to relaxation by cumulative concentrations of isoproterenol or forskolin was apparent in aortic strips from SHR when compared with aortic strips from WKY animals. This decreased cAMP-responsiveness is similar to previous reports in the literature for β-adrenergic stimulation (Cohen and Berkowitz, 1976; Godfraind and Dieu, 1978). However, neither the concentration of soluble cAPK nor the relative distribution of the cAPK isoenzymes were different in aortic smooth muscle (Table 1). Moreover, no major differences in WKY vs SHR in either parameter were evident in other sources of arterial smooth muscle, even though differences related to the arterial source were apparent. The notable exception were branches of the femoral artery, which had significantly less cAPK than similar branches from WKY animals. In further studies, in situ activation of cAPK was assessed during forskolin-mediated relaxation in SHR and WKY aortic strips. No difference in the extent of activation of cAPK was apparent even though the rate and extent (concentration-dependent) of relaxation was markedly less in SHR. These results suggest that decreased relaxation responsiveness to cAMP-increasing vasodilators in aortic smooth muscle from SHR is not related to events proximal to and including activation of cAPK. It seems likely that events distal to activation of cAPK, such as phosphorylation of relevant substrates and subsequent alteration of Ca^{2+} and other ion fluxes (see Webb and Bhalla, 1976; Bhalla et al., 1978; Scheid et al., 1979; Jones et al., 1984), may be primarily responsible for this decreased responsiveness in hypertension.

Table 1

Protein Kinase Activities of Vascular Smooth Muscles
From Normotensive and Hypertensive Rats

Source	Animal[c]	Kinase Activity (pmol/mg/min)[a]		cAPK Isozymes[b] (% I/% II)
		cAPK	MLCK	
Aorta	WKY	300 ± 20	6,467 ± 1,229	55/45
Aorta	SHR	315 ± 10	6,209 ± 898	57/43
Aorta	DOCA	ND	5,823 ± 625	ND
Aorta	SO RHR	ND	10,453 ± 1,500	ND
Aorta	RHR	ND	9,296 ± 1,263	ND
Caudal	WKY	200 ± 20	ND	58/42
	SHR	190 ± 10	ND	56/44
Carotid	WKY	250 ± 30	ND	50/50
	SHR	225 ± 30	ND	51/49
Femoral	WKY	700 ± 30	ND	55/45
	SHR	710 ± 50	ND	53/47
Femoral Branch	WKY	1,070 ± 130	ND	51/49
	SHR	660 ± 80*	ND	51/49

[a] Values are the mean ± S.E. for 6-15 samples/group.

[b] Quantitated by native gel electrophoresis/autoradiography.

[c] WKY - Wistar Kyoto; SHR - Spontaneously Hypertensive rat; DOCA - deoxycorticosterone hypertensive rat; SO RHR - Sham-operated renal hypertensive rat; RHR - renal hypertensive rat.

ND = Not determined.

* = Significantly different from normotensive control ($p < 0.05$; Student-Newman-Keuls multiple comparisons test).

A decreased sensitivity to contractions elicited by $CaCl_2$ was also apparent in aortic smooth muscle from SHR (Table 2). A similar depressed responsiveness was evident with aortic smooth muscle from RHR. This phenomenon was observed with both K^+-depolarized and norepinephrine-contracted muscles, although a further difference between the two models of hypertension was only evident with K^+-depolarized muscles. Many previous investigators, using a wide variety of arterial smooth muscles from different models of hypertension, have demonstrated either increased or decreased responsiveness to $CaCl_2$ (see Webb and Bohr, 1981; Winquist et al., 1982; Mulvaney, 1983 for reviews; also, Moreland et al., 1982; Harris et al., 1984). Another major factor in these types of studies is the methodology used. In our studies, muscle strips were equilibrated with EGTA but were not contracted repeatedly with norepinephrine in the presence of EGTA to deplete intracellular Ca^{2+} storage sites.

Table 2

Comparative Ca^{2+} Sensitivity and Potency of Antagonists in Aortic Smooth Muscles from Normotensive and Hypertensive Rats

Animal[c]	SBP[c] (mm Hg)	EC_{50} - $CaCl_2$ [a]		pA_2 - 100 mM KCl; $CaCl_2$ [b]		
		100 mM KCl	10 μM Norepinephrine	Nifedipine	Diltiazem	W-7
WKY	130 ± 2	0.15	0.07	9.23	6.92	4.6
SHR	202 ± 5	0.31	0.30	9.05	6.96	5.1
RHR	181 ± 4	0.60	0.25	ND	ND	ND

[a]EC_{50} is the concentration of $CaCl_2$ required to produce 50% of the maximum contraction for KCl or norepinephrine-induced contractions. Muscle strips were rinsed in the presence of 0.05 mM EGTA prior to addition of the agonist and subsequent cumulative additions of $CaCl_2$ (0.05 to 7.5 mM).

[b]pA_2 values for antagonists were determined versus $CaCl_2$ (100 mM K^+-depolarized) contractions

[c]WKY -Kyoto Wistar; SHR - Spontaneously hypertensive rat; RHR - Renal hypertensive rat; SBP - Systolic blood pressure.

ND = not determined

In further experiments, the potency of various antagonists for inhibiting $CaCl_2$-dependent - K^+-depolarized force development was assessed in aortic strips from WKY and SHR. No difference in relative potency of the Ca^{2+} channel blockers nifedipine or diltiazem (which binds to a different receptor in the Ca^{2+} channel) were noted between WKY and SHR. This suggests that altered Ca^{2+} sensitivity in SHR is not related to the voltage-dependent Ca^{2+} channel, and is similar to previous findings (Harris et al., 1984). In contrast, the calmodulin antagonist, W-7, was approximately one-half log unit more potent in SHR than in WKY. This does suggest that a calmodulin-mediated regulatory system, such as MLC phosphorylation, might be responsible for this decreased Ca^{2+} sensitivity.

Initial studies examining the MLC phosphorylation system have quantitated the concentration of MLCK in aortic smooth muscles from various models of hypertension (Table 1). No major differences in hypertensive models vs. normotensive controls were evident although RHR did appear to have higher MLCK activity. No differences in WKY vs SHR aorta were apparent that might be linked to the increased potency of W-7 in intact strip studies. Other possibilities which remain to be explored include altered sensitivity of MLCK to activation by Ca^{2+}-calmodulin, altered MLC phosphatase activity, or other calmodulin, calmodulin-like, or non-calmodulin (such as PKC) related systems which may be affected by W-7.

Pharmacological Modulators of Vascular Contractile Protein Interactions

Direct pharmacological regulation of vascular contractile protein interactions can theoretically occur at several sites. Among these are the sites of regulation on the thin-filament (phosphorylation of 21,000 dalton protein, leiotonin system, caldesmon) or on the thick-filament (Ca^{2+} binding to MLC, phosphorylation of the MLC). Most reported efforts, to date, have focused on developing modulators of MLC phosphorylation as an approach to pharmacologically modulate vascular contractile protein interactions.

Direct alteration of MLC phosphorylation can conceivably occur by four distinct modes; inhibition of Ca^{2+} binding to calmodulin, inhibition of Ca^{2+}-calmodulin activation of MLCK, direct inhibition of MLCK catalytic activity, or direct stimulation of MLC phosphatase activity. Organic agents which relax vascular smooth muscle and function exclusively by stimulating phosphatase activity have not been identified. Synthesis of peptide analogs of smooth muscle MLC (Kemp and Pearson, 1985) and even more recent identification of a 3 Kda internal inhibitory peptide fragment of MLCK which may be shifted during Ca^{2+}-calmodulin binding (Strebinska et al., 1986) offer possible avenues for future drug development of specific modulators of MLCK activity. Also, the disclosure of isoquinolinesulfonamide and short-chain naphthalenesulfonamides, which directly inhibit most protein kinases (Hidaka et al., 1984; Inagaki et al., 1986), also demonstrates that direct inhibition of MLCK catalytic activity can occur, and offers the possibility for future development of specific agents.

Calmodulin antagonism is currently the most popular method for inhibiting smooth muscle MLC phosphorylation. By far, the largest group of calmodulin antagonists are those agents which compete with the regulated enzyme for the Ca^{2+}-calmodulin complex. Agents which directly modify Ca^{2+} binding to calmodulin have not been as numerous; most of these are various metal ions, such as Hg, Zn, Co and Sr (See Prozialeck, 1983 for review) and the recently reported HT-74 (Tanaka et al., 1986).

Of the many agents which inhibit calmodulin-regulated systems through binding of the Ca^{2+}-calmodulin complex, the most extensively studied have been the phenothiazine antipsychotic agents. However, several other classes of drugs have been shown to bind to calmodulin in a Ca^{2+}-dependent manner, and also inhibit various calmodulin-regulated systems. Among these are diphenylbutylpiperidine and butyrophenone antipsychotics, tricyclic and non-tricyclic antidepressants, various neuropeptides (such as β-endorphin) and insect venoms (such as mastoparan and melittin), benzodiazepine anti-anxiety agents, alpha adrenergic antagonists, naphthalene sulfonamide smooth muscle relaxants and the anti-fungal miconazole analog calmidazolium (see Prozialeck, 1983; Roufogalis, 1985 for some reviews). Separate drug binding sites for some of these agents, which appear to exhibit allosteric modulation of binding, have been reported (Johnson, 1983).

Most vasodilators, such as nitroprusside, hydralazine and diazoxide, and Ca^{2+} entry blockers like nifedipine, and verapamil do not directly inhibit actin-myosin interactions (superprecipitation) or MLC phosphorylation in Triton X-100 purified aortic actomyosin (Silver et al., 1984b). Nitrendipine has been reported to inhibit avian gizzard MLCK activity (Movsesian et al., 1984) but not MLC phosphorylation or actin-myosin interactions in arterial actomyosin (Table

3). Felodipine has been shown to bind to calmodulin (Bostrom et al., 1981) with a dissociation constant in the low micromolar range, and to inhibit avian gizzard MLCK activity (IC_{50} = 10 µM; Movsesian et al., 1984) and arterial actomyosin MLC phosphorylation and actin-myosin interactions (IC_{50} = 200 µM; Table 3). The pharmacological significance of this inhibition, however, is questionable as inhibition of force development in intact smooth muscle occurs in the nanomolar range (Bostrom et al., 1981; Silver et al., 1984a). Lipophilic weakly basic Ca^{2+} antagonists, such as prenylamine, cinnarizine, flunarizine, bepridil, and perhexiline, also bind to calmodulin and inhibit calmodulin-regulated MLC phosphorylation (Table 3), and phosphodiesterase activity (Lugnier et al., 1984; Kubo et al., 1984). These agents comprise the Class III Ca^{2+} blockers as classified by Spedding (1984; 1985).

Table 3

Direct Inhibition of Arterial Actin-Myosin Interactions and Myosin Light Chain Phosphorylation (MLCP) by Ca^{2+} Antagonists

Class[a]	Compound	Inhibition of Superprecipitation (% at 100 µM)	Inhibition of MLCP (% at 100 µM)	IC_{50} (µM)
I	Dihydropyridines			
	Nifedipine	0	0	
	Nitrendipine	0	0	
	Felodipine	42 ± 2	28 ± 2	200
II	Verapamil	0	0	
	Diltiazem	30 ± 1	0	
III	Bepridil	66 ± 4	59 ± 4	38
	Cinnarizine	60 ± 4	58 ± 3	60
	Perhexiline	80 ± 1	73 ± 2	33
?	Hexahydronaphthyridines			
	Wy 46531	70 ± 2	68 ± 3	18
?	Arylsulfonamides			
	Wy 46622	73 ± 1	67 ± 3	26
	Wy 47324	80 ± 2	72 ± 2	18
	Calmodulin Antagonist			
	W-7	65 ± 2	60 ± 2	35
	Trifluoperazine	70 ± 3	62 ± 3	18

[a] As classified by Spedding (1985).

Another Ca^{2+} entry blocker, diltiazem, can directly inhibit smooth muscle contractile protein function. Inhibition of force development at micromolar concentrations has been reported in chemically-skinned smooth muscle (Saida and van Breemen, 1983). Interestingly, diltiazem inhibits Ca^{2+}-dependent actin-myosin interactions in Triton-purified actomyosin independent of any effects on MLC phosphorylation (Silver et al., 1984b). This effect is evident at 0.5 µM diltiazem when Ca^{2+} is submaximal in the assay. Higher concentrations of diltiazem (100 µM) are needed to suppress interactions when Ca^{2+} is maximal. Exogenous calmodulin attenuates diltiazem-mediated inhibition and is independent of effects on MLC phosphorylation at high free Ca^{2+} concentrations. This shows that it is possible for some drugs to directly inhibit arterial actin-myosin interactions by mechanisms independent of MLC phosphorylation, and also supports the concept of additional Ca^{2+}-dependent regulatory mechanisms in vascular smooth muscle. Recent findings suggest that Ca^{2+} dependent latchbridge regulation also involves calmodulin or a calmodulin-like protein (Moreland and Moreland, 1986).

The pharmacological effects of two novel series of agents which structurally resemble the dihydropyridines or verapamil/W-7 hybrids have been described (Table 3). Both series appear to incorporate Ca^{2+} entry blockade and direct inhibition of aortic actin-myosin interactions, analogous to the previously discussed Class III Ca^{2+} entry blockers. Inhibition of actin-myosin interactions is related to concomitant inhibition of MLC phosphorylation; inhibition of MLC phosphorylation is attenuated with exogenous calmodulin with agents of the dihydropyridine-like series (hexahydronaphthyridines) and, for the most part, with agents from the verapamil/W-7 hybrid series (arylsulfonamides). This suggests that these agents are functioning as calmodulin antagonists, although some agents from the arylsulfonamides may have other mechanisms. The effects of these agents in other calmodulin-regulated systems have yet to be determined. Potency of these agents is similar to or greater than most of the aforementioned calmodulin antagonists; IC_{50} values range from 18 to 30 µM.

These agents can be differentiated from standard Ca^{2+} entry blockers in intact muscle systems. That is, these agents are less cardiac depressant and more efficacious versus most mediators (norepinephrine, angiotensin II, histamine, serotonin, leukotrienes, $PGF_{2\alpha}$, thromboxane A_2) of vasoconstriction in rabbit aortic or porcine coronary arterial smooth muscle. These effects are consistent with an intracellular effect at the contractile proteins, since regulation of cardiac contractility is via a different mechanism (troponin C) and vasoconstrictor agents may mobilize pools of Ca^{2+} (intracellular and/or entry through voltage-independent Ca^{2+} channels) which are not amenable to inhibition by standard Ca^{2+} entry blockers. Among the vasoconstrictor agents, leukotriene C_4 appears to be the best agent for differentiating between Ca^{2+} blockade and MLC phosphorylation inhibition, since neither verapamil nor nifedipine inhibit force development while the corresponding Wy agents almost completely inhibit force development.

The Ca^{2+} agonist Bay K 8644 can also be used to distinguish between relaxant mechanisms. Dose-related inhibition of K^+-depolarized force development by nitrendipine and verapamil can be attenuated in a concentration-related manner by Bay K 8644. However, equieffective, MLC phosphorylation inhibitory concentrations of some Wy agents and W-7 are not subject to reversal of inhibition by Bay K 8644.

Utility in other systems is also apparent with some of these agents. Specifically, some of the arylsulfonamides are more potent than standard Ca^{2+} blockers or W-7 at inhibiting platelet aggregation. Again, this would be consistent with an intracellular mechanism of action, as platelet aggregation is thought to be mediated by MLC and/or other phosphoprotein phosphorylation. Interestingly, increasing the length of the intermediate methylene chain in the arylsulfonamides series produced more potent MLC phosphorylation inhibitors (IC_{50} = 3-6 µM), but generally less effective vascular relaxants or platelet aggregation inhibitors. This stresses the importance of developing pharmacological intact cell models to accurately address and support biochemical findings.

Recent studies have shown that structurally diverse agents with differing pharmacological profiles can also directly modulate troponin C-cardiac contractile protein regulatory mechanisms. Within the diverse group of agents previously shown to antagonize calmodulin-regulated enzymes, differential effects (stimulation, inhibition) on cardiac myofibrillar ATPase activity are evident (Fig. 1). Mechanistically, most stimulatory agents increased Ca^{2+} sensitivity with differential effects on maximum ATPase activity. These effects are consistent, for the most part, with an increase in Ca^{2+} binding affinity of troponin C, which has previously been reported for trifluoperazine (Levin and Weiss, 1978).

Development of Ca^{2+} binding protein modulators that function by inhibiting Ca^{2+}-calmodulin activation of smooth muscle MLC phosphorylation or also by stimulation Ca^{2+}-troponin C activation of cardiac myofilaments offers novel cellular approaches to treatment of various diseases, such as hypertension, asthma or congestive heart failure. Future studies and research along the lines discussed in this manuscript and in the following reports presents a bright future for the goals of defining the involvement of these regulatory systems in aberrant, pathophysiological smooth muscle function, and developing specific pharmacological modulators of these key regulatory systems for research and possibly, therapeutic development.

ACKNOWLEDGEMENTS

The author acknowledges the expert technical assistance of Ms. J. Ambrose, Mr. J. Dachiw, Ms. S. Kocmund, Mr. R. Michalak and Ms. P. Pinto (all of Wyeth Labs, Inc.). The excellent secretarial assistance of Ms. Darlene DeFrancesco (Sterling-Winthrop Research Institute) and the discussions and comments of Drs. R.W. Lappe and R.L. Wendt (Wyeth Labs, Inc.) are also appreciated.

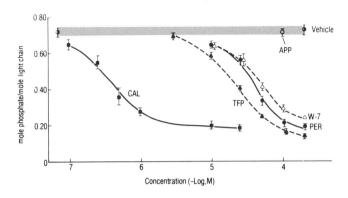
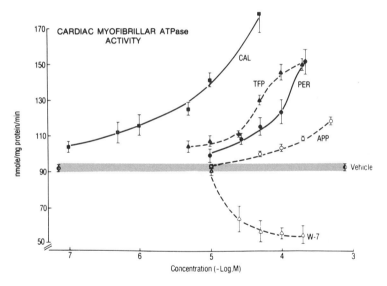

Figure 1

Concentration-related effects of calmodulin-antagonists and the cardiotonic agent APP 201-533 (APP) on arterial actomyosin MLC phosphorylation (top) or cardiac myofibrillar ATPase activity (bottom). Values are the mean ± S.E. for 5-7 preparations. Cal = calmidazolium; TFP = trifluorperazine, PER = perhexiline.

REFERENCES

Bhalla RC, Webb, RC, Singh D, Ashley T, Brock T (1978) Calcium fluxes, calcium binding, and adenosine 3':5'-monophosphate-dependent protein kinase activity in the aorta of spontaneously hypertensive and Kyoto Wistar normotensive rats. Mol Pharmacol 14:468-477.

Bostrom SL, Ljung B, Mardh S, Forsen S, Thulin E (1981) Interaction of the antihypertensive drug felodipine with calmodulin. Nature 292:777-778.

Cohen ML, Berkowitz BA (1976) Decreased vascular relaxation in hypertension. J Pharmacol Exp Ther 196:396-406.

Corbin JD, Keely SL, Soderling TR, Park CR (1975) Hormonal regulation of adenosine 3':5'-monophosphate-dependent protein kinase. Adv Cyclic Nucleotide Res 5:265-279.

Corbin JD, Reimann E (1974) Assay of cyclic AMP-dependent protein kinases. Methods Enzymol 38:287-290.

Corbin J, Soderling T, Park CR (1973) Regulation of adenosine 3':5'-monophosphate-dependent protein kinase. J Biol Chem 248:1821-1831.

Godfraind T, Dieu D (1978) Influence of aging on the isoprenaline relaxation of aortae from normal and hypertensive rats. Arch Int Pharmacodyn Ther 236:300-302.

Harris AL, Swamy VC, Triggle DJ (1984) Calcium reactivity and antagonism in portal veins from spontaneously hypertensive and normotensive rats. Can J Physiol Pharmacol 62:146-150.

Hidaka H, Inagaki M, Kawamoto S, Sasaki Y (1984) Isoquinolinesulfonamides, novel and potent inhibitors of cyclic nucleotide dependent protein kinase and protein kinase C. Biochemistry 23:5036-5041.

Inagaki M, Kawamoto S, Itoh H, Saitoh M, Hagiwara M, Takahashi J, Hidaka H (1986) Naphthasnesulfonamides as calmodulin antagonists and protein kinase inhibitors. Mol Pharmacol 29:577-581.

Johnson JD (1983) Allosteric interactions among drug binding sites on calmodulin. Biochem Biophys Res Commun 112:787-793.

Jones AW, Bylund DB, Forte LR (1984) cAMP-dependent reduction in membrane fluxes during relaxation of arterial smooth muscle. Am J Physiol 246:H306-H311.

Kemp BE, Pearson RB (1985) Spatial requirements for location of basic residues in peptide substrates for smooth muscle myosin light chain kinase. J Biol Chem 260:3355-3359.

Kubo K, Matsuda Y, Kase H, Yamada K (1984) Inhibition of calmodulin-dependent cyclic nucleotide phosphodiesterase by flunarizine, a calcium-entry blocker. Biochem Biophys Res Commun 124:315-321.

Levin RM, Weiss B (1978) Specificity of the binding of trifluoperazine to the calcium-dependent activator of phosphodiesterase and to a series of other calcium-binding proteins. Biochim Biophys Acta 540:197-204.

Lugnier C, Follenius A, Gerard D, Stoclet JC (1984) Bepridil and flunarizine as calmodulin inhibitors. Eur J Pharmacol 98:157-158.

Miller JM, Silver PJ, Stull JT (1983) The role of myosin light chain kinase phosphorylation in beta-adrenergic relaxation of tracheal smooth muscle. Mol Pharmacol 24:235-242.

Moreland RS, Moreland S (1986) Calmodulin antagonists inhibit stress maintenance by latchbridges in skinned arterial fibers. Biophys J 49:69a.

Moreland RS, Webb RC, Bohr DF (1982) Vascular changes in DOCA hypertension. Influence of a low protein diet. Hypertension 4 (Suppl III):III-99-III-107.

Movsesian MA, Swain AL, Adelstein RS (1984) Inhibition of turkey gizzard myosin light chain kinase activity by dihydropyridine calcium antagonists. Biochemical Pharmacol 33:3759-3764.

Mulvaney MJ (1983) Do resistance vessel abnormalities contribute to the elevated blood pressure of spontaneously-hypertensive rats? Blood Vessels 20:1-22.

Palmer WK, McPherson JM, Walsh DA (1980) Critical controls in the evaluation of cAMP-dependent protein kinase activity ratios as indices of hormonal action. J Biol Chem 255:2663-2666.

Prozialeck WC (1983) Structure-activity relationships of calmodulin antagonists. Annu Rep Med Chem 18:203-212.

Roufogalis BD (1985) Calmodulin antagonism. In Marme D (ed.): "Calcium and Cell Physiology" Berlin: Springer-Verlag, pp. 148-168.

Saida K, Van Breeman C (1983) Mechanism of Ca^{++} antagonist-induced vasodilation: Intracellular actions. Circ Res 52:137-142.

Scheid CR, Honeyman TW, Fay FS (1979) Mechanism of β-adrenergic relaxation of smooth muscle. Nature (Lond.) 277:32-36.

Silver PJ, Ambrose JM, Michalak RJ, Daichiw J (1984a) Effects of felodipine, nitrendipine and W-7 on arterial myosin phosphorylation, actin-myosin interactions and contraction. Eur J Pharmacol 104:417-424.

Silver PJ, Dachiw J, Ambrose JA (1984b) Effects of calcium antagonists and vasodilators on arterial myosin phosphorylation and actin-myosin interactions. J Pharmacol Exp Therap 230:141-148.

Silver PJ, Dachiw J, Ambrose JM, Pinto PB (1985a) Effects of the calcium antagonists perhexiline and cinnarizine on vascular and cardiac contractile protein function. J Pharmcol Exp Therap 234:629-635.

Silver PJ, Michalak RJ, Kocmund SM (1985b) Role of cAMP protein kinase in decreased arterial cyclic AMP responsiveness in hypertension. J Pharmacol Exp Therap 232:595-601.

Silver PJ, Schmidt-Silver CJ, DiSalvo J (1982) β-adrenergic relaxation and cAMP kinase activation in coronary artertial smooth muscle. Am J Physiol 242:H177-H184.

Spedding MA (1985) Activators and inactivators of Ca^{++} channels: New perspectives. J Pharmacol (Paris) 16:319-343.

Spedding MA, Cavero I (1984) "Calcium antagonists": a class of drugs with a bright future. Part II - Determination of basic pharmacological properties. Life Sciences 35:575-587.

Strebinska M, Ikebe M, Hartshorne DJ (1986) Tryptic hydrolysis of myosin light chain kinase: Conversion of calmodulin-dependent to calmodulin-independent forms. Biophys J 49:66a.

Tanaka T, Umekawa H, Saitoh M, Ishikawa T, Shin T, Ito M, Itoh H, Kawamatso Y, Sugihara H, Hidaka H (1986) Modulation of calmodulin function and of Ca^{2+}-induced smooth muscle contraction by the calmodulin antagonist, HT-74. Mol Pharmacol 29:264-269.

Webb RC, Bhalla RC (1976) Altered calcium sequestration by subcellular fractions of vascular smooth muscle from spontaneously hypertensive rats. J Mol Cell Cardiol 8:651-661.

Webb RC, Bohr DF (1981) Recent advances in the pathogenesis of hypertension. Consideration of structural, functional, and metabolic vascular abnormalities resulting in elevated arterial resistance. Am Heart J 102:251-264.

Winquist RJ, Webb RC, Bohr DF (1982) Vascular smooth muscle in hypertension. Fed Proc 41:2387-2392.

SELECTIVE INHIBITORS OF PHOSPHORYLATION IN SMOOTH MUSCLE

Masatoshi Hagiwara and Hiroyoshi Hidaka

Department of Molecular and Cellular Pharmacology
Mie University School of Medicine, Edobashi, Tsu,
Mie 514, Japan

INTRODUCTION

The biochemical basis of transduction of extracellular signals into cellular and subcellular motility, including contraction, mitosis, intracellular transport, exo- and endocytosis, and cell surface mobility has long been a subject of great interest. Since the 1960s, ongoing studies have established that many eukaryotic cells share with muscle cells, common contractile proteins, and actin and myosin, compounds which provide the molecular basis for diverse motile activities, a special example of which is smooth muscle contraction.

In 1977, we reported at the USA-Japan joint congress in Hawaii that calmodulin antagonists such as N-(6-aminohexyl)-5-chloro-1-naphthalenesulfonamide (W-7) which we newly synthesized, relaxed vascular strip and inhibited aortic actomyosin superprecipitation, thereby predicting the involvement of calmodulin in vascular smooth muscle contraction (Hidaka, 1977; Hidaka et al., 1978). The next year, Dabrowska et al. identified myosin light chain kinase (MLC-kinase) as a Ca^{2+}-calmodulin dependent and substrate-specific enzyme (Dabrowska et al., 1978).

Although the phosphorylation of myosin by MLC-kinase may play important role in regulation sysmte of actin-myosin interaction, other protein kinases, including cAMP-dependent protein kinase (Noiman, 1980; Walsh et al., 1981), casein kinase I, casein kinase II, phosphorylase kinase (Singh et al., 1983), protease-activated kinase (Tuazon et al., 1982) and epidermal growth factor-stimulated kinase (Gallis et al., 1983) also reportly phosphorylate the isolated myosin light chain. Whether or not they all act on not only the intact cells but also the intact myosin has yet to be determined. In addition to these we proved that protein kinase C phosphorylates the intact myosin in vitro and in vivo (Endo et al., 1982; Naka et al., 1983) and suggested the possibility that protein kinase C catalyzed phosphorylation of 20,000 dalton myosin light chain may function as a secondary modulating system for the Ca^{2+}-calmodulin-dependent regulation, which acts as the primary switch in controlling the actin-myosin interaction. In order to solve the considerable complexity of myosin phosphorylation and distinguish the role of each protein kinase in smooth muscle contraction system, pharmacological methods using specific inhibitors of the enzymes are thought to be invaluable.

MLC-KINASE INHIBITORS

The small acidic protein, calmodulin, is ubiquitous throughout much of plant and animal kingdoms. We previously demonstrated that W-7 and its derivatives bind to calmodulin in Ca^{2+} dependent manner and with high affinity (Hidaka et al., 1979), and that calmodulin may play an important role in vascular smooth muscle and platelet function through Ca^{2+}-calmodulin dependent myosin light chain phosphorylation. Calmodulin, however, regulates various enzyme activities such as cyclic nucleotide phosphodiesterase, adenylate cyclase, $(Ca^{2+} + Mg^{2+})$ATPase and MLC-kinase. All the calmodulin activation of these enzymes is inhibited by W-7. The hydrophobic regions of calmodulin, which are exposed by a conformational change induced by Ca^{2+}-binding to the high affinity sites of calmodulin, are responsible for the activation of Ca^{2+}- calmodulin dependent enzyme (Tanaka and Hidaka, 1980). These postulation based on the findings that the affinity for calmodulin of naphthalenesulfonamide derivatives increased with extention of the length of hydrocarbon chain (C_5 to C_{10}) and depends on the hydrophorbicity of each compound determined as octanol-buffer partition coefficients (Hidaka et al., 1981). Furthermore, a positive correlation between potency in relaxation of vascular strips and affinity for calmodulin was obtained (C_5 to C_{10}). However, a shorter alkyl chain derivative of W-7, N-(6-aminoethyl)-5-chloro-1-naphthalenesulfonamide (A-3) of which affinity for calmodulin was about 5 - 7 times lower than that of W-7 (calmodulin affinity was determined by displacement of [^{3}H]W-7 from calmodulin) produced relaxation of rat and rabbit aortic strips more potently than does W-7. The chalacteristics of vasodilating effect of A-3 was, however, similar to those of W-7. An addition of A-3 in concentrations ranging from 1 - 100 µM elicited a dose-dependent relaxation of rat aortic strips contracted by 20 mM KCl in the presence of phentolamine (1 µM), propranolol (1 µM) and atropine (0.3 µM). The ED_{50} value of A-3 (the concentration of which produced a 50% relaxation of sustained contraction) was 19 µM (N=6), whereas that of W-7 was 28 µM (N=6). To elucidate the reason of this discrepancy between calmodulin affinity and vasodilatory effect, the relaxant effect of A-3 on the aortic strips contracted by various contractile agonists was examined. Phenylephrine, histamine, serotonin, A-23187, angiotensin II and prostagrandin $F_{2\alpha}$ were used as the contractile agonists. The ED_{50} values of the agonists increased significantly by the addition of A-3 and maximum contractile tension of various antagonists was reduced by A-3. 3×10^{-5} M of A-3 reduced 66.0 ± 7.0 % of 3×10^{-6} M phenylephrine-induced contraction of aortic strips elicited in Ca^{2+}-free solution. Furthermore, A-3 relaxed aortic strips previously contracted by A23187, a Ca^{2+} ionophore, whereas nifedipine, a Ca^{2+} channel blocker, did not. These results suggest that A-3 as well as calmodulin antagonist W-7 is a vasorelaxant and exerts its action at intracellular or submembranal levels.

Then we examined the effects of A-3 and W-7 on the phosphorylation of 20,000 dalton myosin light chain in bovine aorta subcellular fractions prepared by the method of Doctrow and Lowenstein (1985). Vascular smooth muscle from bovine aorta was homogenated and centrifuged. The insoluble material was subjected to sucrose density gradient centrifugation and the contractile protein rich fraction (F4) as shown by gel electrophoresis (Fig. 1A) was obtained. 20,000 dalton myosin light chain was a major phosphorylated band when this fraction (1 mg/ml) was incubated with [$\gamma-^{32}P$] ATP. This phosphorylation of myosin light chain was nearly abolished by the addition of 10^{-4}M A-3 as well as 10^{-4}M W-7 (Fig. 1B). Addition cAMP ($10^{-7} - 10^{-5}$) and GTP ($10^{-6} - 10^{-5}$) altered neither the degree of phosphorylation of myosin light chain or the inhibition of its phosphorylation by A-3. These results suggest the inhibition of myosin light chain phosphorylation by A-3 is not mediated via the activation and/or inhibition of cAMP-dependent protein kinase under this

Effect of W-7 and A-3 on phosphorylation of 20KD protein in bovine aorta subcellular fractions

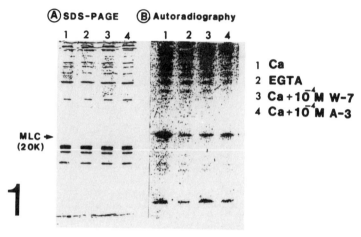

condition. As 2.0 mM EGTA nearly abolished this phosphorylation, myosin light chain was phosphorylated in Ca^{2+} dependent fashion. In two kinds of Ca^{2+} dependent protein phosphorylations, Ca^{2+}-phospholipid-dependent phosphorylation and Ca^{2+}-calmodulin dependent phosphorylation, A-3 has only weak effect on Ca^{2+}-phospholipid dependent activity of protein kinase C. All these evidence suggest that the incorporation of ^{32}P from ATP into myosin light chain in subcellular fraction prepared from aortic smooth muscle was mainly performed by Ca^{2+}-calmodulin dependent phosphorylation and A-3 preferencially inhibited MLC-kinase.

In the purified assay system of MLC-kinase the inhibitory effect of A-3 was further investigated. W-7 induced inhibition of MLC-kinase was overcome by the addition of excess calmodulin but A-3 induced inhibition was not. In proceeding papers, MLC-kinase was reported to be alternatively activated, in an irreversible manner, by limited proteolysis with trypsin or chymotrypsin (Tanaka et al., 1980; Walsh et al., 1982). In this process, the catalytically active fragment produced was entirely independent of Ca^{2+} and calmodulin. A-3 but not W-7 inhibits both the Ca^{2+}-calmodulin dependent and independent activities of MLC-kinase with a similar concentration dependency. In light of these findings, the inhibitory actions of A-3 seems to be results of direct interaction with the active site or the adjacent site of MLC-kinase and not due to effects on the enzyme activating process. Kinetic analysis by double-reciprocal plots revealed that the inhibition of MLC-kinase produced by A-3 was competitive with respect to ATP and Ki value was about 7 µM in the presence or absence of Ca^{2+}-calmodulin complex.

Because A-3 competes with ATP in the MLC-kinase reaction, we investigated the effects of the compound on a wider range of protein kinases and on other ATP or GTP-utilizing enzymes, specifically, adenylate cyclase and guanylate cyclase from human platelets and myosin ATPase from rabbit skeletal muscle. Table 1 summarizes the results obtained with A-3 and W-7. A-3 potently inhibite MLC-kinase, cAMP-dependent and cGMP-dependent protein kinase and casein kinase II with Ki values of $10^{-6} - 10^{-5}$ M, and weakly inhibited protein kinase C and casein kinase I with Ki values of $10^{-5} - 10^{-4}$ M, competitively with respect to ATP or GTP, and not with respect to protein substrates. We already reported that W-7 inhibits Ca^{2+}-dependent protein phosphorylation by MLC-kinase and protein kinase C in a competitive fashion with enzyme activators such as calmodulin and phosphatidylserin (Tanaka et al., 1982). However, inhibitions of W-7 for trypsin treated MLC-kinase and protein kinase C were both in a competitive fashion with respect to ATP, and Ki values of W-7 for Ca^{2+}-calmodulin and Ca^{2+}-phosphatidylserine independent enzyme activities were 110 µM and 340 µM, much higher than those of both enzymes activated by their co-factors. W-7 acted on the other protein kinases at $10^{-4} - 10^{-3}$ M concentration with a mode of action apparently similar to that of A-3. Contrary to the potent abilities of A-3 to inhibit protein kinases, other ATP- or GTP-utilizing enzymes such as adenylate cyclase, guanylate cyclase and myosin ATPase activities were not significantly affected even by 10^{-3} M A-3. Thus, it has been clearly demonstrated that the short chain derivative of naphthalenesulfonamides, A-3, did not work as a simple analogue of ATP and exhibited selective affinity toward the nucleotide-binding sites of protein kinases.

TABLE 1. Effect of A-3 and W-7 on ATP- or GTP-utilizing enzymes

Enzyme	Ki value (µM)	
	A-3	W-7
MLC-kinase (trypsin treatment)	7.0	110
Protein kinase C (trypsin treatment)	47	340
cAMP-dependent protein kinase	4.3	170
cGMP-dependent protein kinase	3.8	130
Casein kinase I	80	1000
Casein kinase II		
(ATP)	5.1	110
(GTP)	5.2	65
Adenylate cyclase	>1000	>1000
Guanylate cyclase	>1000	970
Myosin ATPase		
(Ca^{2+}-ATPase)	>1000	>1000
(K^+, EDTA-ATPase)	>1000	>1000

These studies of naphthalenesulfonamides suggest the possibility that naphthalenesulfonamides or certain derivatives may serve as useful tools with which to study directly protein kinases. However, more specific inhibitors for a certain protein kinase need to be developed before <u>in vivo</u> studies can be performed. After further investigation of naphthalenesulfonamide derivatives, we found that when the alkyl chain of A-3 was replaced by homopiperazine ring, the derivative, named ML-9 only weakly inhibited cAMP-dependent protein kinase, protein kinase C, but increased the potency to inhibit MLC-kinase (Ki value was 3.8 µM). ML-9 also revealed potent vasodilating effect on vascular smooth muscle. These results suggest that the phosphorylation of myosin light chain by MLC-kinase is indispensable process in smooth muscle contraction system and that ML-9 may be an available probe to investigate the function of MLC-kinase <u>in vitro</u> and <u>in vivo</u> as a MLC-kinase specific inhibitor.

cAMP-DEPENDENT PROTEIN KINASE INHIBITOR

Recent studies on smooth muscle cyclic nucleotides have suggested that cyclic nucleotides, especially cAMP, have a regulatory role in smooth muscle contraction (Diamond, 1978; Scheid et al., 1979). ß-adrenergic relaxation of smooth muscle by catecolamines has been associated with elevated levels of cyclic AMP. pharmacologically we reported an inhibitor of cyclic nucleotide phosphodiesterase produced vascular relaxation (Hagiwara et al., 1985). The question arises whether subsequent activation of cyclic AMP-dependent protein kinase has a role in the regulation of smooth muscle contraction. A current popular view holds that cAMP can produce relaxation of smooth muscle by a reduction in the cytoplasmic free Ca^{2+}-concentration and by inhibition of actin-myosin-interaction via an effect on MLC-kinase. ß-Stimulation or cAMP has been suggested to decrease the cytoplasmic Ca^{2+} level by affecting various mechanisms of Ca^{2+} influx, efflux, and intracellular sequestration (Meisheri and Van Breemen, 1982; Itoh et al., 1982). On the other hand, ß-adrenergic stimulation of the heart is thought to increase cardiac muscle contractility by activation of cAMP-dependent protein kinase and concomitant increase in the phosphorylation of certain proteins (Krebs and Beavo, 1979). It has also been suggested in heart muscle that cAMP-dependent phosphorylation of some component of the Ca^{2+} channel increase the amount of Ca^{2+} which enter the cell during depolarization (Cachelin et al., 1983). Adelstein et al. (1978) have shown that the catalytic subunit of cAMP dependent protein kinase play a part in negative regulation, by phosphorylation of MLC-kinase, which results in a decrease in the activity of the kinase in vitro. Moreover, the catalytic subunit of cAMP-dependent protein kinase inhibits Ca^{2+}-activated tension in skinned smooth muscle fiber preparation (Kerric and Hoar, 1981). However in intact vascular smooth muscle the role of cAMP-dependent protein kinase is still elusive because subcellular components including Ca^{2+}-mobilization mechanisms may have been destroyed by chemical skinning procedures of smooth muscle. In order to elucidate the physiological role of cyclic nucleotide-dependent protein kinases in intact smooth muscle, cell membrane permeable and cyclic nucleotide dependent protein kinase specific inhibitor was thought to be indispensable.

As we shown before, A-3 potently inhibits cyclic nucleotide dependent protein kinases. On the contrary to ML-9, the naphthalene ring of A-3 was replaced by isoquinoline, the derivative, H-8 was neither calmodulin antagonist nor MLC-kinase inhibitor, and inhibited potently cyclic nucleotide-dependent protein kinase in competition with ATP (Ki value was 1.2 µM). H-8 also relaxed rabbit aortic strips contracted by various agonists and exters its action at the intracellular or submembranal level (Ishikawa et al., 1985). H-8 is a intracllular Ca^{2+} antagonist of afferent class from Ca^{2+} entry blockers such as nicardipine (Asano and Hidaka, 1985) and excess $CaCl_2$ antagonized the relaxation induced by H-8 in depolarized smooth muscle, whereas W-7 or ML-9-induced relaxation was not reversed by excess $CaCl_2$. This suggests that cyclic nucleotide-dependent protein kinases may interact the Ca^{2+}-mobilization at intracellular level, although it is still unclear at the present time that the vasodilatory effect of H-8 is related to its ability to specifically inhibit cyclic nucleotide-dependent protein kinases.

PROTEIN KINASE C INHIBITOR

In the series of isoquinolinesulfonamides, a derivative with the sulfonylpiperazine residue, 1-(5-isoquinolinyl-sulfonyl)-2-methylpiperazine (H-7) was the most potent in inhibiting protein kinase C (Ki value was 6.0 µM). Recently, we found that, protein kinase C phosphorylates the myosin light chain in the intact as well as in the isolated form (Endo et al., 1982; Naka et al., 1983). The sites of phosphorylation on the 20,000-dalton light chain are different for protein kinase C and for MLC-kinase (Inagaki et al., 1984) and myosin phosphorylated by protein kinase C formed a bent 10S monomer while that phosphorylated by MLC-kinase was unfolded and extended 6S monomer in the present of 0.2 M KCl (Umekawa et al., 1985). Sequential phosphorylation of (H)meromyosin by MLC-kinase and protein kinase C cause a decrease in the actin-activated ATPase activity, as compared to the activity of myosin phosphorylated by MLC-kinase (Nishikawa et al., 1984). In addition protein kinase C phosphorylates MLC-kinase (2 mol of phosphate/molecule) and these sites are different from those utilized by cAMP-dependent protein kinase (Ikebe et al., 1985). However, phosphorylation in both regions results in a reduced affinity of MLC-kinase for calmodulin. The pharmacological effect of a protein kinase C inhibitor, H-7 was investigated using human platelets which is one of protein kinase C rich tissues (Minakuchi et al., 1981). H-7 enhanced serotonin release from human platelets that was induced by the 12-O-tetradecanoyl phorbol 13-acetate (TPA) and correspondingly decreased incorpolation of radioactive phosphate into 20,000-dalton light chain of platelet myosin. Although both protein kinase C and MLC-kinase was involved in its phosphorylation, two dimensional peptide mapping following tryptic hydrolysis revealed that H-7 selectively inhibited the protein kinase C-catalyzed phosphorylation of myosin light chain. This pharmacological evidence suggests that protein kinase C-catalyzed myosin light chain phosphorylation may play an inhibitory role in the release reaction induced by MLC-kinase-catalyzed phosphorylation of myosin in platelets. Although further studies are required to determine the function of protein kinase C in smooth muscle contraction system, this hypothesis may apply to smooth muscle.

CONCLUSION

Although increasing evidence suggests that MLC-kinase catalyzed phosphorylation of 20,000 dalton myosin light chain act as the primary regulation system of actin-myosin interaction in vascular smooth muscle, it is difficult to determine the functions of other protein kinase in secondary modulating system of smooth muscle contraction.

Molecular pharmacological manipulation of phosphorylation in smooth muscle

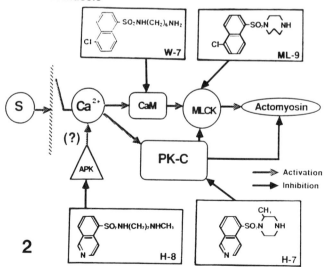

Calmodulin antagonists such as W-7 are useful tools to study Ca^{2+}-calmodulin regulatory system but are not so useful for elucidating the physiological role of each protein kinase. In this article, we presented data on some naphthalenesulfonamide and isoquinolinesulfonamide derivatives of W-7 and Fig. 2 summarizes our molecular pharmacological manipulation of phosphorylation in vascular contraction system. These newly developed compounds will shed light on the physiological significance and molecular mechanism of smooth muscle contraction.

ACKNOWLEDGMENT

We thank Drs. T. Tanaka and M. Inagaki for their kind cooperation. This work was supported in part by Grant-in-Aid for Scientific Research from the Ministry of Education, Science, and Culture, Japan.

REFERENCES

Adelstein RS, Conti MA, Hathaway DR, Klee CB (1978). Phosphorylation of smooth muscle myosin light chain kinase by the catalytic subunit of adenosine 3',5'-monophosphate-dependent protein kinase. J Biol Chem 253: 8347-8350.

Asano T, Hidaka H (1985). Intracellular Ca^{2+} antagonist, HA1004: pharmacological properties different from those of nicardipine. J Pharmacol Exp Ther 233: 454-458.

Cachelin AB, Depeyner JE, Kokubun S, Reuter H (1983). Ca^{++} channel modulation by 8-bromocyclic AMP in cultured heart cells. Nature (Lond) 304: 462-474.

Dabrowska R, Sherry JMF, Aromatorio DK, Hartshorne DJ (1978). Modulator protein as a component of the myosin light chain kinase from chicekn gizzard. Biochemistry 17: 253-258.

Diamond J (1978). Role of cyclic nucleotides in control of smooth muscle contraction. In George WJ, Ignarro LJ (eds): "Adv Cycl Nucleo Res", New York: Raven Press, pp 327-340.

Doctrow SR, Lowenstein JM (1985). Adenosine and 5'-chloro-5'-dexyadenosine inhibit the phosphorylation of phosphatidylinositol and myosin light chain in calf aorta smooth muscle. J Biol Chem 260: 3469-3476.

Endo T, Naka M, Hidaka H (1982). Ca^{2+}-phospholipid dependent phosphorylation of smooth muscle myosin. Biochem Biophys Res Commun 105: 942-948.

Gallis B, Edelman AM, Casnellie JE, Krebs EG (1983).
Epidermal growth factor stimulates tyrosine phosphorylation of the myosin regulatory light chain from smooth muscle. J Biol Chem 258: 13089-13093.

Hagiwara M, Endo T, Kanayama T, Hidaka H (1984). Effect of 1-(3-Chloroamino)-4-phenylphthalazine (MY-5445), a specific inhibitor of cyclic GMP phosphodiesterase, on human platelet aggregation. J Pharmacol Exp Ther 228: 467-471.

Hidaka, H (1977). U.S.A.-Japan cooperative science program on molecular and cellular aspects of vascular smooth muscle in health and disease.

Hidaka H, Yamaki T, Asano M, Totsuka T (1978). Involvement of calcium in cyclic nucleotide metabolism in human vascular smooth muscle. Blood Vessels 15: 55-64.

Hidaka H, Yamaki T, Totsuka T, Asano M (1979). Selective inhibitors of Ca^{2+}-binding modulator of phosphodiesterase produce vascular relaxation and inhibit actin-myosin interaction. Mol Pharmacol 15: 49-59.

Ikebe M, Inagaki M, Kanamaru K, Hidaka H (1985). Phosphorylation of smooth muscle myosin light chain kinase by Ca^{2+}-activated, phospholipid-dependent protein kinase. J Biol Chem 260: 4547-4550.

Inagaki M, Kawamoto S, Hidaka H (1984). Serotonin secretion from human platelets may be modified by Ca^{2+}-activated, phospholipid-dependent myosin phosphorylation. J Biol Chem 259: 14321-14323.

Ishikawa T, Inagaki M, Watanabe M, Hidaka H (1985). Relaxation of vascular smooth muscle by HA-1004, an inhibitor of cyclic nucleotide-dependent protein kinase. J Pharmcol Exp Ther 235: 495-489.

Itoh T, Izumi H, Kuriyama H (1982). Mechanisms of relaxation induced by activation of β-adrenoceptors in smooth muscle cells of the guinea-pig mesenteric artery. J Physiol 236: 475-493.

Kamps MP, Taylor SS, Sefton BM (1984). Direct evidence that oncogenic tyrosine kinases and cyclic ATP-binding sites. Nature (Lond) 310: 589-592.

Kerrick WGL, Hoar PE (1981). Inhibition of smooth muscle tension by cyclic AMP-dependent protein kinase. Nature 292: 253-255.

Krebs EG, Beavo JA (1979). Phosphorylation-dephosphorylation of enzymes. Annu Rev Biochem 48: 923-959.

Meisheri KD, Breemen van C (1982). Effects of Beta-adrenergic stimulation on calcium movements in rabbit aortic smooth muscle: relationship with cyclic AMP. J Physiol 331: 429-441.

Minakuchi R, Takai Y, Binzu YU, Nishizuka Y (1981). Widespread occurrence of calcium-activated, phospholipid dependent protein kinase in mammalian tissues. J Biochem 89: 1651-1654.

Naka M, Nishikawa M, Adelstein RS, Hidaka H (1983). Phorbol ester-induced activation of human platelets is associated with protein kinase C phosphorylation of myosin light chains. Nature 306: 490-492.

Nishikawa M, Sellers JR, Adelstein RS, Hidaka H (1984). Protein kinase C modulates in vitro phosphorylation of the smooth muscle heavy meromyosin by myosin light chain kinase. J Biol Chem 259: 8808-8814.

Noiman ES (1980). Phosphorylation of smooth muscle myosin light chains by cAMP-dependent protein kinase. J Biol Chem 255: 11067-11070.

Scheid CR, Honeyman TW, Fay FS (1979). Mechanism of beta-adrenergic relaxation of smooth muscle. Nature 277: 32-36.

Singh TJ, Akatsuka A, Huang KP (1983). Phosphorylation of smooth muscle myosin light chain by five different kinases. FEBS Lett 159: 217-220.

Tanaka T, Hidaka H (1980). Hydrophobic regions function in calmodulin-enzyme(s) interaction. J Biol Chem 255: 11078.

Tanaka T, Naka M, Hidaka H (1980). Activation of myosin light chain kinase by trypsin. Biochem Biophys Res Commun 92: 313-318.

Tanaka, T, Ohmura T, Yamakado T, Hidaka H (1982). Two types of calcium-dependent protein phosphorylation modulated by calmodulin antagonists: naphthalenesulfonamide derivatives. Mol Pharmacol 22: 408-412.

Tuazon PT, Stull JT, Trangh JA (1982). phosphorylation of myosin light chain by a protease-activated kinase from rabbit skeletal muscle. Biochem Biophys Res Commun 108: 910-197.

Walsh MP, Dabrowska R, hinkins S, Hartshorne DJ (1982). Calcium-independent myosin light chain kinase of smooth muscle. Preparation by limited chymotryptic digestion of the calcium ion dependent enzyme, purification and characterization. Biochemistry 21: 1919-1925.

Walsh MP, Persechini A, Hinkins S, Hartshorne DJ (1981). Is smooth muscle myosin a substrate for the cAMP-dependent protein kianse? FEBS Lett 126: 107-110.

REGULATION OF cAMP CONTENT AND cAMP-DEPENDENT PROTEIN KINASE ACTIVITY IN AIRWAY SMOOTH MUSCLE

Theodore J. Torphy, Miriam Burman, Lisa B.F. Huang, Stephen Horohonich and Lenora B. Cieslinski

Department of Pharmacology, Smith Kline & French Laboratories, Philadelphia, Pennsylvania 19101

INTRODUCTION

This chapter will focus on two somewhat disparate aspects of the regulation of airway smooth muscle tone. The first area concerns the complex manner in which bronchoconstricting and bronchodilating pathways interact to determine steady state smooth muscle tone. The second area involves the separation, identification and characterization of cyclic nucleotide phosphodiesterase isozymes in airway smooth muscle, and the importance of each isozyme in hydrolyzing cAMP and cGMP in the intact tissue.

INTEGRATION OF CONTRACTILE AND RELAXANT PATHWAYS

The effects of various substances on smooth muscle tone and the biochemical processes involved in regulating contraction and relaxation generally are studied in isolation. Rarely, for example, are the mechanical and biochemical responses to ß-adrenoceptor agonists examined in the presence of different contractile agents or even in the presence of various concentrations of a single contractile agent. The uncomplicated approach has its advantages: unraveling the biochemical pathways regulating smooth muscle function is difficult enough without introducing additional experimental complexities! Nevertheless, it is unlikely that smooth muscle tone *in vivo* is regulated entirely by one neurohumoral factor or by a single biochemical pathway. Instead, smooth muscle tone is likely to be determined by multiple – and often opposing – neural and humoral inputs. In asthma, for example, the effects of bronchoconstrictors such as histamine, acetylcholine and the peptidoleukotrienes are counterbalanced by endogenous bronchodilators such as epinephrine, prostaglandin E_2 and prostacyclin.

The aphorism "functional antagonism" is used to describe an interaction between two or more agonists that act through separate receptors to exert opposite effects on a common effector system (Offermeier and Van den Brink, 1974). A fundamental tenet of functional antagonism is that it is not unidirectional (Offermeier and Van den Brink, 1974). Thus, in the case of airway smooth muscle, a bronchorelaxant can reduce the response to a bronchoconstrictor and, hypothetically, a bronchoconstrictor can inhibit the response to a bronchorelaxant. The latter phenomenon – the ability of contractile agents to inhibit the response to relaxants – is discussed below.

Effect of Muscarinic Tone on the Mechanical and Biochemical Responses to ß-Adrenoceptor Agonists

The first comprehensive experiments demonstrating an inverse relationship between the degree of bronchoconstriction and the efficacy of bronchorelaxants was conducted by Van den Brink (1973). In this study, contracting bovine tracheal strips with increasing concentrations of muscarinic agonists markedly increased isoproterenol EC_{50} values and decreased the maximal isoproterenol-induced relaxation. We began to explore the biochemical basis for the inhibitory interaction between bronchoconstricting and bronchodilating pathways using canine trachealis as a model.

The effect of contracting canine isolated tracheal strips with increasing concentrations of methacholine on the relaxant response to isoproterenol is shown in Figure 1. The inhibitory effect of methacholine is striking: contracting tracheal strips with increasing concentrations of methacholine progressively decreases, and eventually abolishes, the relaxant response to isoproterenol.

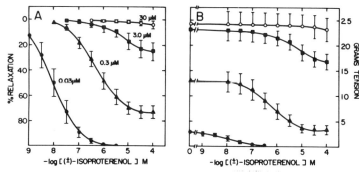

Figure 1. Typical pattern of functional antagonism in airway smooth muscle. Canine tracheal strips were first contracted with various concentrations of methacholine (shown to the right of the concentration-response curves in panel A) before being relaxed by the cumulative addition of isoproterenol. Data are expressed as percentage of relaxation to resting state (A) or absolute relaxation (B). See Torphy et al. (1983) for details. Reprinted with permission. © by the American Society for Pharmacology and Experimental Therapeutics.

Although increasing the degree of muscarinic tone could inhibit isoproterenol-induced relaxation by a number of mechanisms, we chose to examine the effect of methacholine on the ability of isoproterenol to increase cAMP content and activate cAMP-dependent protein kinase (cAMP-PK), crucial molecular events thought to mediate smooth muscle relaxation in response to ß-adrenoceptor stimulation (Hardman, 1981). Methacholine had no effect on basal cAMP content, but did inhibit isoproterenol-stimulated cAMP accumulation (Torphy et al., 1983). However, maximal inhibition of isoproterenol-stimulated cAMP accumulation was produced by 0.3 µM methacholine; 3.0 and 30 µM methacholine had no additional effect. Thus, although methacholine reduced both isoproterenol-induced relaxation and cAMP accumulation, the concentration-dependence of these two phenomena was different. Nevertheless, the activation state of cAMP-PK was assessed in the same tissue samples (Fig. 2). As with cAMP accumulation, methacholine had no effect on basal cAMP-PK activity ratios. But, more importantly, methacholine inhibited isoproterenol-stimulated cAMP-PK activity in a concentration-dependent manner, with concentrations above 0.3 µM eliciting a graded inhibition of kinase activity (Fig. 2). In other experiments, methacholine suppressed hormone-stimulated cAMP-PK activation in response to a full range of isoproterenol concentrations (Torphy et al., 1983). Thus, there is a strong correlation between the inhibition of isoproterenol-stimulated cAMP-PK activity and the ability of methacholine to reduce the mechanical response to isoproterenol. The correlation with cAMP content is less convincing.

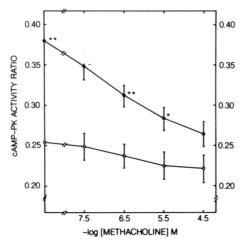

Figure 2. Effect of methacholine on basal (O) and isoproterenol-stimulated (●) cAMP-dependent protein kinase (PK) activity ratios in canine tracheal smooth muscle. Values were obtained in the same tissues as those used in Figure 1. Significantly different from the activity ratio in the presence of methacholine alone (*P<0.025; ** P<0.005). See Torphy et al. (1983) for details. Reprinted with permission. © by the American Society for Pharmacology and Experimental Therapeutics.

Specificity of Functional Antagonism: Effects of Various Relaxants

It was of interest to know whether increasing muscarinic tone functionally antagonized the response to smooth muscle relaxants other than ß-adrenoceptor agonists. To determine whether muscarinic receptor stimulation inhibits the mechanical and biochemical responses to other agents that activate adenylate cyclase, a series of experiments similar to those described above were conducted using prostaglandin E_2 and forskolin as the relaxants (Torphy et al., 1985). As observed with isoproterenol, contracting tracheal strips with increasing concentrations of methacholine markedly reduced the relaxant response to both prostaglandin E_2 and forskolin. Once again, the maximum inhibition of drug-stimulated cAMP accumulation was elicited by 0.3 µM methacholine, a concentration lower than that producing maximum inhibition of the relaxant response (Torphy et al., 1985). The ability of these two agents to activate cAMP-PK also was decreased by 0.3 µM methacholine but in contrast to the results with cAMP, 3.0 µM methacholine produced a further decrement in drug-stimulated kinase activity (Torphy et al., 1985). A detailed discussion of the apparent discrepancy between cAMP accumulation and cAMP-PK activity appears in Torphy et al. (1985). Regardless of this discrepancy, the data suggest that the ability of muscarinic agonists to antagonize functionally the relaxant response to activators of adenylate cyclase is due, at least in part, to an inhibition of drug-stimulated cAMP-PK activity.

If a decrease in protein kinase activation is indeed responsible for a portion of the methacholine-induced inhibition to the relaxant response to isoproterenol, prostaglandin E_2 and forskolin, then increases in muscarinic tone should have less of an inhibitory effect on bronchorelaxants that do not act via the adenylate cyclase/ protein kinase cascade. This is the case when sodium nitroprusside and 8-bromo-cGMP are used as the bronchorelaxants (Torphy et al., 1985). Similarly, recent experiments have shown that increasing the concentration of methacholine over a 100-fold range has a marked inhibitory effect on isoproterenol-induced relaxation but only a negligible effect on the relaxation induced by papaverine, a nonselective phosphodiesterase inhibitor, or trimethoxybenzoate (TMB-8), an intracellular Ca^{2+} antagonist (data not shown). Thus, the degree of functional antagonism produced by methacholine varies substantially with the type of bronchorelaxant used.

Specificity of Functional Antagonism: Effects of Various Contractile Agents

Muscarinic agonists have been used in nearly all studies on functional antagonism in airway smooth muscle. An important question to answer is whether all contractile agents antagonize the response to ß-adrenoceptor agonists in a similar manner. Because of the intense interest in the actions of the peptidoleukotrienes (LTs) on airway smooth, a comparison was made between the inhibitory effects of LTD_4

versus methacholine on isoproterenol-induced relaxation of the opossum trachea (Torphy et al., 1986). The opossum trachea was used in these studies because it responds to LTD_4, unlike the canine trachea, and because enough smooth muscle is present to conduct biochemical analyses. Tissues were contracted with three concentrations of methacholine or three concentrations of LTD_4, each of which produced the same degree of contraction as the corresponding equieffective concentration of methacholine. After plateau tone was reached, cumulative isoproterenol concentration-response curves were constructed (Fig. 3). Although the relaxant response to isoproterenol was reduced as tissues were contracted with higher concentrations of either methacholine or LTD_4, the inhibitory effect of methacholine (Fig. 3A) was substantially greater than the inhibitory effect of the leukotriene (Fig. 3B). Similar results were obtained with the guinea-pig trachea (Torphy, 1984). Thus, the sensitivity of airway smooth muscle to ß-adrenoceptor agonists is influenced both by the amplitude of the preexisting contraction and, perhaps more importantly, by the contractile agent used (Torphy et al., 1986).

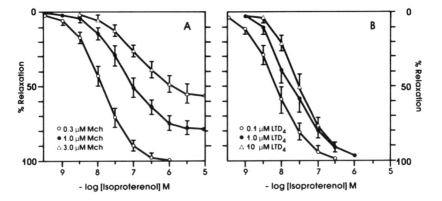

Figure 3. Differential inhibitory effects of contractile agents on isoproterenol-induced relaxation of the opossum trachea. Cumulative isoproterenol concentration-response curves were constructed with tracheal strips contracted with three concentrations of methacholine (Mch, panel A) or three corresponding equieffective concentrations of LTD_4 (panel B). See Torphy et al. (1986) for details. Reprinted with permission. © by the American Society for Pharmacology and Experimental Therapeutics.

DISCUSSION

Evidence is emerging which suggests that airway smooth muscle tone in vivo is determined by a complex integration of bronchoconstricting and bronchodilating inputs. In this context, it is important to recognize not only that bronchodilators (e.g., ß-adrenoceptor agonists, prostaglandin E_2) can inhibit the response to bronchoconstrictors but, conversely, bronchoconstrictors can inhibit the response to bronchodilators. Moreover, the response of airway smooth muscle to bronchorelaxants is influenced by at least two factors: the initial contractile state of the tissue and the agent used to induce tone.

As suggested previously (Torphy et al., 1983), inhibitory interactions between bronchoconstricting and bronchodilating pathways may contribute to the development of "catecholamine resistance" during severe asthmatic episodes; as bronchoconstriction becomes more intense, the bronchodilator efficacy of ß-adrenoceptor agonists declines. In vivo, evidence for an inverse relationship between the degree of airway obstruction and the bronchorelaxant response to sympathomimetics has been obtained in both animals (Jenne et al., 1985) and man (Barnes and Pride, 1983). Obviously, the clinical importance of this phenomenon may vary depending on the mediator(s) responsible for the bronchoconstriction. In view of this, it is noteworthy that the isoproterenol-resistant airway obstruction present in a subpopulation of chronic bronchitics is highly responsive to muscarinic antagonists (Marini and Lakshminarayan, 1980). In these patients, perhaps both the airway obstruction and its refractoriness to ß-adrenoceptor agonists is a manifestation of muscarinic receptor stimulation through elevated vagus nerve activity.

IDENTITY AND ROLE OF PHOSPHODIESTERASE ISOZYMES IN CANINE TRACHEALIS

The term "cyclic nucleotide phosphodiesterase (PDE)" refers to a family of enzymes that inactivate cAMP or cGMP by catalyzing the hydrolysis of the 3',5'-phosphodiester bond to form the corresponding 5'-nucleotide monophosphate. The existence of multiple forms of PDEs has been known for nearly two decades. These forms differ with respect to their substrate (cAMP or cGMP) specificity, kinetic characteristics, intracellular location and endogenous activators (Weishaar et al., 1985). But perhaps the most important feature of these isozymes is that their presence and overall role in cyclic nucleotide hydrolysis varies substantially form one tissue to another (Weishaar et al., 1985). Furthermore, PDE inhibitors are available that possess considerable isozyme-selectivity (Weishaar et al., 1985). This not only will permit researchers to assess the role of different PDE isozymes in intact tissues, but these compounds may represent the foundation for new therapeutic classes of PDE inhibitors that can be targeted for specific cells or organs. The remainder of this section is devoted to the results of preliminary experiments concerning the presence and role of PDE isozymes in airway smooth muscle.

Identification and Characterization of PDE Isozymes in the Canine Trachealis

To separate PDE isozymes in canine tracheal smooth muscle, the supernatant fraction of trachealis homogenates (prepared in the presence of 0.1% Triton X-100 and sonicated to remove particulate-bound enzymes) was applied to a DEAE-sepharose anion-exchange column. Isozymes were eluted with a linear sodium acetate gradient (0.05-0.8M) and fractions were assayed for cAMP and cGMP PDE activity. A typical chromatograph is shown in Figure 4. Cyclic nucleotide PDE activity eluted in two distinct peaks, tentatively designated PDE I and PDE III (see Weishaar et al., 1985 for explanation of nomenclature). When assayed at a "physiologic" cyclic nucleotide substrate concentration (1 µM), PDE I appeared to account for all of the cGMP PDE activity and 85% of the cAMP PDE activity (Fig. 4). PDE III accounted for approximately 15% of the total cAMP hydrolysis and did not hydrolyze cGMP (Fig. 4).

Fractions corresponding to the two peak PDE activities were pooled and analyzed for kinetic characteristics, activation by Ca^{2+}•calmodulin and effects of specific PDE isozyme inhibitors. PDE I, often called "cGMP PDE", hydrolyzes both cAMP and cGMP with the same K_m (1-2 µM) and V_{max}. The hydrolysis of both cAMP and cGMP is stimulated severalfold by Ca^{2+}• calmodulin and, based on relative IC_{50}s, M&B 22,948 (zaprinast) is approximately 20-fold selective for PDE I (IC_{50}=8 µM) versus PDE III.

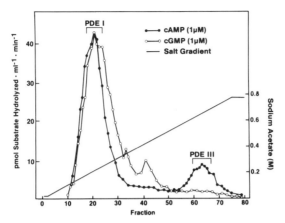

Figure 4. Separation of PDE isozymes in canine tracheal smooth muscle using DEAE-sepharose anion-exchange chromatography. Column fractions (1 ml) were assayed directly for PDE activity using either 1 µM [^{3}H]cAMP (●) or [^{3}H]cGMP (○) as a substrate. Recovery of total homogenate PDE activity from ion-exchange columns was consistently 85-90%.

Anamolous kinetics and partial inhibition by isozyme-specific enzyme inhibitors suggest that the second peak of PDE activity (PDE III) actually contains two distinct isozymes, PDE IIIa and PDE IIIb. PDE IIIa has a low K_m for cAMP (0.3 µM), does not hydrolyze cGMP and is inhibited specifically by SK&F 94120 [5-(4-acetamidophenyl)-pyrazin-2(1H)-one] with an IC_{50} of 6 µM. PDE IIIb has a higher K_m (10 µM) and V_{max} for cAMP than does PDE IIIa. Like PDE IIIa, PDE IIIb does not hydrolyze cGMP. PDE IIIb accounts for two-thirds, and PDE IIIa one-third, of the total PDE III activity when assayed in the presence of 1 µM cAMP.

An important observation from these studies is that PDE I accounts for all of the measurable cGMP hydrolytic activity and, under what may be considered a "physiologic" substrate concentration (1 µM), approximately 85% of the total cAMP hydrolytic activity in the canine trachealis (see Fig. 4). In contrast, neither PDE IIIa nor IIIb hydrolyzes cGMP. Moreover, these isozymes account for only 5% and 10%, respectively, of the total cAMP PDE activity when measured in the presence of the standard 1 μM substrate concentration. Thus, from these preliminary biochemical analyses, it appears that most of the total cellular PDE activity for both cyclic nucleotides is attributable to PDE I.

Role of PDE Isozymes in Intact Canine Trachealis

Biochemical analyses of partially purified enzymes provide information concerning the types and relative amounts of PDEs present, along with their kinetic characteristics and inhibitor profiles. Such studies may not, however, accurately predict the importance of individual isozymes in regulating cyclic nucleotide content in the intact tissue. This problem was approached by attempting to potentiate the biochemical and mechanical responses to isoproterenol (a relaxant that elevates cAMP) or sodium nitroprusside (a relaxant that elevates cGMP) by using isozyme-selective PDE inhibitors. In these studies, canine trachealis strips were first contracted with 3.0 μM methacholine, which eliminated the intrinsic relaxant response to the isozyme-selective PDE inhibitors, before being treated 5 min later with vehicle or various concentrations of M&B 22,948 or SK&F 94120. After an additional 10-min incubation period, tissues were untreated (controls) or relaxed by the cumulative addition of sodium nitroprusside or isoproterenol. When the maximum relaxant response was attained, tissues were flash-frozen and assayed for cAMP and cGMP content.

The effect of M&B 22,948 on sodium nitroprusside-induced relaxation and cGMP accumulation is shown in Figure 5. M&B 22,948 potentiated both the maximum relaxant response to sodium nitroprusside (Fig. 5A) and sodium nitroprusside-stimulated cGMP accumulation (Fig. 5B). These data suggest that PDE I is indeed a major contributor to cGMP hydrolysis in the intact tissue. In contrast, the same concentrations of M&B 22,948 had no effect on isoproterenol-induced relaxation or cAMP accumulation (data not shown). This observation is significant for two reasons. First, eventhough PDE I accounts for 85% of the total cAMP hydrolytic activity in trachealis homogenates, it does not appear to be important for regulating cAMP content in the intact tissue. Second, at the concentrations used, M&B 22,948 appears not to inhibit PDE III (a or b) in the intact tissue.

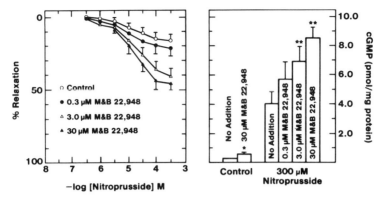

Figure 5. Effect of various concentrations of M&B 22,948 on sodium nitroprusside-induced relaxation (left) and cGMP accumulation (right) in the canine trachealis. See text for details. *Significantly greater than control in the absence of M&B 22,948 ($P<0.05$). **Significantly greater than sodium nitroprusside-induced cGMP accumulation in the absence of M&B 22,948 ($P<0.05$).

A complementary series of experiments were conducted using SK&F 94120. SK&F 94120 produced a concentration-dependent potentiation of both the relaxant response to isoproterenol, as manifested by an increase in maximum relaxation and a shift to the left of the concentration-response curve (Fig. 6A), and isoproterenol-induced cAMP accumulation (Fig. 6B). These data suggest that although PDE IIIa accounts for only 5% of the total cAMP PDE activity in broken cell preparations, it is important for the regulation of cAMP content in the intact tissue. SK&F 94120 (100 μM) had no effect on the mechanical and biochemical responses to sodium nitroprusside (data not shown), suggesting that SK&F 94120 does not inhibit PDE I in the intact tissue.

DISCUSSION

The inability of PDE I inhibition to potentiate the biochemical and mechanical responses to isoproterenol is intriguing. Excluding a spurious inflation of the relative activity of PDE I due to artifacts arising from the isolation procedure, the data indicate that this isozyme accounts for 85% of the total cAMP hydrolysis in tissue homogenates (when activity is assayed in the presence of a "physiologic" substrate concentration) but does not appear to modulate basal or hormone-stimulated cAMP accumulation in the intact tissue. In contrast, PDE IIIa contributes only 5% the total cAMP PDE activity when assessed in vitro but seems to be an important regulator of cAMP content in the intact trachealis. Perhaps the most important concept reinforced by these studies is that sweeping conclusions concerning the regulation of cell function should not be based solely on data derived from biochemical experiments on broken cell preparations. Whenever possible, such information should be confirmed with appropriate studies using intact tissues.

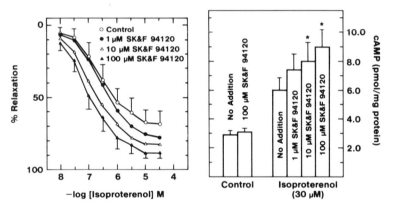

Figure 6. Effect of various concentrations of SK&F 94120 on isoproterenol-induced relaxation (left) and cAMP accumulation (right) in the canine trachealis. See text for details. *Significantly greater than isoproterenol-stimulated cAMP content in the absence of SK&F 94120 ($P<0.05$).

REFERENCES

1. Barnes PJ, Pride NB (1983). Dose-response curves to inhaled ß-adrenoceptor agonists in normal and asthmatic subjects. Br J Clin Pharmacol 15:677-682.
2. Hardman JG (1981). Cyclic nucleotides and smooth muscle contraction: Some conceptual and experimental considerations. In Bulbring E, Brading AF, Jones AW and Tomita T (eds.): Smooth Muscle: An Assessment

of Current Knowledge, pp. 249-262, University of Texas Press, Austin.
3. Jenne JW, Shaughnessy TK, Manfredi CJ, Drug WS (1985). Bronchodilation by isoproterenol depends on both dose and nature of the constricting stimulus in the dog. Am Rev Resp Dis 131 (suppl.):A280.
4. Marini JJ, Lakshminarayan S (1980). The effect of atropine inhalation in "irreversible" chronic bronchitis. Chest 77:591-596.
5. Offermeier J and Van den Brink FG (1974). The antagonism between cholinomimetic agonists and ß-adrenoceptor stimulants. The differentiation between functional and metaffinoid antagonism. Eur J Pharmacol 27:206-213.
6. Torphy TJ, Freese, WB, Rinard GA, Brunton LL, Mayer SE (1982). Cyclic nucleotide-dependent protein kinases in airway smooth muscle. J Biol Chem 257:11609-11616.
7. Torphy TJ, Rinard GA, Rietow MG, Mayer SE (1983). Functional antagonism in canine tracheal smooth muscle: Inhibition by methacholine of the mechanical and biochemical responses to isoproterenol. J Pharmacol Exp Ther 227:694-699.
8. Torphy TJ (1984). Differential relaxant effects of isoproterenol on methacholine- versus leukotriene D_4-induced contraction in the guinea-pig trachea. Eur J Pharmacol 102:549-553.
9. Torphy TJ, Zheng C, Peterson SM, Fiscus RR, Rinard GA, Mayer SE (1985). Inhibitory effect of methacholine on drug-induced relaxation, cyclic AMP accumulation, and cyclic AMP-dependent protein kinase activation in canine tracheal smooth muscle. J Pharmacol Exp Ther 233:409-417.
10. Torphy TJ, Burman M, Schwartz LW, Wasserman MA (1986). Differential effects of methacholine and leukotriene D_4 on cyclic nucleotide content and isoproterenol-induced relaxation in the opossum trachea. J Pharmacol Exp Ther 237:332-340.
11. Van den Brink FG (1973). The model of functional interaction. II. Experimental verification of a new model: The antagonism of ß-adrenoceptor stimulants and other agonists. Eur J Pharmacol 22:279-286, 1973.
12. Weishaar RE, Cain MH, Bristol JA (1985). A new generation of phosphodiesterase inhibitors: Multiple molecular forms of phoshodiesterase and the potential for drug selectivity. J Med Chem 28:537-545.

MOLECULAR MECHANISMS UNDERLYING INCREASED CONTRACTILITY TO NOREPINEPHRINE STIMULATION IN SHR VASCULAR SMOOTH MUSCLE.

R.C. Bhalla, R.V. Sharma and M.B. Aqel

Department of Anatomy, The University of Iowa, Iowa City, IA 52242.

INTRODUCTION

It is now well accepted that arterial smooth muscle of spontaneously hypertensive rat (SHR) is hyperreactive to norepinephrine (NE) stimulation (Webb, 1984). NE regulates vascular smooth muscle function by interacting with both α-and β-adrenoceptors. Stimulation of β-adrenoceptor produces vascular smooth muscle relaxation by i) decreasing free intracellular Ca^{2+} concentration (Bhalla et al., 1978a; Brockbank & England, 1980; Saida & van Breeman, 1984) and ii) by decreasing the sensitivity of myosin light chain kinase for Ca^{2+}-calmodulin (Adelstein et al., 1982; Bhalla et al., 1982; Gerthoffer et al., 1984; Hathaway et al., 1985). Conversely, stimulation of α-adrenoceptor causes vascular smooth muscle contraction by: i) increasing release of Ca^{2+} from non-mitochondrial intracellular stores, presumably sarcoplasmic reticulum (Bond et al., 1984); ii) increasing the influx of extracellular Ca^{2+} (Cauvin et al., 1984). An alteration in any one of these mechanisms would lead to a decrease in relaxation and an increase in contraction of vascular smooth muscle of SHR. In this paper we will discuss alterations in α- and β-adrenoceptor mediated mechanisms in relation to pathophysiology of vascular smooth muscle cell of SHR.

β-adrenoceptor mediated responses in vasulcar smooth muscle of SHR.

Vascular smooth muscle strips from hypertensive rats show reduced relaxation compared with normotensive control rats after treatment with dibutyryl cAMP or with isoproterenol and theophyline suggesting an alteration in the cAMP mediated mechanisms (Triner et al., 1975; Cohen & Berkowitz, 1976; Bhalla et al., 1979). However, conflicting reports exist in the literature regarding the differences in the basal and hormonal stimulated adenylate cyclase activity in vascular smooth muscle membranes of WKY and SHR (Triner et al., 1975; Amer et al., 1975). We have observed that GTP was necessary for hormonal stimulation of vascular smooth muscle adenylate cyclase activity (Bhalla & Sharma, 1982). The guanine nucleotide-, isoproterenol- and fluoride-stimulated enzyme activity was significantly decreased ($p<.05$) in SHR vascular smooth muscle as compared to WKY (Bhalla and Sharma, 1982). The discrepancies in the literature with regard to hormonal stimulation of adenylate cyclase in the vascular smooth muscle of SHR could be due to lack of GTP in the assays in earlier studies. In addition, we also observed that cAMP-phosphodiesterase activity was significantly ($p<.05$) increased in the particulate fraction of SHR vascular smooth muscle as compared to WKY (Sharma & Bhalla, 1978). Our results would provide the molecular basis for reduced cAMP levels observed in the vascular smooth muscle of SHR (Amer et al., 1975).

We have also observed that cAMP-dependent protein kinase activity was decreased approximately 30-40% in both soluble and particulate fractions (Bhalla et al., 1978b; Bhalla et al., 1980). Two major peaks of isozymes I and II of soluble cAMP-dependent protein kinase activity could be separated by DEAE cellulose chromatography. The distribution of isozymes I and II was 40% and 60% respectively in WKY as compared to 26% and 74% in SHR (Gupta et al., 1982). The enzyme activity under the peak of isozyme I was reduced by approximately 55% in SHR as compared to WKY. The apparent Km values for cAMP, ATP and Mg^{2+} for isozyme I and II of SHR were comparable to WKY. Similar to our observations, Coquill and Hamet (1980) also documented reduced cAMP-dependent protein kinase activity in aorta of SHR as compared to WKY. On the other hand, Silver et al. (1985) observed a decreased cAMP-dependent protein kinase activity only in the branches of femoral artery of SHR and not in other blood vessels tested. However, they observed that isoproterenol and forskolin mediated relaxation of aortic strips was significantly diminished in SHR. Moreover, these authors also found a significant correlation between rate of relaxation mediated by forskolin and the activation state of cAMP-dependent protein kinase both in SHR and WKY.

The decreased cAMP-dependent protein kinase activity could lead to decreased relaxation of SHR vascular smooth muscle by the following mechanisms: i) decreased Ca^{2+} sequestration into sarcoplasmic reticulum, ii) decreased Ca^{2+} efflux through plasma membrane iii) increased myosin light chain kinase activity due to reduced down regulation of the enzyme activity by cAMP-dependent protein kinase mediated phosphorylation. Thus, it has been shown that cAMP-dependent protein kinase mediated phosphorylation of aortic plasma membrane fraction and microsomes increased energy-dependent Ca^{2+} uptake (Bhalla et al., 1978a; Brockbank & England, 1980). Similarly, Saida and van Breeman (1984) using saponin permeabilized aortic strips have shown that cAMP increases Ca^{2+} uptake into sarcoplasmic reticulum. Moreover, the interaction of Ca^{2+}-calmodulin with MLCK is weakened by cAMP-dependent protein kinase mediated phosphorylation resulting in an inhibition of MLCK activity (Adelstein et al., 1982; Bhalla et al., 1982; Hathaway et al., 1985) with a concomitant inhibition of actomyosin ATPase (Silver et al., 1981). Using chemically skinned vascular smooth muscle preparations it has been shown that the isometric tension development could be partly inhibited by the addition of cAMP-dependent protein kinase (Pfitzer et al., 1984). These studies suggest that changes in vascular contractility cannot be described solely in terms of changes in cytoplasmic Ca^{2+}, and that changes in the sensitivity of the contractile proteins to a given Ca^{2+} concentration are also potential mechanisms for the regulation of contractile response of vascular smooth muscle.

Decreased microsomal phosphorylation with a concomitant reduction in Ca^{2+} uptake has been demonstrated in SHR vascular smooth muscle as compared to WKY (Webb, 1984; Bhalla et al., 1978b, 1980). Similarly, plasma membrane enriched fractions isolated from caudal and mesenteric artery of SHR have reduced energy-dependent Ca^{2+} uptake as compared to WKY (Kwan, 1985). Intracellular free Ca^{2+} concentration is increased in platelets, lymphocytes and smooth muscle cells of SHR and in patients with essential hypertension (Erne et al., 1984; Bruschi et al, 1984; Nabika et al., 1985). Patients with borderline hypertension have mildly elevated Ca^{2+} levels which fall in the intermediate range of normotensive and hypertensive subjects (Erne et al., 1984). These results would strongly suggest that increased cytoplasmic Ca^{2+} concentration could be responsible for the increased peripheral resistance in the hypertensive subjects.

Evidence presented above suggests that, only at low concentrations of free Ca^{2+} can MLCK be inhibited by cAMP-dependent protein kinase. In SHR, Ca^{2+} uptake is decreased in both S.R. and S.L., which may not allow cytoplasmic Ca^{2+} levels to decrease to such low levels that cAMP-dependent protein kinase mediated phosphorylation of MLCK could sufficiently inhibit the enzyme activity. Furthermore, cAMP-dependent protein kinase activity has also been reported to be decreased in vascular smooth muscle of SHR which may further result in increased MLCK activity due to decreased phosphorylation leading to increased muscle contraction.

α-Adrenoceptor mediated responses in vascular smooth muscle of SHR.

Observations made in our laboratory (Aqel et al., 1986) and other laboratories (Mulvany & Nyborg, 1980) have demonstrated an increased Ca^{2+} sensitivity to NE activation in the vascular smooth muscle of SHR. We have further demonstrated that the increased Ca^{2+} sensitivity to NE stimulation is specifically due to alterations in α_1-adrenoceptor mediated mechanisms (Aqel et al., 1986). In contrast to our observations, it has been shown that increased contraction of SHR caudal artery to NE stimulation is due to an increase in the α_2-adrenoceptor mediated mechanisms (Hicks et al., 1984). Recently it has been shown that in the proximal part of the rat caudal artery, α_1-adrenoceptors play a predominant role, while in the distal part of the caudal artery both α_1- and α_2-adrenoceptors may be involved in vasoconstriction (Medgett, 1985). Therefore, it is likely that the differences between our study and the earlier studies could be due to the different parts of caudal arteries used.

Stimulation of α-receptor produces vascular smooth muscle contraction by increasing cytosolic Ca^{2+} concentration through one or both of the following pathways: i) by causing release of Ca^{2+} from the intracellular stores

(Bond et al., 1984); ii) by opening Ca^{2+} channels to allow influx of extracellular Ca^{2+} into the cell (Cauvin et al., 1984). ^{45}Ca influx is increased in SHR caudal arteries as compared to WKY in response to NE and methoxamine (Bhalla et al., 1986). Similarly, it has been demonstrated that NE stimulated ^{45}Ca influx is increased in SHR mesenteric arteries as compared to WKY (Cauvin & van Breeman, 1985).

Recently Bond et al (1984) have demonstrated that isolated vascular smooth muscle preparations can be repeatedly contracted by NE and caffeine in the absence of extracellular Ca^{2+}. We have compared the intracellular Ca^{2+} pool of WKY and SHR caudal arteries by measuring the contractile responses to NE, and caffeine in Ca^{2+} free solution (3mm). Segments of proximal part of caudal arteries were denervated with 6-hydroxydopamine (Aqel et al., 1986). Following denervation, the contractions were measured according to Bond et al. (1984) with slight modification. Briefly, the rings were first stimulated with high-K^+ solution and the contractile response was recorded until a plateau phase of contraction was achieved (2-3 min). The high-K^+ solution, containing 2 mM Ca^{2+} was then replaced with a high K^+, Ca^{2+}-free solution containing 10 mM $LaCl_3$ for 9 minutes. Then the rings were stimulated by adding different doses of NE or 25 mM caffeine to the high K^+, Ca^{2+}-free solution and the contractile response was recorded. Data are expressed as a percent of maximum tension obtained by high K^+ solution containing 2 mM Ca^{2+}. Addition of 30 µM NE to Ca^{2+}-free, high K^+ solution evoked 80-90% of the contraction produced by high K^+ in the presence of 2 mM Ca^{2+} (Fig. 1).

Using the experimental protocol given in Fig. 1, the effect of different concentrations of NE on the contractile responses of SHR and WKY caudal artery rings was examined. Data given in Fig. 2 show that at all concentrations of NE used between $5 \times 10^{-6}M$ to $10^{-4}M$, the contractile response of SHR caudal artery was significantly greater ($p<0.05$) as compared to WKY. In addition to α-adrenoceptor stimulation, 25 mM caffeine in Ca^{2+}-free solution also produced significantly greater ($p<0.05$) contraction in SHR caudal arteries as compared to WKY. These results would suggest that intracellular Ca^{2+} pool (presumably SR Ca^{2+} pool) is increased in SHR vascular smooth muscle which may account at least in part for the increased contractile response to NE stimulation observed in this study.

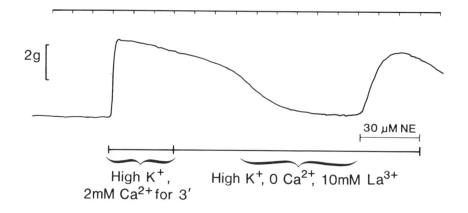

Fig. 1. NE-induced contraction of SHR caudal artery rings in Ca^{2+}-free solution. The arterial rings after denervation and equilibration in PSS were first contracted in a high K^+, 2 mM Ca^{2+} solution for 3 min, followed by high K^+, 0 Ca^{2+}, 10 mM La^{3+} solution for 9 min before addition of 30 µM NE in the same solution.

It has been demonstrated that in vascular smooth muscle phosphoinositide metabolism is increased during hormonal stimulation and inositol-(1,4,5)-trisphosphate (IP_3) is rapidly accumulated (Baron et al., 1984; Alexander et al., 1985). Recently it has also been demonstrated that addition of exogenous IP_3 into permeabilized vascular smooth muscle cells and arterial strips produced repeated contractions and also Ca^{2+} release from intracellular stores, presumably S.R. (Somlyo et al., 1985). It is therefore logical to assume that increased contraction produced by NE in Ca^{2+}-free solution in SHR vascular smooth muscle could be due to either increased production of IP_3, or increased capacity of

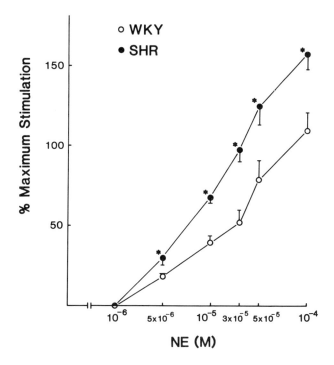

Fig. 2. NE-induced contracton in absence of extracellular Ca^{2+}. Effect of different concentrations of NE on the contractile response of WKY and SHR caudal arteries in Ca^{2+}-free solution. Experimental protocol was similar to that given in Fig. 1. The values are mean ± S.E.M. of 10-12 arterial rings in both WKY and SHR. *Significantly different ($p<0.05$) as compared to WKY.

intracellular Ca^{2+} pool. We have investigated α_1-adrenoceptor number and affinity by ^{3}H-prazosin binding to WKY and SHR caudal artery membranes. The plasma membranes were prepared by the procedure developed in our laboratory with slight modifications (Sharma & Bhalla, 1986). We did not observe any differences in the dissociation constant (K_d) and maximum number of binding sites (B_{max}) for [^{3}H]-prazosin binding in caudal artery membranes of WKY and SHR. Similarly the displacement of [^{3}H]-Prazosin by NE in competitive experiments did not reveal any differences between WKY and SHR in the affinity for NE (Sharma et al., 1986).

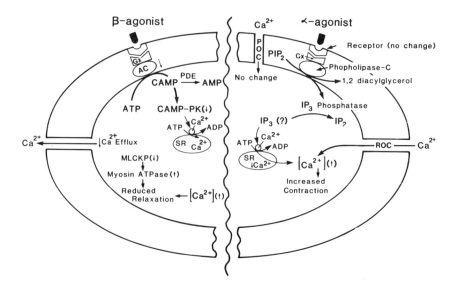

Fig. 3. Schematic representation of possible alterations in α and β-adrenoceptor mediated stimulus-response coupling mechanisms in vascular smooth muscle of SHR. Gx, guanine nucleotide binding protein, MLCK-P, myosin light chain kinase phosphorylation, S.R., sarcoplasmic reticulum; PIP_2, phosphatidyl inositol 4,5 bis phosphate; IP_3, inositol 1,4,5 tris phosphate; IP_2, inositol bis phosphate; POC, potential operated Ca^{2+} channels; ROC, receptor operated Ca^{2+} channels; (↓), denotes decrease in SHR; (↑), denotes increase in SHR.

CONCLUSIONS

There is an overwhelming acceptance that β-adrenoceptor mediated relaxation of SHR vascular smooth muscle are attenuated. However, the molecular mechanisms underlying this defect are much debated. The observations made in our laboratory support the contention that the entire cascade of events starting from hormonal-stimulation of adenylate cyclase to the target-protein phosphorylation are diminished in such a way that would lead to a decreased Ca^{2+} removal from the cytoplasm leading to defective relaxation of VSM in SHR. In addition, the existing information would support the contention that α-adrenoceptor mediated responses are augmented in the vascular smooth muscle of SHR. The increased contraction in response to α-adrenoceptor stimulation appears to be due to both increased Ca^{2+} influx into the cell through receptor-operated channels and increased Ca^{2+} release from the sarcoplasmic reticulum. Thus, it would appear that alterations in β- and α-adrenoceptors have occurred in the vascular smooth muscle of SHR in such a way that it will lead to increased vascular smooth muscle tone.

Acknowledgements:

This work was supported by National Institute of Health grants HL 19027, HL 35682, HL 14388, AM 34986 and a grant-in-aid from American Heart Association with funds contributed by Iowa Heart Association.

REFERENCES

Adelstein RS, Sellers JR, Centi MA, Pato MD, de Lanerolle P (1982). Regulation smooth muscle contractile proteins by calmodulin and cyclic AMP. Fed Proc 41:2873-2878.

Alexander RW, Brock TA Gimbrone MA Jr, Rittenhouse SE (1985). Angiotensin increases inositol trisphosphate and calcium in vascular muscle. Hypertension 7:447-451.

Amer MS, Doba N, Ries DJ (1975). Changes in cyclic nucleotide metabolism in aorta and heart of neurogenically hypertensive rats: possible trigger mechanisms of hypertension. Proc Natl Acad Sci USA 72:2135-2139.

Aqel MB, Sharma RV, Bhalla RC (1986). Increased Ca^{2+} sensitivity of $α_1$-adrenoceptor stimulated contraction in SHR caudal artery. Am J Physiol 250:C275-C282.

Baron CB, Cunningham M Strauss JF, Coburn RF (1984). Pharmacomechanical coupling in smooth muscle may involve phosphatidylinositol metabolism. Proc Natl 81:6899-6903.

Bhalla RC, Sharma RV (1982). Characteristics of hormone-stimulated adenylate cyclase in vascular smooth muscle: altered activity in spontaneously hypertensive rat. Blood Vessels 19:109-116.

Bhalla RC, Aqel MB, Sharma RV (1986) α_1-adrenoceptor-mediated responses in the vascular smooth muscle of SHR. J Hypertension 4:(Suppl. 3) In Press.

Bhalla RC, Sharma RV and Gupta RC (1982). Isolation of two myosin light chain kinases from bovine carotid and their regulation by phosphorylation mediated by cyclic AMP-dependent protein kinase. Biochem J 203: 583-592.

Bhalla RC, Sharma RV, Ramanathan S, (1980). Possible role of phosphorylation-dephosphorylation in the regulation of calcium metabolism in cardiovascular tissues of SHR. Hypertension 2:207-214.

Bhalla, RC, Sharma RV, Webb, RC (1979). Possible role of cyclic AMP and calcium in the pathogenesis of hypertension. Jap Heart J 20:(Suppl. 1) 222-224.

Bhalla RC, Webb RC, Singh D, Brock, T (1978a). Role of cAMP in rat aortic microsomal phosphorylation and calcium uptake. Am J Physiol 234:H508-H514.

Bhalla RC, Webb RC, Singh D, Ashly T, Brock T (1978b). Calcium fluxes, calcium binding, and adenosine cyclic 3':5'-monophosphate dependent protein kinase activity in the aorta of spontaneously hypertensive and Kyoto Wistar normotensive rats. Mol Pharmacol 14:468-477.

Bond M, Kitazawa T, Somlyo AP, Somlyo AV (1984). Release and recycling of calcium by the sarcoplasmic reticulum in guinea-pig portal vein smooth muscle. J Physiol 355:677-695.

Brockbank KJ, England PJ (1980). A rapid method for the preparation of sarcolemmal vesicles from rat aorta and the stimulation of calcium uptake into the vesicles by cyclic AMP dependent protrin kinase. FEBS Lett 122:67-71.

Bruschi G, Bruschi ME, Caroppo M, Orlandini G, Pavarani C, Cavatorta A (1984). Intracellular free [Ca^{2+}] in circulating lymphocytes of spontaneously hypertensive rats. Life Sci 35:535-542.

Cauvin C, van Breeman C (1985) Altered ^{45}Ca fluxes in isolated mesenteric resistance vessels from SHR. Fed Proc 44:1008A.

Cauvin C, Saida K, van Breeman C (1984) Extracellular Ca^{2+} dependence and diltiazem inhibition of contraction in rabbit conduit arteries and mesenteric resistance vessels. Blood Vessels 21:23-31.

Cohen ML, Berkowitz BA (1976). Decreased vascular relaxation in hypertension. J Pharmacol Exp Ther 196:396-406.

Coquil JF, Hamet P (1980). Activity of cyclic AMP-dependent protein kinase in heart and aorta of spontaneously hypertensive rat. Proc Soc Exp Biol Med 164:569-575.

Erne P, Bolli P, Burgisser E, Buhler FR (1984). Correlation of platelet calcium with blood pressure. New Engl J Med 310(17):1084-1088.

Gerthoffer WT, Trevethick MA, Murphy RA (1984). Myosin phosphorylation and cyclic adenosine 3,5'-monophosphate in relaxation of arterial smooth muscle by vasodilators. Circ Res 54:83-89.

Gupta RC, Bhalla RC, Sharma RV (1982). Altered distribution and properties of cAMP-dependent protein kinase isozymes in spontanously hypertensive rat aorta. Biochem Pharmacol 31:1837-1841.

Hathaway DR, Konicki MV, Coolican SA (1985). Phosphorylation of myosin light chain kinase from vascular smooth muscle by cAMP and cGMP-dependent protein kinase. J Mol Cell Cardiol 17:841-850.

Hicks PE, Medgett IC, Langer SZ (1984). Postsynaptic α_2-adrenergic receptor mediated vasoconstriction in SHR tail arteries in vitro. Hypertension 6 Suppl 1:12-18.

Kwan CY (1985). Dysfunction of calcium handling by smooth muscle in hypertension. Can J Physiol Pharmacol 63:366-374.

Mulvany MJ, Nyborg N (1980). An increased calcium sensitivity of mesenteric resistance vessels in young and adult spontaneously hypertensive rats. Br J Pharmacol 71:585-596.

Medgett IC (1985). α_2-adrenoceptors mediate sympathetic vasoconstriction in distal segments of rat tail artery. Eur J Pharmacol 108:281-287.

Nabika T, Velletri PA, Beaven MA, Endo J, Lovenberg W (1985). Vasopressin-induced calcium increases in smooth muscle cells from spontaneously hypertensive rats. Life Sciences 37:579-584.

Pfitzer G, Hoffman P, DiSalvo J, Rüegg JC (1984). cGMP and cAMP inhibit tension development in skinned coronary arteries. Phlugers Arch 401:277-280.

Saida K, van Breeman C (1984). Cyclic AMP modulation of adrenoceptor-mediated arterial smooth muscle contraction. J Gen Physiol 84:307-318.

Sharma RV, Bhalla RC (1978). Cyclic nucleotide phosphodiesterase in heart and aorta of spontaneously hypertensive rats. Biochim Biophys Acta 526:479-488.

Sharma RV, Bhalla RC (1986). Isolation and characterization of plasma membranes from bovine carotid arteries. Am J Physiol 250:(cell Physiol 19)C65-C75.

Sharma RV, Aqel MB, Butters C, McEldoon J, Bhalla RC (1986) Molecular mechanisms of increased vascular smooth muscle contraction in SHR. Fed Proc 45:460A.

Silver PJ, Holroyde MJ, Soloro RJ, DiSalvo J (1981). Ca^{2+}, calmodulin and cyclic AMP-dependent modulation of actin-myosin interaction in aorta. Biochem Biophys Acta 674:65-70.

Silver PJ, Michalak RJ and Kocmund SM (1985). Role of cyclic AMP protein kinase in decreased arterial cyclic AMP responsiveness in hypertension. J Pharmacol Exp Ther 232:595-601.

Somlyo AV, Bond M, Somlyo AP (1985). Inositol triphosphateinduced calcium release and contraction in vascular smooth muscle. Proc Natl Acad Sci USA 82:5231-5235.

Triner L, Vulliemoz Y, Verosky M, Manger M (1975). Cyclic adenosine monophosphate and vascular reactivity in spontaneously hypertensive rats. Biochem Pharmacol 24:743-745.

Webb RC (1984) Vascular changes in hypertension. Cardiovascular pharmacology (Antonaccio M. ed.) pp. 215-255, Raven Press, New York.

SLOWING OF CROSSBRIDGE CYCLING RATE IN MAMMALIAN SMOOTH MUSCLE OCCURS WITHOUT EVIDENCE OF AN INCREASE IN INTERNAL LOAD

T.M. Butler, M.J. Siegman and S.U. Mooers

Department of Physiology
Jefferson Medical College
Thomas Jefferson University
Philadelphia, Pennsylvania 19107

INTRODUCTION

Smooth muscle has the remarkable property of being able to vary both the maximum velocity of shortening and the energetic cost of isometric force maintenance. Most investigators interpret these results to mean that the number of attached crossbridges which give rise to isometric force production by the muscle is regulated independently of actin-activated myosin ATPase activity or crossbridge cycling rate.

On the basis of biochemical studies suggesting that the myosin light chain must be phosphorylated for actin-activation of myosin ATPase (see Hartshorne, 1982, for review) and the observation that force could be maintained under conditions of a low degree of myosin light chain phosphorylation, Murphy and colleagues (see Murphy et al., 1983, for review) hypothesized that there are two classes of crossbridges that interact with actin. The first type consists of phosphorylated crossbridges that maintain force and have a relatively fast cycling rate. The second is a "latchbridge" that results from dephosphorylation of the crossbridges, and has the very special property of interacting with actin to maintain force while either not cycling or cycling very slowly.

Interpretation of the energetics and mechanical properties of smooth muscle can be made in terms of the above model involving two types of force-generating crossbridges. Under isometric conditions, force output would result from all crossbridges that are in the positive force-bearing mode, and would include both phosphorylated crossbridges and latchbridges. However, the energy usage during isometric force maintenance would be determined by the number of phosphorylated crossbridges, since these would consume ATP at a very much higher rate than the latchbridges. The cycling rate of each phosphorylated crossbridge would presumbably be independent of the presence or absence of latchbridges. Variations in the energy cost of isometric force maintenance would result from changes in the relative contribution of cycling and latchbridges to the force output.

In the original latchbridge model (Dillon et al, 1981; Murphy et al, 1983), the slowing of maximum velocity of shortening was postulated to result from an internal load presented by latchbridges against which phosphorylated cycling crossbridges would have to interact. In other words, the cycling bridges cause relative filament sliding, but the latchbridges, because of their slow detachment rate, would be pulled into a negative force-bearing position and provide a load on the more rapidly cycling bridges. As a result, the cycling rate of the phosphorylated, rapidly cycling bridges would slow according to their force-velocity relationship. If a large fraction of the force-bearing bridges are latchbridges under isometric conditions, then when allowed to shorten, velocity would be very slow and most of the mechanical energy derived from ATP splitting would be dissipated internally against the negative force-producing latchbridges. In this model, the maximum velocity of shortening is determined by the relative ratio of the number of phosphorylated cycling crossbridges to the number of latchbridges. Maximum velocity of shortening occurs when the negative force provided by latchbridges equals the positive force generated by the phosphorylated cycling bridges. In this model it is not possible to postulate that latchbridges detach and provide no internal load upon shortening, because under such conditions, there would then be no change in the maximum velocity of shortening. The cycling rate of the phosphorylated crossbridges, upon which Vmax would depend, would be constant in such a mechanism.

We have performed experiments designed to test certain aspects of this two-state model. The results to be described are **inconsistent** with the model in two major respects. The first is that the number of phosphorylated crossbridges is not the major determinant of energy usage under isometric conditions. The second is that we find no energetic evidence for the dissipation of work against an internal load during shortenings under conditions when most of the force would be expected to be generated by latchbridges.

RESULTS AND DISCUSSION

All of the experiments to be described have been performed on the rabbit taenia coli at $18^\circ C$. The chemical energy usage in the muscle has been determined by direct measurement of changes in high energy phosphate contents in muscles in which ATP resynthesis from respiration and glycolysis has been inhibited. The details of this method are described in previous publications (Butler et al, 1978; Siegman et al, 1980). The maximum velocity of shortening has been determined from slack test measurements (Edman, 1979; Siegman et al, 1984), and the degree of myosin light chain phosphorylation was determined by IEF-SDS two-dimensional electrophoresis (Driska et al, 1982; Butler et al, 1983).

Figure 1 shows the energetics and mechanical parameters as well as the degree of myosin light chain phosphorylation during activation of the taenia coli under isometric conditions. In muscles bathed in a solution containing 1.9 mM Ca^{++}, the average rate of energy usage is 4-fold higher during the first 25 sec of stimulation when force is being developed, than during the next 35 sec when force is being maintained. In 4.5 mM Ca^{++}, there is no increase in the active force output of the muscle, but there is a doubling of the energy usage. Interestingly, the 4-fold difference in the energy cost during force development compared to force maintenance is still apparent. These data suggest that there are large changes in the average rate of crossbridge cycling as a function of both duration of stimulation and calcium concentration of the bathing solution. This is supported by the observed variations in the maximum velocity of shortening also shown in Fig. 1.

There is no significant difference in the degree of myosin light chain phosphorylation at the two calcium concentrations even though there are large differences in the average crossbridge cycling rate.

TABLE 1

ATP Turnover Rate of Phosphorylated Myosin S1 Under Isometric Conditions in the Rabbit Taenia Coli

Design	Turnover Rate (sec^{-1})[a]	
	Based on Total Phosphorylation	Based on Suprabasal Phosphorylation
1.9 mM Calcium		
0 - 25 sec First tetanus	0.99	1.63
25 - 60 sec First tetanus	0.27	0.48
4.5 mM Calcium		
0 - 25 sec First tetanus	1.89	3.12
25 - 60 sec First tetanus	0.45	0.80
1.9 mM Calcium		
0 - 25 sec Second tetanus[b]	0.30	0.77

a Based on a myosin S1 content of 0.08 μ mole/g (Siegman et al, 1980).
b In this design, the muscles were stimulated for 25 sec and allowed to relax for 30 sec prior to the test stimulus
Data from Siegman et al (1984) and Butler et al (1986)

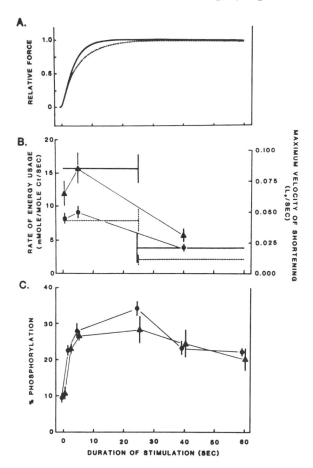

Figure 1. Time course of force output, chemical energy usage, maximum velocity of shortening, and degree of myosin light chain phosphorylation during an isometric tetanus in the rabbit taenia coli.
A. Active force output from muscles bathed in a Krebs solution containing either 1.9 mM Ca^{++} (dotted line) or 4.5 mM Ca^{++} (solid line).
B. Average rate of energy usage in 1.9 mM Ca^{++} (dotted line) and 4.5 mM Ca^{++} (solid line). Also shown are values for Vmax in 1.9 mM (●) and 4.5 mM (▲) Ca^{++}.
C. Myosin light chain phosphorylation as a function of duration of stimulation. Symbols are the same as in B. (From Siegman et al, 1984).

If it is assumed that all of the suprabasal energy consumption is due to actin-activated myosin ATPase activity of phosphorylated myosin, then it is possible to estimate the ATP turnover rates of phosphorylated myosin S1 under the various mechanical conditions. These are shown in Table 1 based on total myosin phosphorylation as well as after correction for basal light chain phosphorylation. There is a seven-fold variation in the calculated cycling rate of phosphorylated myosin under the conditions described. This strongly suggests that the ATPase rate of phosphorylated myosin is regulated under isometric conditions. The simple hypothesis that changes in the economy of force maintenance are caused by a decrease in the number of phosphorylated crossbridges with the transition of some into the latch state is, therefore, not tenable. Any model of the regulation of crossbridge function in smooth muscle must account for the observed changes in average crossbridge cycling rate under isometric conditions independent of changes in myosin light chain phosphorylation.

Another design that we have used to probe the regulation of crossbridge cycling is shown in Figure 2. In this design the muscle is stimulated for 25 sec and allowed to relax for 30 sec before restimulation. The top panel shows the isometric force output. During the second stimulation, there is a rather slow redevelopment of isometric force to $90 \pm 4\%$ of the original value. There is however, a striking change in the light chain phosphorylation response (Fig. 2, lower panel) upon restimulation; there is a significant attenuation of myosin phosphorylation even though force redevelopment is nearly maximum. This supports previous work on other muscles which suggests that there can be large force outputs with very low levels of myosin light chain phosphorylation. (See Kamm and Stull, 1985 for review.)

Measurements of the maximum velocity of shortening show that at 20 sec after the initiation of stimulation, Vmax in the first tetanus is almost two-fold higher (1.82 ± 0.21) than during the second, even though the active force output is not significantly different. Table 2 shows the suprabasal rate of energy usage under the various conditions. High-energy phosphate utilization during the development of force is very low during the second stimulation compared to the first, and is not significantly different from the energy usage during the force maintenance

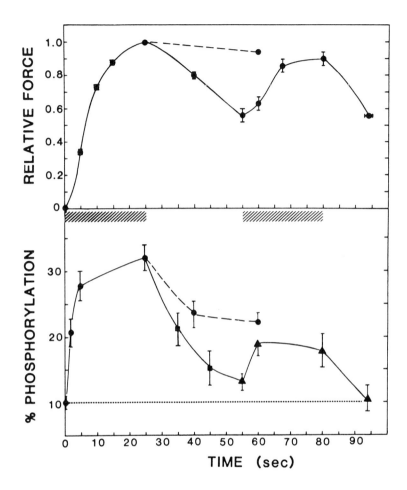

Figure 2. Force output and myosin light chain phosphorylation during stimulation under isometric conditions in the rabbit taenia coli. The upper panel shows active force in response to a 25 sec stimulation followed by either continued stimulation (dashed line) or by a 30 sec period of relaxation and another 25 sec stimulation (solid line). The hatched areas show the times during which the muscle was stimulated. The lower panel shows myosin light chain phosphorylation under the same conditions. The dotted line shows the light chain phosphorylation under unstimulated conditions (From Butler et al, 1986).

phase of the first contraction. Interestingly, the turnover rate of phosphorylated myosin S1 during the second stimulation is about the same as that during the force maintenance phase of the initial stimulation (see Table 1).

TABLE 2

Average Suprabasal Rate of Chemical Energy Usage Under Isometric Conditions

Design	Average Rate of Energy Usage $(mmol/mol\ C_t/sec)^a$
1.9 mM Calcium	
0 - 25 sec First tetanus	8.2 ± 0.8 N = 15
25 - 60 sec First tetanus	2.0 ± 0.6 N = 18
0 - 25 sec Second tetanus[b]	1.6 ± 1.2 N = 11
4.5 mM Calcium	
0 - 25 sec First tetanus	15.7 ± 2.0 N = 9
25 - 60 sec First tetanus	4.1 ± 1.9 N = 9

a C_t = total creatine = 2.7 μmole/g wet wt.
b   In this design the muscles were stimulated for 25 sec and allowed to relax for 30 sec prior to the test stimulus
    Data from Siegman et al (1984) and Butler et al (1986)

When consecutive contractions occur, it seems that some process is initiated during the first stimulation and relaxation period that alters the characteristics of the second contraction, such that the average crossbridge cycling rate is slowed, there is a low degree of myosin light chain phosphorylation, but no change in the maximum force generation. Presumably, this is the same process that has given rise to a decrease in the cycling rate of phosphorylated crossbridges in a single tetanus and is maintained even during the relaxation period. Other experiments (Butler et al, 1986) suggest that the effect of this "inhibitory process" is long-lived, in that recovery of Vmax between contractions takes approximately one minute while recovery of the original phosphorylation response takes longer.

The data described above concerning the ATPase rate of phosphorylated myosin under isometric conditions rules out the possibility that changes in energy usage are due only to the loss of cycling crossbridges by dephosphorylation to latchbridges during continuous stimulation. Some mechanism must control the cycling rate of phosphorylated myosin under isometric conditions. This raises the interesting possibility that the slowing in maximum velocity of shortening during the second tetanus might result from this same long-lived inhibitory process that can change the inherent cycling rate of the crossbridge. This would differ considerably from the latchbridge hypothesis, in which the slowing in maximum velocity of shortening was the result of dephosphorylated latchbridges providing an internal load against which phosphorylated cycling crossbridges must shorten. Since we have some evidence that there might be a mechanism allowing for the slowing of velocity by direct control of cycling rate of phosphorylated crossbridges rather than the internal load model, we thought that it would be helpful to test one of the predictions of the internal load model for slowing of velocity.

In order for latchbridges to slow velocity by providing an internal load, some of the work produced by the cycling crossbridges would have to be dissipated internally against the non-cycling crossbridges. More of the energy derived from ATP splitting would thus be degraded into heat if there were a large internal load presented by a class of latchbridges. Energetic evidence for the presence of an internal load would be a decrease in the quantity of external work produced per mol of ATP which is utilized. In other words, the apparent efficiency of work production would decrease with the presence of an internal load.

TABLE 3

Chemical Energy Cost of Work Production During Isovelocity Shortening in the Rabbit Taenia Coli

Design	Chemical Energy Cost of Work Production (mol/kJ)
1. 0 - 35 sec First tetanus	0.26 ± 0.04 N = 7
2. 20 - 55 sec First tetanus	0.29 ± 0.07 N = 11
3. 0 - 25 sec Second tetanus[a]	0.18 ± 0.12 N = 13

[a] In this design the muscles were stimulated for 25 sec and allowed to relax for 30 sec prior to the test stimulus.
All muscles were bathed in Krebs solution containing 1.9 mM Calcium.
Data from Butler et al, 1983, and Butler et al, 1986.

Table 3 shows the results from energetics experiments designed to determine the energy cost of active external work production. The periods and conditions chosen for determination of the energy cost of work production during shortening were the following: 1) during the initial force development phase when isometric energy usage, Vmax and degree of phosphorylation are at their highest; 2) during the subsequent period of isometric force maintenance when both the isometric energy usage and Vmax are low, but phosphorylation is only slightly lower than initially; and

3) during the redevelopment of force according to the design shown in Fig. 2, when the isometric energy usage, Vmax and degree of phosphorylation are all very low. There is no significant difference in the energy cost of work production in any of these three designs. In fact, in the restimulation case in which the latchbridge model would predict the largest internal load, the energy cost of work production is smallest.

The force-velocity relation for the rabbit taenia coli (Gordon and Siegman, 1971) shows that a load of about 20% Po is required for a two-fold decrease in shortening velocity. From this relation, we can calculate the quantity of work that should be dissipated for an internal load of 20% Po. If it is assumed that this work has the same energetic cost as the total external work produced during the initial stimulation, then it is possible to calculate an expected energy usage due to work production in the restimulation design. We found that the predicted energy usage was significantly higher than the observed (Butler et al, 1986). In other words, there wasn't enough chemical energy used to account for the observed external work plus the internal work dissipation. These experiments thus present energetic evidence against the idea that changes in velocity of shortening are mediated by changes in internal loading.

CONCLUSION

Our results question the validity of the two-state model for regulation of crossbridge cycling in smooth muscle in two respects. The first is that the number of phosphorylated crossbridges is not the major determinant of energy usage under isometric conditions. The ATPase rate of phosphorylated myosin S1 is likely to vary by almost an order of magnitude. The most likely explanation for this finding is that there are regulatory systems present which either act independently of myosin light chain phosphorylation or at least modify the cycling rate of phosphorylated myosin. These could be leiotonin (Ebashi et al, 1987), caldesmon (both phosphorylated and unphosphorylated) (Ngai et al, 1984, 1985; Marston et al, 1985), and direct calcium binding to myosin (Chacko and Rosenfeld, 1982; Nag and Seidel, 1983). With all of these potential regulatory systems, it might be surprising to find a simple relationship between ATP usage and the number of phosphorylated myosin S1 units.

The second problem which we have identified concerning the two-state crossbridge model is potentially more difficult to fit into any model. This is the absence of energetic evidence that would suggest that changes in shortening velocity reflect changes in internal loading. Vmax would be determined by the fastest cycling crossbridges if there were no change in internal load. Our data suggest that there are regulatory systems which, singly or in concert, regulate the fastest cycling crossbridges in a graded manner, and that maximum velocity of shortening is governed by this mechanism rather than by a variable resistance to shortening imposed by a variable number of slowly cycling latchbridges.

ACKNOWLEDGEMENTS

This work was supported by HL15385 to the Pennsylvania Muscle Institute. TMB is the recipient of a Research Career Development Award AM 00873.

REFERENCES

Butler TM, Siegman MJ, and Mooers SU (1983). Chemical energy usage during shortening and work production in mammalian smooth muscle. Am J Physiol 244: C234-C242.

Butler TM, Siegman MJ, and Mooers SU (1986). Slowing of cross-bridge cycling in smooth muscle without evidence of an internal load. Am J Physiol 251: C945-C950.

Butler TM, Siegman MJ, Mooers SU, and Davies RE (1978). Chemical energetics of single isometric tetani in mammalian smooth muscle. Am J Physiol 235: C1-C7.

Chacko S, and Rosenfeld A (1982). Regulation of actin-activated ATP hydrolysis by arterial myosin. Proc Natl Acad Sci USA 79: 292-296.

Dillon PF, Aksoy MO, Driska SP, and Murphy RA (1981). Myosin phosphorylation and the crossbridge cycle in arterial smooth muscle. Science 211: 495-497.

Driska SP, Aksoy MO, and Murphy RA (1981). Myosin light chain phosphorylation associated with contraction in arterial smooth muscle. Am J Physiol 240: C222-C233.

Ebashi S, Mikawa T, Kuwayama H, Suzuki M, Ikemoto H, Ishizaki Y, and Koga R (1987). Ca^{2+} regulation in smooth muscle: Dissociation of myosin light chain kinase activity from activation of actin-myosin interaction. In Siegman MJ, Stephens NL, Somlyo AP (eds): "Regulation and Contraction of Smooth Muscle," New York: Alan R. Liss, this volume.

Edman, KAP (1979). The velocity of unloaded shortening and its relation to sarcomere length and isometric force in vertebrate muscle fibers. J Physiol Lond 291: 143-159.

Gordon, AR, and Siegman, MJ (1971). Mechanical properties of smooth muscle. I. Length-tension and force-velocity relations. Am J Physiol 221: 1243-1249.

Hartshorne, DJ (1982). Phosphorylation of myosin and the regulation of smooth muscle actomyosin. In Dowben RM, Shay JW (eds): "Cell and Muscle Motility," New York: Plenum, pp. 185-220.

Kamm KE, and Stull JT (1985). The function of myosin and myosin light chain kinase phosphorylation in smooth muscle. Ann Rev Pharmacol Toxicol 25: 593-620.

Marston SB, and Smith CWJ (1985). The thin filaments of smooth muscles. J Muscle Res & Cell Motil 6: 669-708.

Murphy RA, Aksoy MO, Dillon PF, Gerthoffer WT, and Kamm KE (1983). The role of myosin light chain phosphorylation in regulation of the crossbridge cycle. Fed Proc 42: 51-56.

Nag S, and Seidel JC (1983). Dependence on Ca^{+2} and tropomyosin of the actin-activated ATPase activity of phosphorylated gizzard myosin in the presence of low concentrations of Mg^{+2}. J Biol Chem 258: 6444-6449.

Ngai PK, and Walsh M (1984). Inhibition of smooth muscle actin-activated myosin Mg^{2+}-ATPase activity by caldesmon. J Biol Chem 259: 13656-13659.

Ngai PK, and Walsh M (1985). Properties of caldesmon isolated from chicken gizzard. Biochem J 230: 695-707.

Siegman MJ, Butler TM, Mooers SU, and Davies RE (1980). Chemical energetics of force development, force maintenance and relaxation in mammalian smooth muscle. J Gen Physiol 76: 609-629.

Siegman MJ, Butler TM, Mooers SU, and Michalek A (1984). Ca^{+2} can affect Vmax without changes in myosin light chain phosphorylation in smooth muscle. Pflugers Archiv 401: 385-390.

FORCE:VELOCITY RELATIONSHIP AND HELICAL SHORTENING IN SINGLE SMOOTH MUSCLE CELLS

David Warshaw, Whitney McBride, and Steven Work

Department of Physiology & Biophysics
University of Vermont
Burlington, Vermont 05405

INTRODUCTION

In 1938 A.V. Hill (Hill, 1938) presented the relationship of active force production with shortening velocity in whole skeletal muscle as a rectangular hyperbola. In smooth muscle, a similar force versus velocity relationship (F:V) has been described (Murphy, 1980; Fay et al., 1981). It was this similarity in F:V shape between the two muscle types that served as evidence for a qualitatively similar crossbridge mechanism in smooth and skeletal muscle. However, to date all studies of smooth muscle F:V have been limited to multicellular tissue preparations in which it is difficult to assess the dependence of the F:V on possible inhomogeneities in cellular activation and mechanical responses. With the ability to isolate single smooth muscle cells and to measure their mechanical responses (Warshaw and Fay, 1983; Warshaw et al., 1986), it is possible to obtain single smooth muscle cell velocity data. Isotonic release data were used to construct F:V in single smooth muscle cells. The maximum shortening velocity (Vmax) estimated from F:V data compared favorably with values obtained from the slack test procedure. However, Vmax determined from mechanical measurements in which a single cell is fixed at both ends to a recording device may prevent the cell from expressing its true shortening capabilities. Video images of freely shortening cells revealed that single cells shorten helically in a corkscrew-like fashion. Thus the cell's inherent Vmax may be significantly underestimated using standard methods for determining mechanical performance.

METHODS

Single Cell Isolation and Preparation

The procedures for isolation and preparation of single smooth muscle cells from the toad (Bufo marinus) stomach muscularis have been previously described in detail (Fay et al., 1982; Warshaw and Fay, 1983). In brief, single smooth muscle cells (SMCs) were obtained by enzymatically digesting tissue slices of stomach muscularis. Cells were then transferred to an inverted microscope on a glass slide containing amphibian physiological saline. Cells were attached at one end to a force transducer (200 Hz natural frequency; 0.01uN resolution, 0.1 um/uN compliance) and at the other end to a piezoelectric length displacement device (1KHz natural frequency; ±10um displacement) by means of specially designed glass micropipettes. Knots were tied in each end of the cell and thus prevented cell slippage upon contraction. Cells were stimulated by transverse electric field stimulation through platinum paddles at $20^{o}C$. Once stimulated to contract, cells generated active force isometrically. At the peak of contraction cells were subjected to experimental protocols in which either muscle force (i.e. isotonic release) or cell length (i.e. isometric release for slack test procedure) was controlled (see Figures 1,2).

In studies of freely shortening cells, cells were decorated with 1 μm diameter anion exchange resin beads. These charged beads would adhere to the cell sulface and thus act as markers for cell motion. A single cell, decorated with beads, was then attached at one end only and allowed to shorten freely following electrical stimulation. Video images of contracting cells were digitized by computer every 500 ms and the digitized image digitally filtered to enhance the appearance of the beads (see Fig. 6).

Protocols

<u>Isotonic Releases</u>. In order to determine the F:V in a single SMC it was necessary to measure cell shortening velocity at various force levels. Therefore the following

computer controlled protocol was used. The tension signal was digitized both at rest and peak of isometric force. By subtracting the resting force value from that at the peak of contraction, a value for maximum active force (Fmax) was determined. At this point the cell was subjected to a series of isotonic releases (each release complete in 30-65 ms) of varying magnitudes (0.10-0.90Fmax). Each force step was maintained for 500 ms after which time the force was returned gradually to Fmax (see Fig. 1).

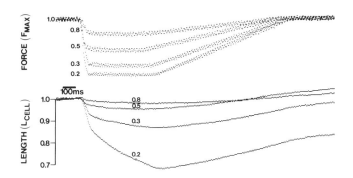

Figure 1. Isotonic releases.

Since the entire isotonic release protocol was completed within a single contraction (12.5s), it was necessary to control for any slowing in shortening velocity that may have occurred during the protocol. Therefore an isotonic release protocol was performed with the magnitude of the isotonic releases held constant at 0.5 Fmax. Therefore, any variations in the observed shortening velocity would thus characterize any dependence of shortening velocity with duration of contraction.

After a detailed mathematical analysis of the length response following completion of the force step, the length response was best fit to a single exponential function by nonlinear regression (R^2=0.94). A value for velocity of shortening normalized to cell length was determined from the derivative of the exponential fit of cell length and time at a point immediately after completion of the isotonic release.

The isotonic release protocol provided a series of active muscle force steps and their resultant length changes which were used to construct the F:V in a single cell. Hill (Hill, 1938) described the dependence of shortening velocity on load as a rectangular hyperbola of the form: $(F/Fmax+a/Fmax)(V+b)=(1+a/Fmax)b$ with constants $a/Fmax$ and b.

<u>Slack Test Method</u>. In addition to determining the maximum shortening velocity (Vmax) from the force:velocity relationship, Vmax can also be estimated from the slack test procedure (Edman, 1979). This procedure is based upon a cell's ability to take up its slack immediately following a length change which is sufficient to make the cell go slack. Thus at peak isometric force, a series of length changes between 0.10 and 0.30 Lcell was imposed (see Fig. 2). Each length change was complete in 5 ms after which length was maintained constant for 1 s and then returned to the original length over a 1 s period.

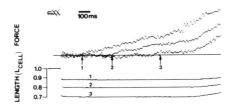

Figure 2. Slack test data.

The slack test data were analyzed in the following manner. Once the cell length was released to a slack length, the time to take up this slack was determined as the time from the beginning of the release to that time when the redevelopment of force was just detected (see Figure 2). Since the load on the cell is zero while slack, the amount of slack cell length taken up by the cell divided by the time to take up this slack is thus an estimate of unloaded shortening velocity. To be certain that the amount of slack to be taken up by the cell had no effect on unloaded shortening, the magnitude of varying length changes were plotted against the time to take up different amounts of slack as described above (see Figure 5). The slope of the line fit by linear regression to these points was then taken as an estimate of Vmax.

RESULTS

Isotonic Releases and Force:Velocity Relationship

A single smooth muscle cell's ability to shorten under various loads was studied following a sudden change in force. The timecourse of the length change in response to a force step (see Figure 1) was biphasic. The initial phase is a rapid decrease in cell length that is coincident with the sudden drop in force and reflects the cell's elastic response. Following the force step, for the entire period (500 ms) that cell force was maintained at a new level, cell length decreased with a single time constant ranging between 75 and 600 ms.

The velocities of shortening estimated from the isotonic release protocol were used to construct the

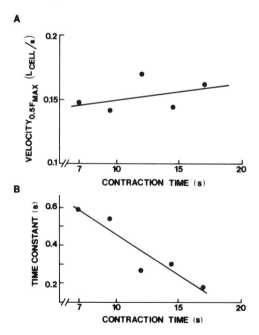

Figure 3. Shortening velocity and time constant for slowing of velocity with time of contraction

relationship between active cell force and shortening velocity. Since these isotonic releases were obtained in a single contraction, it was important to control for any variation in shortening velocity with duration of contraction. Figure 3A shows the lack of any such dependence of shortening velocity at a constant load (0.5 Fmax). Thus the estimates for shortening velocity at various loads were used without correction to construct the force versus velocity relationship in single smooth muscle cells (see Figure 4). It is interesting to note though that the time constant for slowing of velocity at a given load did decrease with duration of contraction (see Figure 3B). Therefore the observed slowing of shortening velocity after completion of an isotonic release would be greater at a point later in the contraction.

The relationship between active cell force and shortening velocity was best fit by the Hill hyperbola (R^2=0.94). The force and velocity data from four experiments were grouped and the resultant hyperbolic constants were 0.268 for (a/Fmax) and 0.163 Lcell/s for b (see Figure 4). From the fitted F:V equation, an estimate for maximum velocity of shortening was determined at zero force and equalled 0.608 Lcell/s.

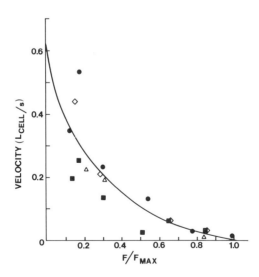

Figure 4. Force:velocity relationship of single smooth muscle cell.

Vmax: Slack Test

When cells were subjected to large rapid releases to slack lengths, the time to take up the slack was dependent upon the magnitude of the length change being longer for longer releases (see Figure 2). From six experiments the average slope of the regression line (i.e. Vmax) through the slack test data equalled 0.583 Lcell/s (see Figure 5) with individual cells having Vmax that ranged between 0.26 and 0.89 Lcell/s. The slack test estimate of Vmax thus agrees with the Vmax estimate obtained from the F:V. However, note that in 2 experiments in which 3

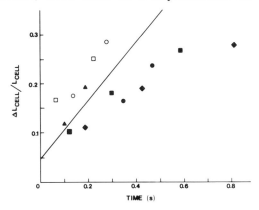

Figure 5. Amount of imposed slack length vs. time to take up slack.

releases were performed, the relationship between the length change and time to take up the slack became non-linear for larger releases. If this trend existed for all length changes, then the slack test may in fact underestimate Vmax. This nonlinearity may relate to similar phenomena reported by Gunst (Gunst, 1986) in which force redevelopment following a release is slower for larger releases.

Contraction of Freely Shortening Cells

Time lapse digitized video images of freely shortening cells are seen in Figure 6. As the cell shortens it is apparent that beads which are used to decorate the cell surface in order to detect motion begin to rotate around the cell surface. Since the cell remains within the plane of focus, the observed changes in the beads spatial relationships can not be attributed to oscillations of the cell within the focal plane. The bead movements suggest that the cells shorten in a corkscrew-like fashion with the maximum rotation for a given bead being as much as 180 degrees from its initial position on the cell membrane. After viewing 15 cells, there does not appear to be any preferred direction of rotation with cells helically contracting in both a clockwise and counter-clockwise manner. Notice in Figure 6 that the position of beads on the upper membrane surface appear on the lower membrane surface after 2.5 s of helical contraction as the cell shortens.

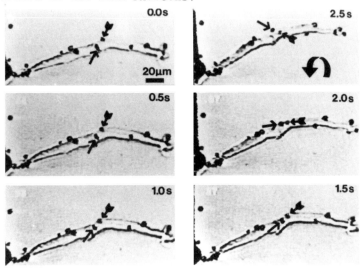

Figure 6. Rotation of freely shortening cell.

DISCUSSION

Since Hill (Hill, 1938) first described the hyperbolic shape of the F:V in whole skeletal muscle, similar hyperbolic functions have been reported in single skeletal muscle fibers (Julian, 1971; Edman, 1979), cardiac (Meiss and Sonnenblick, 1974) and smooth muscle tissue preparations (Murphy, 1980; Fay et al., 1981). Since smooth muscle F:V have only been obtained in multicellular tissue preparations, it may be difficult to assess the extent to which the F:V shape is dependent upon inhomogeneities in the individual cellular responses. In the present study, the ability to obtain F:V in a single smooth muscle cell has been demonstrated. Since single smooth muscle cell F:V are also hyperbolic in shape, these data support the idea that a qualitatively similar crossbridge mechanism exists in smooth muscle as compared to that in striated muscle. In addition the similarity in hyperbolic constants that describe F:V from smooth muscle tissue preparations and single smooth muscle cells supports what to date has been assumed, that tissue F:Vs are a reasonable estimate of their constituent cells.

Isotonic Releases

F:V in single smooth muscle cells were constructed from individual length responses following a sudden change in cell force. In the past, investigators studying the timecourse of these length responses (i.e. isotonic transients) in single skeletal muscle fibers (Civan and Podolsky, 1966) have interpreted the multiphasic length responses as reflecting the crossbridge mechanical response (i.e. the initial shortening phase that coincides with the force step), followed by the transitions of a relatively synchronized population of crossbridges passing through a series of steps in the crossbridge cycle before reaching steady state shortening. The length transients upon completion of the force step in skeletal muscle are characterized by an initial rapid shortening that lasts for 2-5 ms which is followed by a slowing or hesitation in shortening lasting 50 ms and a third and final phase of steady state shortening (Civan and Podolsky, 1966). The length responses from single SMC are also characterized by an initial length change that coincides with the force change. Estimates of cell compliance (approximately 1.9% Lcell) from this initial length response agree with previous estimates from single toad stomach smooth muscle cells (Warshaw and Fay, 1983) as well as estimates from other smooth muscle tissue preparations (Pfitzer et al., 1982).

During the 500ms following the sudden change in force, cell shortening does not exhibit a multiphasic length response that is characteristic of fast skeletal muscle. The SMC shortening velocity is monophasic and slows with a time constant of 75-600 ms. The apparent lack of an initial rapid velocity transient (observed in smooth muscle tissue preparations (Hellstrand and

Johannson, 1979; Mulvany, 1979; Dillon and Murphy, 1982)) and subsequent hesitation of shortening in the single cell is related most likely to the amount of time required to impose the change in force (i.e. 65 ms). Therefore any attempt to analyze the kinetics of the length responses from SMC in terms of specific steps in the crossbridge cycle would be premature at this time.

Following an isotonic release, the slowing of cell shortening velocity observed in single smooth muscle cells is similar to that reported in both smooth muscle tissue preparations (Hellstrand and Johannson, 1979; Mulvany, 1979; Dillon and Murphy, 1982). The slowing of velocity during the maintained force step may result from: 1) shortening deactivation; 2) internal compressive forces that resist further cell shortening; 3) an ongoing process that slows crossbridge cycling as a function of time. If shortening deactivation exists in smooth muscle, then as the cell shortens following an isotonic release, it may enter a range of lengths where velocity is length dependent and thus slows. Compression of internal structures within the cell may also cause the shortening velocity to slow during isotonic releases. Evidence for internal compressive forces is supported by the ability of single toad stomach SMCs to fully reextend immediately following a contraction to 30% of resting cell length (Fay and Delise, 1973). Although the structures responsible for cell reextension are unknown, their compression during shortening would thus act as an internal load to further shortening. The final possibility is the existence of some process that results in a continual slowing of crossbridge cycling with time of contraction. Possible processes are the appearance during a contraction of non- or slowly cycling crossbridges (i.e. latchbridges (Dillon and Murphy, 1982)) which act as an internal load to the normally cycling bridges or actin-binding proteins which may hinder the crossbridge interaction with actin (e.g. caldesmon (Nagai and Walsh, 1985)). Eventhough, over the duration of the experimental protocol (i.e. 12.5 s), the velocities at a constant load (determined immediately after completion of the force step) did not slow with time (see Figure 3A), the slowing of velocity following the force step did become more pronounced with time of contraction (see Figure 3B). These data may be consistent with the latchbridge hypothesis only if during the force step latchbridges detach and then reattach following completion of the force step. Therefore the time constant might reflect the reattachment of latchbridges which hinder the normally cycling bridges. In addition, if the latchbridge population increased with time then in subsequent isotonic releases the greater number of latchbridges that would reattach following completion of the force step could result in the observed slowing of velocity during the step with time of contraction. This explanation is highly speculative and requires that the attachment of latchbridges be highly sensitive to changes in load. Thus the exact cause of the slowing during individual length responses remains to be investigated.

Form of Single SMC Force:Velocity Relationship

From the best fit of single SMC F:V data to the Hill hyperbola, Vmax (0.61 Lcell/s) was calculated at zero force. This value agrees with Vmax determined by the slack test procedure in this laboratory and recently in a preliminary report by Yagi and Fay (1985) in the same preparation. No F:V data are available from tissue from which these cells are derived. However comparisons to other smooth muscle tissue preparations indicate that the cells' average hyperbolic constants and Vmax are in the range of values reported for tissue whose cellular orientations are well defined. Thus by mathematically averaging individual cell F:V, the resultant cell F:V is similar to tissue F:V that arises from mechanically averaging individual cell F:Vs.

Crossbridge Cycle Kinetics

Since the F:V is a basic property of muscle, investigators have attempted to correlate the F:V shape with specific events in the crossbridge cycle. Huxley (Huxley, 1957) proposed a model of muscle contraction in which rate constants for "side piece" (i.e. crossbridge) attachment and detachment could be derived from the F:V constants. Using this model two aspects of the crossbridge cycle in smooth muscle as compared to fast striated muscle can be predicted: 1) slower rate of crossbridge detachment; 2) greater percentage of attached crossbridges. The crossbridge model proposed by Huxley (1957) predicts that as the muscle shortens, an attached population of crossbridges shift to positions relative to the actin binding site that results in these bridges producing negative force. These crossbridges would oppose further shortening and thus decrease velocity. In smooth muscle then it is possible that detachment of these bridges is slower than in fast skeletal muscle and would retard shortening velocity to a greater extent accounting for the slower Vmax in smooth muscle. With regard to the percentage of attached crossbridges, the Huxley model predicts that multiplying the hyperbolic constants (i.e. ab) equals the ratio, g_1/f_1, where g_1 and f_1 are the crossbridge detachment and attachment rates respectively, in the positional region where myosin crossbridges will attach to actin. Since the Huxley model consists of only one attached and one detached crossbridge state, the ratio f_1/g_1 (i.e. 1/(ab)) is an estimate of the equilibrium constant between detached and attached states or the relative proportion of attached to detached crossbridges. Using this as a first approximation, f_1/g_1 for SMC ranges between 12 and 23 as compared to 12 for fast frog skeletal muscle at $0^{\circ}C$ (Hill, 1938). Therefore SMC may have twice as many attached crossbridges than fast striated muscle in the isometric steady state. It is interesting to note that both slow amphibian (Lannergren, 1978) and tortoise (Woledge, 1968) skeletal muscles have a and b values which predict similar increased numbers of attached crossbridges. It is possible that all slowly contracting muscles have crossbridge cycle kinetics that result in a greater percentage of attached bridges relative to fast striated muscle. An increased number of attached crossbridges may help to explain smooth muscle's ability to generate comparable or greater force per cross sectional area as in striated muscle with far less of the contractile protein, myosin (Cohen and Murphy, 1979).

Vmax of Freely Shortening Cells

Most determinations of Vmax require attaching both ends of either a tissue or single cell preparation to a mechanical recording device. Thus Vmax is determined from end-to-end measurements of preparation length. However when single smooth muscle cells are allowed to freely shorten while being held at one end only, we observed motion of marker beads on the cell surface which indicated that cells shorten in a corkscrew-like fashion. Thus if one takes into account the true distance traveled by a point on the cell membrane during cell shortening, the actual velocity could be 1.5 times greater than that measured by end-to-end shortening. To what extent this phenomenum occurs in vivo is a matter of speculation. However for arterial constrictions in which diameters change by as much as 50%, the resultant cell shortening within the wall could be sufficient to demonstrate a helical contraction.

Does the cell's corkscrew-like shortening suggest a specific arrangement of the contractile apparatus within the cell? If contractile units were arranged in series and lie parallel to the cells long axis then one would not expect to see the beads on the cell surface rotate as the cell shortens. Bagby (1983) in a recent review, discusses the various models that have been proposed for the arrangement of the contractile apparatus in smooth muscle. References to undulating and helical appearance of the contractile apparatus have been reported in freely shortening isolated cells (Fisher and Bagby, 1977) and triton treated taenia (Small, 1977). Therefore the observed rotation of beads on the cell surface must be associated with a helical arrangement of the contractile apparatus.

REFERENCES

Bagby RM (1983). Organization of contractile/cytoskeletal elements. In Stephens NL, (ed): "Biochemistry of Smooth Muscle Vol. I," Boca
Raton, Florida: CRC Press, pp 1-84.
Civan MM, Podolsky RJ (1966). Contraction kinetics of striated muscle fibers following quick changes in load. J Physiol. 184:511-534.
Cohen DM, Murphy RA (1979). Cellular thin filament protein contents and force generation in porcine arteries and veins. Circ Res. 45:661-665.
Dillon PF, Murphy RA (1982). Tonic force maintenance with reduced shortening velocity in arterial smooth muscle. Am J Physiol. 242:C102-C108.
Edman KAP (1979). The velocity of unloaded shortening and its relation to sarcomere length and isometric force in vertebrate muscle fibers. J Physiol. 291:143-159.
Fay FS, Delise CM (1973). Contraction of isolated smooth muscle cells-structural changes. PNAS 70:641-645.
Fay FS, Hoffman R, Leclair S, Merriam P (1982).
Preparation of individual smooth muscle cells from the

stomach of Bufo marinus. Meth Enzymol. 85:284-292.
Fay FS, Rees DD, Warshaw DM (1981). The contractile mechanism in smooth muscle. In Bitar EE (ed): "Membrane Structure and Function," New York: John Wiley & Sons, pp 79-130.
Fisher BA, Bagby RM (1977). Reorientation of myofilaments during contraction of a vertebrate smooth muscle. Am J Physiol. Cell 1:C5-C14.
Gunst SJ (1986). Effect of Length History on Contractile Behavior of Canine Tracheal Smooth Muscle. Am J Physiol. 250:C146-C154.
Hellstrand P, Johansson B (1979). Analysis of the length response to a force step in smooth muscle from rabbit urinary bladder. Acta Physiol Scand. 106:231-238.
Hill AV (1938). The heat of shortening and the dynamic constants of muscle. Proc Roy Soc Lond. Ser B: Biol Sci. 126:136-195.
Huxley AF (1957). Muscle structure and theories of contraction. Prog Biophys Biophysic Chem. 7:255-318.
Julian FJ (1971). The effect of calcium on the force: velocity relation of briefly glycerinated frog muscle fibers. J Physiol. 218:117-145.
Lannergren J (1978). The froce-velocity relation of isolated twitch and slow muscle fibers of Xenpous laevis. J Physiol. 283:501-521.
Meiss RA, Sonnenblick EH (1974). Dynamic elasticity of cardiac muscle as measured by controlled length changes. Am J Physiol. 226:1370-1381.
Mulvany MJ (1979). The undamped and damped series elastic components of a vascular smooth muscle. Biophys J. 26:401-414.
Murphy RA (1980). The mechanics of vascular smooth muscle. In Bohr DF, Somlyo AP, Sparks HV, (eds): "Handbook of Physiology: Section 2 The Cardiovascular System," Bethesda: Am Physiol Soc, pp 325-351.
Nagai PK, Walsh MP (1985). Detection of caldesmon in muscle and non-muscle tissues of the chicken using polyclonal antibodies. Biochem Biophys Res Comm. 127:533-539.
Pfitzer G, Peterson JW, Ruegg JC (1982). Length dependence of calcium activated isometric force and immediate stiffness in living and glycerol extracted vascular smooth muscle. Pflug Arch. 361:174-181.
Small JV (1977). Studies on isolated smooth muscle cells. The contractile apparatus. J Cell Sci 24:327-349.

Warshaw DM, Fay FS (1983). Cross-bridge elasticity in single smooth muscle cells. J Gen Physiol. 82:157-199.

Warshaw DM, Szarek JL, Hubbard MS, Evans JN (1986). Pharmacology and force development of single freshly isolated bovine carotid artery smooth muscle cells. Circ Res. 58:399-406

Woeledge RC (1968). The energetics of tortoise muscle. J Physiol. 197:685-707.

Yagi S, Fay FS (1985). Measurement of maximum unloaded shortening velocity in single isolated smooth muscle cells. Biophys J. 47: 296a.

THE EFFECTS OF CALCIUM ON SMOOTH MUSCLE MECHANICS AND ENERGETICS

Richard J. Paul, John D. Strauss and Joseph M. Krisanda

Department of Physiology and Biophysics
College of Medicine
University of Cincinnati
Cincinnati, OH 45267-0576

INTRODUCTION

Our understanding of the mechanisms by which calcium controls smooth muscle contractility has undergone rapid evolution. In just over 10 years, the consensus has shifted from a troponin-tropomyosin system, dismissed for the lack of troponin in smooth muscle (Driska and Hartshorne, 1975), to the recent, highly volatile state in which both thick and thin filament linked regulatory systems are candidates. Recent work has implicated effector systems including phosphorylation of the 20 kD myosin light chains, (Sobieszek and Small, 1977; DeLanerolle et al., 1983) the binding of calcium to the thick filament, and thin-filament systems involving leiotonin (Nonomura and Ebashi, 1980) and caldesmon (Ngai and Walsh, 1984; Moody et al., 1986). Several current studies suggest that a calmodulin-dependent mechanism independent of myosin phosphorylation (Pfitzer et al., 1985; Rüegg, this volume) and a mechanism involving protein kinase C (Foster and Chatterjee, 1986; this volume) may also be involved. Which mechanism, if any, predominates, and the relations between these various hypotheses to say the least, are poorly understood.

On the other hand, as the literature has rapidly expanded, studies on the effects of calcium per se on the mechanical properties and energetics of smooth muscle appear to be more consistent than in the past and a general consensus for both "chemically skinned" and intact smooth muscle preparations seems to be arising. We would like to briefly review the evidence for this consensus, trying to highlight the areas of both agreement and disagreement, with the goal of consolidating the experimental bases to which the various theories of calcium regulation must conform.

MECHANICS

a. Isometric Force

Probably the most oft shared observation is that following various Ca^{2+}-depletion protocols, the addition of Ca^{2+} to the bathing medium elicits an increase of force in intact, K^+-depolarized smooth muscle. Fig. 1 is an example of such an experiment. Treatment of intact preparations with detergents, glycerol or other agents is sometimes referred to as "chemical skinning" and operationally the effects are defined by a substantial increase in cell permeability, lack of response to their normal pharmacological agonists and caffeine. Importantly, for these skinned preparations the dependence of isometric force on Ca^{2+} is about three orders of magnitude lower than that of the depolarized, intact preparation (Meisheri et al., 1985). An example of this is shown in Fig. 2. The ED_{50} for calcium of these "skinned" preparations is on the order of $10^{-6}M$ and, by analogy to similar results for the actomyosin ATPase, is thought to represent the calcium sensitivity of the contractile apparatus. There is considerable variation in the reported calcium sensitivity of the skinned preparations. A major factor in this reported variability is that the apparently more permeable preparations, generally those exposed to long term incubations in glycerol containing solutions, lose calmodulin. Addition of exogenous calmodulin to the media bathing these skinned fibers can markedly shift their calcium sensitivity (Sparrow et al., 1981; Rüegg and Paul, 1982). Moreover the addition of other exogenous factors such as cAMP, the catalytic subunit of cAMP-dependent protein kinase (Cassidy et al., 1981; Rüegg and Paul, 1982; Meisheri and Rüegg, 1983) and myosin light chain phosphatase (Bialojan et al., 1985) have also been reported to shift the calcium sensitivity of skinned fibers. Though there is general agreement in terms of the calcium sensitivity of isometric force, there are yet areas of controversy. In particular, Chatterjee and Murphy (1983) reported the sensitivity of isometric force to calcium was greater when the calcium concentration was decreased from the optimal value than when it was increased to the optimum. Such hysteresis of isometric force to calcium has not been observed in other preparations (Tanner et al., 1986; Strauss and Paul, unpublished observations; Fig. 3). This may be attributable to differences in muscles, skinning procedures or different kinetics. Resolution of this difference

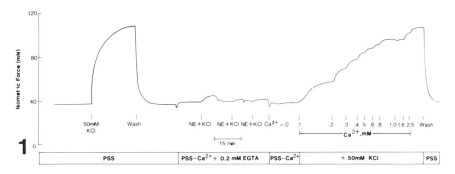

Figure 1. Protocol used to measure calcium sensitivity in response to 50 mM KCl in thoracic aorta from a control rat. Following a test contracture to 50 mM KCl, intracellular stores of calcium were first depleted by repeated exposure to 1 μM norepinephrine and 50 mM KCl in Ca^{++}-free solution containing 0.2 mM EGTA. Calcium was then added back cumulatively in the presence of 50 mM KCl to determine the calcium dose-response relationship. From McMahon and Paul, 1985.

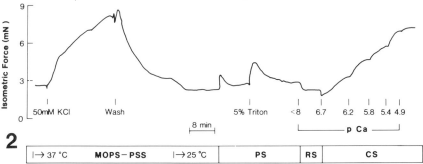

Figure 2. Protocol for measuring calcium sensitivity in Triton X-100 skinned rings of thoracic aorta from a rat. Before skinning, the isometric tension response to the addition of 50 mM KCl was measured in the intact tissue at 37°C. After the bath temperature was lowered to 25°C, the ring was exposed to a 5% Triton solution for 10 minutes, then rinsed briefly in a calcium-free solution. The isometric tension response to increasing concentrations of free Ca^{++} was then recorded to obtain the Ca^{++} dose-response relationship. From McMahon and Paul, 1985.

could play an important role in our understanding of the mechanisms underlying the calcium sensitivity of isometric force.

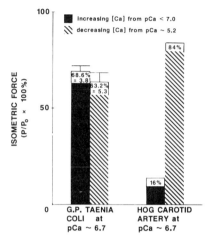

Figure 3. Isometric force (P/P_0) in guinea pig taenia coli and porcine carotid artery at pCa $\simeq$ 6.7. Solid bars indicate force when free calcium in increased to pCa $\simeq$ 6.7 from pCa > 7.0. Hatched bars indicate force when free calcium is decreased from pCa < 5.2 (sufficient for maximum force generation) to pCa $\simeq$ 6.7. (Porcine carotid artery data adapted from Chatterjee and Murphy, 1983. Taenia coli data are unpublished observations.)

b. Shortening Velocity

In contrast to skeletal muscle, shortening velocity in skinned smooth muscle shows a marked dependence on calcium (Arner, 1983; Arner and Hellstrand, 1983; Paul et al., 1983). Fig. 4a shows the plots of the length changes vs. the time required for redevelopment of isometric force, used to determine unloaded shortening velocity (V_{us}) in the "slack test" protocol, as a function of media calcium in skinned guinea pig taenia coli smooth muscle. Fig. 4b shows that the calcium dependence of V_{us} nearly parallels that of isometric force. We have also shown however, that V_{us} can be further increased by the addition of either calcium or calmodulin after force has achieved its maximal value (Fig. 5). The generality of this finding (Arner and Hellstrand, 1985) and its underlying mechanism (Barsotti et al., 1985) remain open questions. In contrast to shortening velocity, neither the magnitude of the series elastic component (Paul et al., 1983; Siegman et al., 1984) nor the apparent stiffness (Arner, 1983; Arner and Hellstrand, 1985) appear to be substantially dependent on calcium.

For the intact preparation, there is less agreement on a calcium dependence of shortening velocity. In earlier works (Hellstrand et al., 1972; Dillon et al., 1981) little effects of calcium were reported. However, in more recent works (Paul et al., 1984; Siegman and Butler, 1984; Aksoy et al., 1983) convincing

Effects of Calcium on Energetics and Mechanics / 323

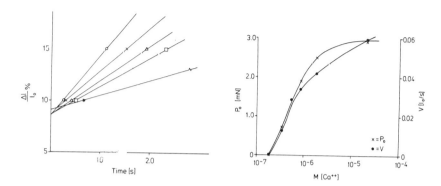

Figure 4. A (Left Panel) *The relation between the imposed step shortening and the time at zero force as a function of Ca⁺⁺ concentration for a guinea pig Taenia coli fiber, 25°C. Isometric force records following steps of 10 and 15% l_i at Ca⁺⁺ concentration of 0.34 (*), 0.53, ( ), 0.8 (Δ), 1.8 (X), and 21.5 (O) µM. From Paul et al., 1983.* **B (Right Panel)** *Relation between isometric force (F_O X's), unloaded shortening velocity (V_{us}, O's), and the Ca⁺⁺ concentration (M). Data from experiment on guinea pig taenia coli fiber are presented in Figure 4a. From Paul et al., 1983.*

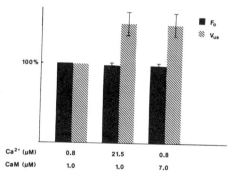

Figure 5. *Dependence of isometric force (F_O) and unloaded shortening velocity (V_{us}) on the concentrations of Ca²⁺ and calmodulin (CaM). Following the attainment of maximum force, velocities can be further increased by increasing either Ca²⁺ or CaM concentrations. Adapted from Paul et al., 1983.*

evidence for a strong dependence on calcium has been presented. Similar to the observations in skinned fibers, these studies on intact smooth muscle have shown that velocity can be increased by about two-fold after isometric force has plateaued by the addition of calcium to non-physiological levels (4-7.5 mM).

In contrast to skinned fibers, in which V_{us} monotonically achieves a maximal value following the increase in calcium (Paul et al., 1983), a most striking feature of V_{us} in intact preparations is its biphasic time course. As first detailed by Murphy and colleagues (Dillon et al., 1981) shortening velocity rapidly attains a maximal value, and then decreases some 2-4 fold although isometric force is increasing or maintained over the same time frame. Examples of this difference are shown in Fig. 6 a & b. The time course of velocity and its calcium dependence have been and will continue to be a major driving force underlying the theories put forth to explain the calcium regulation of smooth muscle contractility. To this end it is of interest to note that the maximum velocities observed in contractions induced by thiophosphorylation (Paul et al., 1983) or the calcium independent myosin light chain kinase (Mrwa et al., 1985; Walsh et al., 1983) in skinned fibers in the absence of calcium are similar to those found in contractions elicited in the presence of optimal calcium concentrations.

ENERGETICS

a. Isometric Tension Cost

One of the outstanding characteristics of smooth muscle is its ability to maintain isometric force with little expenditure of energy (Rüegg, 1971). The tension cost, i.e., the ATP utilization per unit isometric force maintained, may be as much as 1000-fold lower than skeletal muscle (Paul, 1980; 1986). The underlying mechanism(s) has recently received much attention in view of the apparent ability of smooth muscle to slow down its cross bridge cycle rate while maintaining high levels of isometric force as inferred from the decrease in V_{us} with time after stimulation.

In hog carotid artery, the high energy phosphagen content shows little change following stimulation (Krisanda and Paul, 1983). ATP utilization is thus closely matched with ATP synthesis. This allows non-tissue destructive techniques, such as polarographic measurement of tissue O_2 consumption, to continuously monitor ATP utilization. We have shown that isometric force is a major determinant of the steady state suprabasal J_{O_2} in stimula-

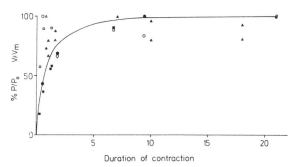

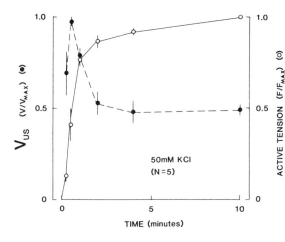

Figure 6. A (**Top Panel**) Dependence of active tension (T/F_{max}) (filled symbols, solid line) and unloaded shortening velocity (V_{us}) (V/V_{max}) (open symbols) on the duration of contraction. Each symbol refers to measurements on a single "chemically skinned" taenia coli fiber, 21°C, 1 μM calmodulin. Both parameters are normalized to the maximal value measured for each fiber and expressed as percent. Solid line represents the average time course of isometric force for three fibers at 12.5 μM Ca^{2+}. From Paul et al., 1983. B (**Bottom Panel**) Plot of unloaded shortening velocity (V_{us}) (V/V_{max}) and active tension (F/F_{max}) vs. time for porcine carotid artery stimulated with 50 mM KCl. V_{us} time course was estimated from a single-step decrease in length ($< 0.3\ L_i$). V_{us} data at 10 min were not significantly different from data collected 20 min after stimulation. Data are means ± SE. From Krisanda and Paul, 1984.

ted vascular smooth muscle with about 80% being associated with the in situ actomyosin ATPase (Paul, 1980; 1986). In recent work we have focussed our attention on the time course and Ca^{2+}-dependence of J_{ATP} under isometric conditions in order to provide an estimate of the cross bridge cycle rate (and tension cost) which is independent of velocity measurements. As shown in Fig. 7, the time course of J_{ATP} also shows a biphasic change after KCl depolarization, decreasing some 2-3 fold from its peak value while isometric force remains constant. We have also reported similar results for rat portal vein (Hellstrand and Paul, 1983) and similar trends can be seen in the work of Stephens and Skoog (1974) who measured oxygen consumption in canine tracheal smooth muscle. A similar decrease in energy utilization with time following stimulation has been reported by Siegman et al. (1980) who measured the decrease in high energy phosphagen content in metabolically inhibited rabbit taenia coli. The time courses of V_{us} and J_{O_2} in hog carotid artery were similar, though not superimposible, and it is likely that both reflect a slowing of the cross bridge cycle rate while isometric force remains constant.

In a recent series of experiments, we investigated the calcium dependence of J_{ATP} under isometric conditions (Paul et al., 1984). In a manner similar to that for V_{us}, J_{ATP} increased with bath Ca^{2+} in the intact hog carotid. In the range of calcium from .15 to 1.6 mM both isometric force (F_0) and J_{ATP} increased proportionately. In this range J_{O_2} and F_0 were linearly related, a relationship that has been nearly universally observed in smooth muscle preparations (Paul and Hellstrand, 1984). However, when the extracellular calcium concentration was increased to 7.5 mM, there was little increase in force with nearly a doubling in J_{ATP}.

The results for V_{us}, J_{ATP} and F_0 in these studies are summarized in Fig. 8. Similar results can be seen in the work of Arner and Hellstrand (1983) for J_{O_2} in portal vein, although the divergence in the dependencies of F_0 and J_{ATP} at high calcium is most clearly seen in the skinned portal vein preparation. Siegman et al. (1984) have also reported that changes in the high energy phosphagen levels in metabolically inhibited taenia coli are calcium dependent with isometric force showing saturation before the chemical energy utilization rate. Thus there is reasonable agreement that F_0, which is often taken as an index of cross bridge number, and V_{us} and J_{ATP}, taken as indices of cross bridge cycling rate, are strong functions of calcium and that V_{us} and J_{ATP} can be varied independently of force. The time-dependent

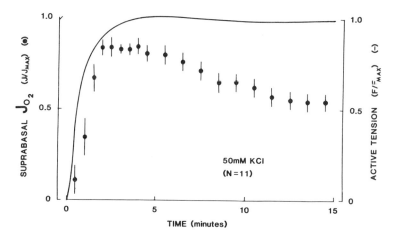

Figure 7. A plot of suprabasal ATP consumption (J_{ATP}) (J/J_{max}) and isometric tension (F/F_{max}) vs. time for tissues stimulated with 50 mM KCl. Suprabasal J_{ATP}, unlike isometric tension, attains a peak rate during early phases of contraction and subsequently declines to a stable steady-state rate. Bars are means ± 1 SE. Bars for tension were not included for figure clarity. Tension data did not vary by more than 5% of mean for KCl stimulation. From Krisanda and Paul, 1984.

decrease in V_{us} and tension cost following stimulation may thus be directly related to the change in intracellular calcium in intact preparations, which has been shown to have a similar biphasic time course (Morgan and Morgan, 1982). While these changes are likely related to those of Ca^{2+}, they may not all necessarily reflect the same universal mechanism. To this end it is of interest to note that changes in the activity of phosphorylase a, a key calcium-dependent enzyme in the glycogenolytic cascade shows a similar time course (Galvas et al., 1985).

The time dependent changes in ATP utilization following stimulation during periods of maintained isometric force appear to account for changes in tension cost in the range of 2-4 fold. While not inconsequential in physiological terms, a modulation of the tension cost of this magnitude is not sufficient to be the sole mechanism underlying the difference between smooth and skeletal muscle (Krisanda and Paul, 1984).

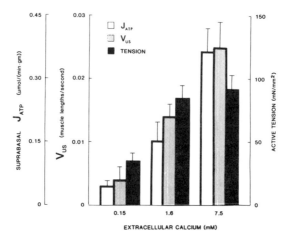

Figure 8. *Plot of the suprabasal rate of ATP utilization (J_{ATP}) (n = 6), the unloaded shortening velocity (V_{us}) (0.15 mM Ca^{2+}, n = 4; 1.6 and 7.5 mM Ca^{2+}, n = 15), and the active isometric tension as a function of the extracellular calcium concentration for porcine carotid artery stimulated with 109 mM KCl substituted for NaCl at 37°C. Data represent stimulations of long duration (20 min), for which the measured parameters are in a steady state. The J_{ATP} was calculated from measurements of tissue oxygen (J_{O_2}), using a stoichiometric ratio of 6.42 mol ATP/mol O_2. Data are presented as means ± SEM. From Paul et al., 1984.*

b. Efficiency

The efficiency of muscle during work producing contractions, i.e., the work produced per unit free energy change, is an energetic parameter closely tied to our understanding of the mechanism of mechanochemical transduction. Although the tension cost in smooth muscle is substantially less than that of skeletal muscle, this does not appear to be the case in terms of efficiency. As estimates of the free energy of hydrolysis of ATP, which range from 35 to 50 kJ/mol, depend on several assumptions, comparisons of efficiency are often given in terms of work per unit high energy phosphagen breakdown. For skeletal muscle this number is on the order of 20 kJ/mol. Butler and Siegman (1984) reported that for working contractions in rabbit taenia coli the efficiency of working contractions was some 5-fold less than that of skeletal muscle. We (Krisanda and Paul, 1984) reported similar findings for hog carotid artery. It should be noted that in both reports a Fenn effect was found, that is, the chemical energy breakdown in work producing contractions was greater than that observed under isometric conditions. In a recent study, we (Krisanda and Paul, 1986) investigated the effects of calcium on the efficiency of hog carotid artery in contractions in which work was performed after the attainment of a steady state. At low calcium concentration (0.15 mM), V_{us} and F_o were lower than that observed at 2.5 mM as previously reported. No Fenn effect was found at the low calcium concentration. However, and perhaps more important, little change in the efficiency was seen in comparison to the contraction at 2.5 mM calcium or when compared to contractions elicited by both KCl and histamine, conditions designed to maximize the intracellular calcium levels. As the efficiency would be anticipated to be decreased by the presence of an "internal load", these results suggest that the decrease in velocity observed at low extracellular concentrations is not likely to be due to the presence of such an internal load. Butler and Siegman (this Symposium) also presented data which indicated that under conditions designed to optimize the effect of an internal load, no change in efficiency was detected in rabbit taenia coli. Thus it is likely that the low efficiency of smooth muscle and the dependence of V_{us} on calcium are not due to the presence of an internal load.

SUMMARY AND CONCLUSIONS

In smooth muscle isometric force, shortening velocity, the ATP utilization under isometric conditions, and tension cost are all functions of calcium. This suggests that both cross bridge number and cycle rate are dependent on calcium. On the other hand, the series elastic component does not exhibit a significant dependence on calcium under conditions in which V_{us} can be increased with little change in F_o. And although the data is not as extensive, the efficiency in working producing contractions is not a strong function of calcium. Our challenge is to fit these mechanical and energetic data into a mechanism which fits the ever growing and controversial body of evidence on the biochemical basis for the action of calcium on the contractile apparatus. At present the most striking feature is the observation that calcium can control the velocity in a manner which is independent of isometric force and efficiency. This would appear to favor a mechanism in which calcium directly affected cross bridge cycle rate in the absence of an internal load.

ACKNOWLEDGEMENTS:

Supported in part by NIH 23240 and an Established Investigatorship of the American Heart Association (RJP).

REFERENCES

Aksoy, M.O., Mras, S., Kamm, K.E. and Murphy, R.A. (1983) Ca^{2+}, cAMP, and changes in myosin phosphorylation during contraction of smooth muscle. Am. J. Physiol. 245 (Cell Physiol. 14):C255-C270.

Arner, A. (1983) Force-velocity relation in chemically skinned rat portal vein: Effects of Ca^{2+} and Mg^{2+}. Pflugers Arch. 397:6-12.

Arner, A. and Hellstrand, P. (1983) Activation of contraction and ATPase activity in intact and chemically skinned smooth muscle of rat portal vein. Circ. Res. 53:695-702.

Arner, A. and Hellstrand, P. (1985) Effects of calcium and substrate on force velocity-relation and energy turnover in skinned smooth muscle of the guinea pig. J. Physiol. 360:347-365.

Barsotti, R., Ikebe, M., and Hartshorne, D. (1985) Effects of Ca^{2+} and M^{2+} on myosin and shortening velocity of gizzard smooth muscle. Biophysical J. 47:299a.

Bialojan, C., Rüegg, J.C., and Di Salvo, J. (1985) Phosphatase mediated modulation of actin-myosin interaction in bovine aortic actomyosin and skinned porcine carotid artery. Proc. Soc. Exp. Biol. Med. 178:36-45.

Butler, T.M. and Siegman, M.J. (1983) Chemical energy usage and myosin light chain phosphorylation in mammalian smooth muscle. Federation Proc. 42:57-61.

Butler, T.M., Siegman, M.J., and Mooers, S.U. (1984) Chemical energy usage during stimulation and stretch of mammalian smooth muscle. Pflugers Archiv. 401:391-395.

Cassidy, P.S., Kerrick, W.G.L., Hoar, P.E. and Malencik, D.A. (1981) Exogenous calmodulin increases Ca^{2+} sensitivity of isometric tension activation and myosin phosphorylation in skinned smooth muscle. Pflugers Arch. 392:115-120.

Chatterjee, M. and Murphy, R.A. (1983) Calcium-dependent stress maintenance without myosin phosphorylation in skinned smooth muscle. Science (Washington, D.C.) 221:464-466.

DeLanerolle, P., Condit, J.R., Tannenbaum, M. and Adelstein, R.S. (1983) Myosin phosphorylation, agonist concentration and contraction of tracheal smooth muscle. Nature (London) 298:871-874.

Dillon, P.F., Aksoy, M.O., Driska, S.P., and Murphy, R.A. (1981) Myosin phosphorylation and the cross-bridge cycle in arterial muscle. Science 211:495-497.

Driska, S. and Hartshorne, D.J. (1975) The contractile proteins of smooth muscle. Properties and components of a Ca^{2+}-sensitive actomyosin from chicken gizzard. Arch. Biochem. Biophys. 167:203-212.

Foster, C.J. and Chatterjee, M. (1986) A 25,000 dalton protein is phosphorylated in response to phorbol ester stimulation in skinned carotid artery. Biophys. J. 49:73a.

Galvas, P.E., Kuettner, C., Paul, R.J., and Di Salvo, J. (1985) Temporal relationships among isometric force, phosphorylase and protein kinase activities in vascular smooth muscle. Proc. Soc. Expt. Biol. Med. 178:254-260.

Hellstrand, P., Johansson, B. and Ringberg, A. (1972) Influence of extracellular calcium on isometric force and velocity of shortening in depolarized venous smooth muscle. Acta Physiol. Scand. 84:528-537.

Hellstrand, P. and Paul, R.J. (1983) Phosphagen content, breakdown during contraction and oxygen consumption in rat portal vein. Am. J. Physiol. 244:C250-C258.

Krisanda, J.M. and Paul, R.J. (1983) High energy phosphate and metabolite content during isometric contraction in porcine carotid artery. Am. J. Physiol. 244:C385-C390.

Krisanda, J.M. and Paul, R.J. (1984) Energetics of isometric contraction in porcine carotid artery. Am. J. Physiol. 246:C510-C519.

Krisanda, J.M. and Paul, R.J. (1986) The effect of changes in active work production on the efficiency of porcine carotid artery. Fed. Proc. 45:766.

McMahon, E. Garwitz and Paul, R.J. (1985) Calcium sensitivity of isometric force in intact and chemically skinned aortas during the development of aldosterone-salt hypertension in the rat. Circ. Res. 56 No. 3:427-435.

Meisheri, K.D. and Rüegg, J.C. (1983) Dependence of cyclic AMP induced relaxation on Ca^{2+} and calmodulin in skinned smooth muscle of guinea pig taenia coli. Pflugers Archiv. 399:315-320.

Meisheri, K.D., Rüegg, J.C. and Paul, R.J. (1985) Smooth muscle contractility: Studies on skinned fiber preparations. In: Daniels, E.E. and Grover, A.K. Calcium and Smooth Muscle Contractility. Human Press, New Jersey, 191-224.

Moody, C.J., Lehman, W. and Marston, S. (1986) The structural basis for caldesmon regulation of vascular smooth muscle contraction. J. Muscle Res. Cell. Motil. (in press).

Morgan, J.P. and Morgan, K.G. (1982) Vascular smooth muscle: the first recorded Ca^{2+} transients. Pflugers Arch. 395:75-77.

Mrwa, U., Guth, K., Rüegg, J.C., Paul, R.J., Barsotti, R. and Hartshorne, D. (1985) Mechanical and biochemical characterization of the contraction elicited by a calcium-independent myosin light chain kinase in chemically skinned smooth muscle. Experientia 41:1002-1006.

Ngai, P.K. and Walsh, M.P. (1984) Inhibition of smooth muscle actin-activated myosin Mg^{2+}-ATPase activity by caldesmon. J. Biol. Chem. 259:13656-13659.

Nonomura, Y. and Ebashi, S. (1980) Calcium regulatory mechanism in vertebrate smooth muscle. Biomed. Res. 1:1-14.

Paul, R.J. (1980) The chemical energetics of vascular smooth muscle. Intermediary metabolism and its relation to contractility. In: Handbook of Physiology, Section on Circulation II, edited by D.F. Bohr, A.P. Somlyo, and H.V. Sparks, American Physiological Society, Bethesda, MD, pp. 201-236.

Paul, R.J. (1986) Smooth muscle mechanochemical energy conversion: Relations between metabolism and contractility. In "Physiology of the Gastrointestinal Tract, Second Edition", L.R. Johnson, editor, Raven Press, New York, pp. 1401-1424.

Paul, R.J., Doerman, G., Zeugner, C. and Rüegg, J.C. (1983) The dependence of unloaded shortening velocity on Ca^{2+}, calmodulin and the duration of

contraction in "chemically-skinned" smooth muscle. Circ. Res. 53(3):342-351.
Paul, R.J. and Hellstrand, P. (1984) Phosphorylase a activation and contractile activity in rat portal vein. Acta Physiol. Scand. 121:23-30.
Paul, R.J., Krisanda, J.M. and Lynch, R.M. (1984) Vascular smooth muscle energetics. J. Cardiovasc. Pharm. (Suppl. 2) 5320-5327.
Pfitzer, G., Wagner, J. and Rüegg, J.C. (1985) A change in smooth muscle contraction is not always associated with a change in light chain phosphorylation. Fed. Proceed. 44 (No. 3):457.
Rüegg, J.C. (1971) Smooth muscle tone. Physiol. Rev. 51:201-248.
Rüegg, J.C. and Paul, R.J. (1982) Vascular smooth muscle, calmodulin and cyclic AMP dependent protein kinase alter calcium sensitivity in porcine carotid skinned fibers. Circ. Res. 50:394-399.
Siegman, M.J., Butler, T.M., Mooers, S.U., and Davies, R.E. (1980) Chemical energetics of force development, force maintenance and relaxation in mammalian smooth muscle. J. Gen. Physiol. 76:609-629.
Siegman, M.J., Butler, T.M., Mooers, S.U., and Michalek, A. (1984) Ca^{2+} can affect V_{max} without changes in myosin light chain phosphorylation in smooth muscle. Pflugers Arch. 401:385-390.
Sobieszek, A. and Small, J.V. (1977) Regulation of the actin-myosin interaction in vertebrate smooth muscle, activation via a myosin light chain kinase and the effect of tropomyosin. J. Mol. Biol. 112:559-576.
Sparrow, M.P., Mrwa, U., Hofmann, F. and Rüegg, J.C. (1981) Calmodulin is essential for smooth muscle contraction. FEBS Lett. 125:141-145.
Stephens, N.L., and Skoog, C.M. (1974) Tracheal smooth muscle and rate of oxygen uptake. Am. J. Physiol. 226:1462-1467.
Tanner, J.A., Haeberle, J.R. and Meiss, R.A. (1986) The relationship between isometric force and stiffness and the stoichiometry of myosin phosphorylation during steady-state contraction of skinned smooth muscle. Biophys. J. 49:72a.
Walsh, M.P., Bridenbaugh, R., Hartshorne, D.J. and Kerrick, W.G.L. (1983) Gizzard Ca^{2+}-independent myosin light-chain kinase; evidence in favor of the phosphorylation theory. Fed. Proc. 42:45-50.

STIFFNESS AND THE ENERGETICS OF ACTIVE SHORTENING IN CHEMICALLY SKINNED SMOOTH MUSCLE.

Per Hellstrand, Håkan Arheden, Lars Sjölin and Anders Arner

Department of Physiology and Biophysics
University of Lund, S-223 62 Lund, Sweden

INTRODUCTION

An important consideration in experiments designed to reveal cross-bridge properties in muscle is how the macroscopic mechanical properties of the intact muscle are related to the fundamental dimensions of the contractile system. In skeletal muscle, with its well-defined sarcomeric structure, cross-bridge compliance may be referred to the length of a half-sarcomere (about 1.1 µm at optimal overlap in frog skeletal muscle), which is a repeating unit involving only one cross-bridge connection (although there are several cross-bridges acting in parallel). Ford et al. (1977) found that for very rapid length steps (0.2 ms) the force-extension relation is approximately linear, and that by extrapolation the force of the muscle would be completely unloaded in a shortening step of 4 nm/half-sarcomere. In subsequent experiments, Stienen and Blangé (1985) have arrived at a similar estimate of cross-bridge compliance in skinned frog muscle.

In smooth muscle, stiffness measurements have generally been performed with a lower time resolution than that of the skeletal muscle experiments, and the reported values indicate a larger compliance. Pfitzer et al. (1982), using 1.5 ms length steps on chemically skinned swine carotid artery reported basically similar force-extension characteristics as those described in frog muscle, but with a clearly larger compliance. Force was found to be unloaded for a step of about 1.2 % of the muscle length, which by this basis of measurement is about three times the value found in skeletal muscle. In isolated single cells of bullfrog stomach, Warshaw & Fay (1983) arrived at value of about 1.5%. Their conclusion was that the cross-bridges in the smooth muscle are more compliant than those of skeletal muscle, since it is not likely that the equivalent "half-sarcomere" of smooth muscle would be smaller. In fact, the thick filaments in rabbit portal-anterior mesenteric vein have been shown to be about 2.2 µm long (Ashton et al. 1975), which is longer than the 1.6 µm of skeletal muscle thick filaments. This suggest that the contractile units in the smooth muscle are rather longer than the half-sarcomere of skeletal muscle, which would require that based on muscle length the compliance of the smooth muscle should be smaller if cross-bridges have similar properties and represent the only compliance measured. However, the structural arrangement of the contractile units in smooth muscle is not known with certainty and it is possible that a quite different system will be found, involving e.g. a side-polar arrangement of myosin filaments rather than the bipolar filaments of skeletal muscle (Somlyo et al. 1984).

STIFFNESS MEASUREMENTS IN CHEMICALLY SKINNED GUINEA PIG TAENIA COLI.

For a valid comparison of mechanical properties in the different kinds of muscle it is essential that experiments are carried out with a similar degree of resolution. We have therefore investigated the force response of chemically skinned guinea pig taenia coli smooth muscle to length steps complete in 0.2-0.6 ms. The spatial homogeneity of the length step was investigated by observing the movement of markers spaced out along the fibre.

Taenia coli was chemically skinned as described by Arner & Hellstrand (1985). Fiber bundles with a length of 2.5-6 mm and a thickness of 0.15-0.30 mm were dissected out and mounted in a solution containing 3.2 mM MgATP and an ATP regenerating system of 12 mM phosphocreatine and 15 U/ml creatine kinase. The free-Mg concentration was 2 mM and ionic strength 0.15 M. Relaxing solution contained 10^{-9} M free-Ca (pCa 9) and activating solution $10^{-4.5}$ M (pCa 4.5). The ends of the fiber bundles were wrapped in strips of aluminum foil, which were flattened and glued to the arms of a force transducer and a servo motor, respectively. The force transducer was an AME 801 (A/S Mikroelektronikk, Horten, Norway), which had been trimmed down to increase the natural frequency in air to 30-40 kHz. The servo motor was built in the department and was based on a coil mounted on an axially movable shaft. The static magnetic field was produced by an electromagnet A photoelectric displacement transducer recorded the position of a vane mounted on the shaft.

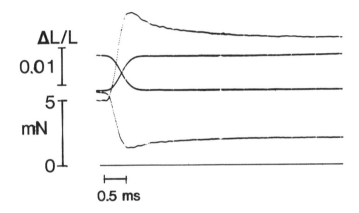

Figure 1. Force response (upper and lower record) to stretch and release (middle records, stretch upwards) of chemically skinned taenia coli. Zero force baseline indicated by straight line.

Fig 1 shows the response of an activated skinned taenia preparation to a stretch and a release of identical magnitude (0.90 % of the muscle length). The length step in this experiment is complete in about 0.6 ms. The response qualitatively resembles that described in skeletal muscle fibres (Huxley & Simmons 1971) and in single smooth muscle cells (Warshaw & Fay 1983). Concomitant with the force step is a momentary change in force ("phase 1") which is followed by a partial recovery ("phase 2") lasting a few milliseconds. The force response then flattens out to an almost constant plateau level ("phase 3") which is followed by a slow recovery to the original force ("phase 4"), which is not seen in the fast time scale of Fig 1. In skeletal muscle phases 3 and 4 are well separated and may in fact show an inflexion of the force record, whereas a monotonous response is found in smooth muscle (Hellstrand & Johansson 1979, Warshaw & Fay 1983). Hence phases 3 and 4 are not clearly distinguishable. For the rather large stretch illustrated in Fig 1 force (F) increases to 2.3 times the isometric force before stretch (F_o), whereas for the release the minimal value of F/F_o is 0.28. Hence the drop in force in the release is not as large as the increase in the stretch, showing that the force response is non-linear. An effect which might influence the response is recovery processes occurring during the step itself, leading to a truncation of the response (Huxley & Simmons 1971). To minimise such effects one needs to use as fast steps as possible.

HOMOGENEITY OF THE LENGTH STEP ALONG THE PREPARATION

A further consideration in interpreting length step experiments is the homogeneity of the response along the preparation. Differences in the relative length change of segments of the fibre bundle would influence the result.

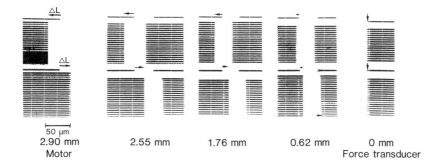

Figure 2. Optical detection of markers placed at indicated positions along fiber bundle. Horizontal arrows show extent of stretch (upper) or release (lower) at each marker position. Separation between sweeps 1 ms (downwards).

To investigate the homogeneity of the length step, in the skinned taenia preparation, an image of the mounted fiber bundle was projected onto a photodiode array allowing a spatial resolution of 1024 points to be obtained with repetitive readings at 1 ms intervals. Transversely through the fiber bundles were inserted short pieces of 35μm platinum wire spaced out along the length of the preparation. The wire pieces served as markers for the optical detection system, which was focused on one marker at a time.

Fig 2 shows data from an experiment where stretches and releases of equal magnitude were applied starting from a constant length of the fiber bundle. The optical signal is displayed as horizontal oscilloscope traces interrupted at points where the light intensity drops below a certain cut-off level, i.e. where the marker obscures the illumination of the photodiode array. The speed of the oscilloscope trace, together with the optical magnification, determines the length calibration. Each new trace is displaced downwards and represents a reading made 1 ms after the preceding trace. The uppermost trace in each half panel was obtained before the release and is separated by a larger (and variable, since it was manually set) vertical distance from the next lower trace, which was triggered 1 ms after the release or stretch itself. The upper half of each panel shows a stretch and the lower half a release. The direction of movement is shown by the horizontal arrows. The markers were placed along the fiber bundle at the positions indicated below each panel. Signals were also obtained from the servo motor and force transducer (vertical arrow). It is seen that the displacement is complete within the first millisecond, and that essentially no movement occurs thereafter. The extent of movement becomes progressively smaller as the force transducer is approached.

Fig 3 shows a plot of the marker displacements against their positions along the fibre bundle. Both for stretches and releases all markers are proportionately displaced, showing that the movement, at this level of resolution, is uniformly distributed along the preparation. The presently used method does not indicate whether the length change is located entirely within the cross-bridge system. Mulvany and Warshaw (1981) have

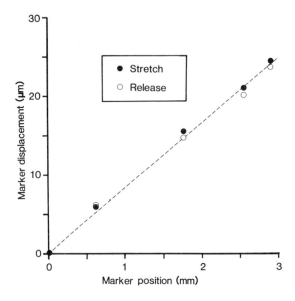

Figure 3. Displacement of markers in experiment shown in Fig 2.

reported evidence for a considerable extracellular compliance in small mesenteric resistance vessels. Their experiments were however done in a slower time scale than that used here and their value for tissue compliance is about 3 times that found in the present study.

THE FORCE-EXTENSION RELATION OF THE ELASTIC COMPONENT

As pointed out above there is need for a high time resolution in stiffness experiments. We therefore tried to optimize conditions for obtaining quick length steps, although at some sacrifice of stability of the system. Fig 4 shows superimposed responses to several length steps of 0.2 ms duration. The force record is disturbed by oscillations with a frequency of about 20 kHz. Note also that the length is not entirely stable after the step. However, the main features of the response are similar to those seen in slower releases.

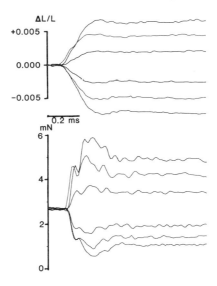

Figure 4. Superimposed stretches and releases on activated taenia coli. Plot of recordings digitized at a sampling frequency of 100 KHz.

To characterize the immediate elastic response the values of F/F_o obtained in the individual stretches and releases were plotted against ΔL. This as shown in Fig 5. The relationship characterizes "phase 1" of the length step response and is commonly referred to as the "T_1-curve" according to the nomenclature used by Huxley & Simmons (1971). Although the relationship is quite curved for large releases ($\Delta L/L<0$) it appears straight for stretches and small releases. As has been found in other studies on smooth muscle (Meiss 1978, Pfitzer et al. 1982, Warshaw & Fay 1983) stiffness decreased when the isometric force was decreased due to varying stimulation level (data not shown). When compared at similar total tension the muscle is therefore stiffer when the intrinsic isometric force is greater, which speaks against a passive element outside the force-generating apparatus as the main source of compliance.

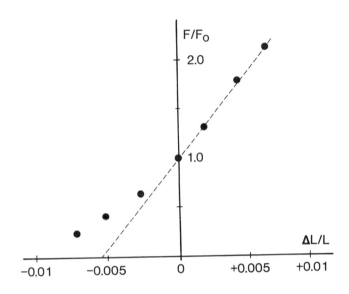

Figure 5. Peak normalized force response (F/F_o) against normalized amount of stretch or release ($\Delta L/L$). Data from Fig 4.

A comparison of absolute stiffness between different kinds of muscle can be done on the basis of the extrapolated linear portion of the T_1 curce, as shown by the dashed line in Fig 5. Assuming linear elasticity the cross-bridges would thus be unloaded for a shortening step of just over 0.5 % of the muscle length. By a similar procedure Ford et al. (1977) arrived at a figure of 4 nm/half sarcomere i.e., 0.36 % of the muscle length in a frog skeletal muscle. They also found that this figure increased with increasing temperature, at 8°C it was about 6 nm/half sarcomere. While we have not investigated different temperatures it is reasonable to conclude that the values found in the skinned smooth muscle at 22°C approach those expected in skeletal muscle.

If the length of the equivalent sarcomere is of about the same magnitude in smooth as in striated muscle then our data suggest that the compliance of the cross-bridge system is not much different. If, however the dimensions of the contractile units turn out to be markedly different this conclusion has to be modified.

BIOCHEMICAL CORRELATES OF CROSS-BRIDGE TURNOVER

While measurements of force and stiffness provide evidence on the instantaneous number of attached cross-bridges, the rate of turnover of cross-bridges can be obtained from measurements of maximal shortening velocity (V_{max}). This index refers to conditions when the filaments are rapidly sliding past each other. Under isometric conditions, on the other hand, the rate of cross-bridge turnover may be obtained from the energetic cost of force maintenance, which is thought to reflect the number of ATP molecules turning over per cross-bridge per unit of time. It is evident that the turnover rate inferred from V_{max} measurements is considerably higher than that given by the isometric tension cost (Hellstrand & Paul 1982). On the other hand the number of simultaneously attached cross-bridges is presumably decreased during shortening, according to current contraction models (reviewed by Woledge et al 1985), although evidence for this is so far lacking in smooth muscle. Thus the overall rate of ATP utilization during shortening as compared to isometric contraction is subject to opposing influences. It should be realized, however, that the actual amount of mechanical power exerted during shortening in a muscle with slow contraction is small, and the identification of a biochemical counterpart of mechanical work is dependent on the proper choice of baseline (see Rall 1982). In rabbit taenia coli an appreciable increase in ATP turnover during shortening was found late in a sustained contracture but not during the initial phase of contraction (Butler et al. 1983). The time difference was not in the energetic cost of work production but in the isometric J_{ATP}, which decreased considerably in the course of contraction. It was therefore of interest to see whether the mechanical conditions of contraction could influence the overall rate of energy turnover in a skinned taenia coli smooth muscle.

Contracting muscles were allowed to shorten for 3 min after a quick release to an isotonic load ∿0.3 F_o. The total amount of shortening during this period amounted to about 40% of the initial muscle length. ATP breakdown was measured during the shortening period and during isometric periods of 3 min immediately before and after the shortening. In these experiments an ATP regenerating system of 5 mM phosphoenolpyruvate (PEP) and 20 U/ml pyruvate kinase was used and the ATP breakdown measured by assay of the amount of pyruvate released into the incubation medium (400 μl aliquots) as described by Arner & Hellstrand (1985). At pCa 4.5 isometric ATP turnover (J_{ATP}) was 1.26 ± 0.06 (n =6) μmol/min g, (mean of rates before and after shortening), and during shortening it was 97 ± 2 % of this isometric rate. Hence there was no difference in ATP turnover rate during isometric contraction and during shortening. In similar experiments performed at a lower Ca^{2+}-concentration, pCa 6.3, which gave 61 ± 7 % (n =6) of the maximal isometric force, J_{ATP} during isometric contraction was 0.56 ± 0.07 μmol/min×g, and the rate of ATP breakdown during shortening was 110 ± 8 % of the isometric rate. Thus when the contractile apparatus is working at submaximal rate due to decreased activation it still does not increase its rate of ATP turnover during shortening. This result is interesting in view of the Ca^{2+} dependence of V_{max} (Arner 1983, Arner & Hellstrand 1985), which implies that not only the number of activated cross-bridges, but also their intrinsic rate of turnover, is decreased at suboptimal activation.

To improve the time resolution of experiments involving shortening of the skinned taenia an alternative experimental approach was developed. Since diffusion limitation could be expected to be significant the need for equilibration between the muscle strip and the surrounding medium limited the time resolution obtainable by serial incubation. However, by arresting the inflow of fresh substrate the fall of high-energy phosphates with time can be followed by analysis of the whole muscle strip. Such an arrest of inflow was achieved by lifting the muscle up from its incubation medium and freezing it after various amounts of time. Comparison was made to control muscles frozen immediately after being lifted up. The results from periods of 20 s and 40 s are shown in Table 1.

TABLE 1. Change in contents of ATP + phosphoenolpyruvate (PEP) in skinned taenia incubated in air under isometric conditions or shortening at $F/F_o \sim 0.4$.

Time (s)	Isometric (μmol/g)	Isotonic (μmol/g)
20	-0.46 ± 0.36 (11)	-0.68 ± 0.27 (10)
40	-0.89 ± 0.38 (11)	-0.62 ± 0.35 (9)

Analysis was done by isotachophoresis (Hellstrand & Paul 1983). The contents of ATP and PEP of each muscle extract was corrected for that in the incubation medium, using EGTA as internal standard for the amount of medium penetrating into the strip. Number of experiments within parentheses.

The isometric rate of ATP breakdown that can be obtained from these figures is in good agreement with those measured by serial incubation (cf.above). Although the rate of ATP breakdown during isotonic shortening seems to be larger than the isometric at 20 s it is smaller at 40 s. The scatter in the data does however not allow a firm conclusion on this point, although overall the data seem to exclude a major deviation of the ATP turnover during shortening as compared to isometric contraction.

CONCLUSIONS

The present study has shown that the stiffness of an activated skinned smooth muscle preparation approaches that of skeletal muscle fibers, when compared on the basis of muscle length. This suggests that the compliance of attached cross-bridges may not be appreciably different in the two kinds of muscle. No change in rate of ATP turnover was found in muscles undergoing shortening as compared to isometric contraction. Thus a decrease in number of simultaneously attached cross-bridges during shortening is matched by an increase in the specific rate of cross-bridge turnover. Further studies combining mechanical and energetic measurements are expected to give more insight into the regulation of cross-bridge attachment and cycling during shortening.

ACKNOWLEDGEMENTS

The present study was supported by the Swedish Medical Research Council (project 14x-28), the Medical Faculty, University of Lund, and AB Hässle, Mölndal. We thank Monica Lundahl, Monica Heidenholm and Ina Nordström for able technical assistance.

REFERENCES

Arner A (1983). Force-velocity relation in chemically skinned rat portal vein: effects of Ca^{2+} and Mg^{2+}. Pflüg Arch 397:6-12.

Arner A, Hellstrand P (1985). Effects of calcium and substrate on force-velocity relation and energy turnover in skinned smooth muscle of the guinea-pig. J Physiol 360:347-365.

Ashton FT, Somlyo AV, Somlyo AP (1975). The contractile apparatus of vascular smooth muscle: intermediate high voltage stereo electron microscopy. J Mol Biol 98:17-29.

Butler TM, Siegman MJ, Mooers SU (1983). Chemical energy usage during shortening and work production in mammalian smooth muscle. Am J Physiol 244:C234-C242.

Ford LE, Huxley AF, Simmons RM (1977). Tension responses to sudden length change in stimulated frog muscle fibres near slack length. J Physiol 269:441-515.

Hellstrand P, Johansson B (1979). Analysis of the length response to a force step in smooth muscle from rabbit urinary bladder. Acta Physiol Scand 106:221-238.

Hellstrand P, Paul RJ (1982). Vascular smooth muscle: relations between energy metabolism and mechanics. In: Crass, III MF, Barnes CD (eds) Vascular Smooth Muscle: Metabolic, Ionic, and Contractile Mechanisms, Ch.1. Academic Press, New York, pp 1-36.

Hellstrand P, Paul RJ (1983). Phosphagen content, breakdown during contraction, and O_2 consumption in rat portal vein. Am J Physiol 244:C250-C258.

Huxley AF, Simmons RM (1971). Proposed mechanism of force generation in striated muscle. Nature 233:533-538.

Meiss RA (1978). Dynamic stiffness of rabbit mesotubarium smooth muscle: effect of isometric length. Am J Physiol 234:C14-C26.

Mulvany MJ, Warshaw DM (1981). The anatomical location of the series elastic component in rat vascular smooth muscle. J Physiol 314:321-330.

Pfitzer G, Peterson JW, Rüegg JC (1982). Length dependence of calcium activated isometric force and immediate stiffness in living and glycerol extracted vascular smooth muscle. Pflüg Arch 394:174-181.

Rall JA (1982). Sense and nonsense about the Fenn effect. Am J Physiol 242:H1-H6.

Somlyo AV, Bond M, Berner PF, Ashton FT, Holtzer H, Somlyo AP (1984). The contractile apparatus of smooth muscle: an update. In: Stephens NL (ed) Smooth Muscle Contraction. Dekker, New York, pop 1-20.

Stienen GJM, Blange T (1985). Tension responses to rapid length changes in skinned muscle fibres of the frog. Pflüg Arch 405:5-11.

Warshaw DM, Fay FS (1983). Cross-bridge elasticity in single smooth muscle cells. J Gen Physiol 82:157-199.

Woledge RC, Curtin NA, Homsher E (1985). Energetic Aspects of Muscle Contraction. Academic Press, London, pp 1-357.

CROSSBRIDGE PROPERTIES STUDIED DURING FORCED ELONGATION OF ACTIVE SMOOTH MUSCLE

Richard A. Meiss

Departments of Physiology/Biophysics and
OB/GYN, Indiana University School of Medicine
Indianapolis, Indiana   46223

INTRODUCTION

The large dimensional changes which take place when smooth muscle shortens isotonically are most likely the result of the relative motion of the thick and thin myofilaments. The magnitude of such changes would be sufficiently great that the length change must be accomplished by many cycles of each crossbridge, given that the filament-based ultrastructure effectively places many crossbridges in parallel with each other. Even at a constant afterload, however, shortening velocity continuously decreases as factors related to activation and tissue geometry come into play. This can introduce uncertainty into measurements of the time-dependent properties of the crossbridges.

In order to produce conditions which may involve relative myofilament shearing at a constant rate, experiments involving constant-velocity stretch (rather than shortening) have been performed in this laboratory for a number of years. Under such conditions, the force response, although not assuming a constant value, does follow a consistent pattern (Meiss, 1982). Because of the large forced changes in muscle length and the apparent yielding behavior, crossbridges must undergo forced detachment during the stretch; the increase in force during long stretches (over length ranges where parallel connective tissue contributions are small) indicates that re-attachment of crossbridges must also take place.

In many striated muscle studies (cf. Kawai, 1982;

Barden, 1981) very small sinusoidal length perturbations have been used as an index of crossbridge activity. The resulting force perturbations bear a proportional relation to active muscle force and stiffness. In those studies, as in the present one using smooth muscle, the stiffness revealed by using small perturbations has been assumed to reflect the number of attached crossbridges, while the muscle force represents a summation of the forces in all stressed crossbridges. A direct proportionality between these two parameters is usually observed during isometric contraction.

Stiffness measurements using oscillatory length perturbations have been carried out in this study during isometric contraction and during forced length changes produced by constant-velocity stretch of isometrically-contracting muscle. The studies were done in order to gain insight into stiffness-related crossbridge mechanisms, especially those involving re-attachment, which may operate under conditions of constant-velocity myofilament shearing.

Because the length and force oscillations are cyclic and periodic, they may interact with steps in the crossbridge cycle in such a way as to reveal the time-dependence inherent in these processes. Such measurements have been used to determine rate constants for crossbridge activities in a number of skeletal muscles (Barden, 1981; Kawai and Brandt, 1980). While a complete frequency analysis of smooth muscle stiffness behavior has not been attempted in the present study, some indications of frequency-sensitive response to imposed mechanical conditions have been observed and are analyzed within the context of hypothetical crossbridge mechanisms.

METHODS

This study used strips of cat <u>duodenal circular-layer muscle</u>, rabbit <u>mesotubarium superius</u>, and rabbit <u>ovarian ligament</u>. Results from among these muscle types were quite similar. The muscles were mounted horizontally in a chamber through which gas-bubbled (95% oxygen, 5% carbon dioxide) Ringer's solution was circulated. The floor of the chamber was made from a feedback-controlled thermoelectric device (Cambion). A wide range of temperatures could be set and maintained. Except where the temperature was an experimental variable, all procedures were carried out at 25 C. Alternating-current square-wave stimuli of optimal voltage, frequency, and duration were applied between silver electrodes lying on either side of the muscle but not in contact with it. One end of the muscle was attached, via a crimped aluminum cylinder, to a photoelectric force transducer with a resonant frequency of 1000 Hz and a compliance of 2.5 um/gm-wt. The other end was attached to the end of a modified penmotor driving arm with an integral position transducer (Gould 2000 series). A sinewave generator and a linear ramp generator supplied the perturbation signals to the length-control circuitry driving the penmotor. Typical sinusoidal perturbation amplitudes were <0.2% of the muscle length, while the ramp stretches were from 10% to 25% of the muscle length. The force and length signals were fed to a system of tunable analog active filters. Two identical bandpass filters and amplitude demodulators extracted the superimposed sinusoidal information, and a phasemeter circuit (referenced to the length signal) provided a calibrated phase output. Notch filters removed the sinusoidal component from the force and length signals. Frequency and phase response over the range of 10 to 120 Hz, as measured by substituting a metal spring for the muscle, were essentially flat. Data traces representing muscle length (L),

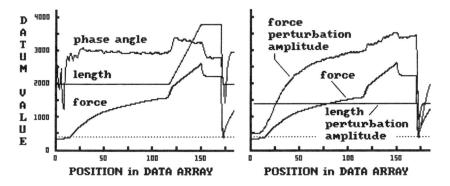

Figure 1. Raw data traces, reproduced from stored data.

length perturbation amplitude (dL), muscle force (F), force perturbation amplitude (dF), and phase angle were displayed on a chart recorder and were also converted into digital form by a microcomputer system using a 12-bit analog-to-digital converter. The conversion rate was 100 or 200 5-trace sample sets per second. Digitized data were stored on flexible diskettes for subsequent analysis. Typical uncalibrated experimental signals are shown in Figure 1. These traces have been reproduced from stored data and are shown in two frames for clarity. The sequence of events (i.e., stimulus, onset of stretch, etc.) was also determined by the microcomputer system. To insure accurate determination of the dF and dL signals, the length perturbation was halted briefly before each contraction, and zero-level reference data were taken.

Figure 1 shows the method of initial data analysis. Using the stored data, the best linear fit between dF/dL and F was determined during the rise of isometric force ('CONTROL REGION'). During the stretch portion ('TEST REGION') the actual stiffness (dF/dL) was determined at each sample point, and the ratio of measured stiffness to the predicted stiffness for that instantaneous force (on the basis of the just-prior isometric determination) was computed. This value was termed the stiffness ratio; values less than 1.0 meant that the muscle was less stiff than would be expected at that force. [Over the control region the stiffness ratio must have an average value of 1.0; see Figure 2, left.]

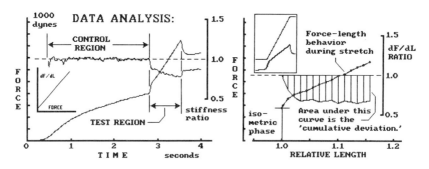

Figure 2. Steps in transforming data for analysis.

The duration of the actual stretch portion depended upon the stretch rate, since the total stretch length was kept constant. To facilitate comparison between contractions in which the stretch rates were different, stretch data were replotted so that the force and the stiffness ratio were functions of the instantaneous muscle length (see Figure 2, right). The difference between the computed stiffness ratio and the value 1.0 was integrated over the stretch length by summing a series of trapezoids determined by the individual data points; negative values were assigned for downward deviations. This procedure produced a single value, the "cumulative deviation", which was an index of the overall change in expected stiffness due to the effects of the stretch.

The stored phase angle data allowed the total (unresolved) stiffness, dF/dL, to be resolved trigonometrically into its in-phase (=elastic) and quadrature (=viscous) components. The above "stiffness ratio" procedure could then be carried out on the resolved stiffness components as well.

RESULTS

Isometric Stiffness

As has been previously reported (e.g., Warshaw and Fay, 1983; Meiss, 1978), the relationship between isometric force and stiffness during the rise of force was linear. This is shown in the present case by the lack of deviation from the 1.0 stiffness ratio over the control region (Fig. 2). The measured isometric stiffness increased somewhat with increasing perturbation frequency; the average increase was 126% (±27% SD) over the range of 30 Hz to 100 Hz (4 determinations on 1 mesotubarium and 3 ovarian ligaments.) As an example, the upper line in Figure 5 shows the unresolved isometric stiffness at peak developed force. The slopes ('k') of the force-stiffness relationships (cf. Meiss, 1978) were not significantly changed with frequency, while the ordinate intercept ('C') increased sufficiently to account for the significant increase of stiffness with increasing perturbation frequency. The components of the resolved stiffness (measured at the maximal developed force level) both showed a similar frequency dependence. The phase angle increased by approximately 1 degree (from 17.5 to 18.5 degrees) over the frequency range.

Effects of Stretch

The principal findings were that a constant-rate stretch produced an increase in force which was linear after an initial step; and, subsequently, that the muscle stiffness was reduced below expected levels as a function of the stretch rate. The change (increase) in phase angle increased during stretch in proportion to the stretch rate. There was a strong interaction between the stretch rate and the perturbation frequency in producing these effects, as outlined below.

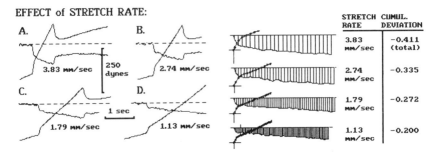

Figure 3. Effects of different stretch rates on stiffness ratio and cumulative deviation.

Figure 3 shows the effects of increasing stretch rate on the cumulative deviation of the stiffness ratio. The perturbation frequency was 80 Hz. Data from another such

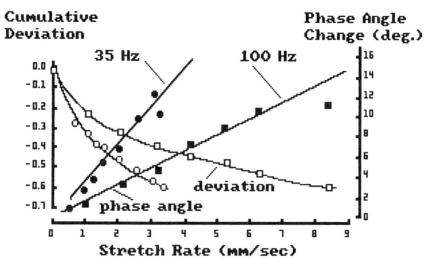

Figure 4. Interaction between stretch rate and perturbation frequency as they affect cumulative deviation and phase angle during stretch.

experiment are shown in Figure 4. Here two perturbation frequencies used were (30 Hz - round symbols; 100 Hz - square symbols). At a given perturbation frequency, increasing the stretch rate produced a linear increase in the phase angle change during stretch and produced a greater cumulative deviation (note the negative ordinate values.) The effects were more pronounced when lower perturbation frequencies were used; equivalent phase angle changes and cumulative deviations were reached with less than half of the stretch rate at the lower perturbation frequency (comparing 30 Hz data with 100 Hz data). Resolution of the cumulative deviation into its stiffness components showed that the stretch-rate sensitivity lay in the elastic component, while the viscous component was not greatly affected by the rate of stretch (Meiss, 1986, unpublished observations). In a related series of experiments done at 35 C and 20 C, it was found that cooling the muscle significantly increased the cumulative deviation during constant-velocity stretch and decreased the rate of rise of isometric force, while it had little effect on the isometric stiffness.

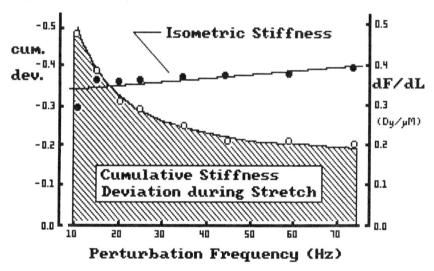

Figure 5. Effects of perturbation frequency on cumulative deviation during constant-velocity stretch, and on unresolved isometric stiffness.

At a constant stretch rate (Figure 5), progressively less cumulative deviation was produced as the perturbation frequency was increased. As previously indicated, the isometric stiffness at peak force increased somewhat with increasing perturbation frequency. Over the range of frequencies used, no maxima or minima were observed for any of the quantities measured.

The effects of perturbation amplitude were also measured as functions of the sinusoidal frequency. Previous work (Meiss, 1984) has shown that the force amplitude response to sinusoidal stretching is diminished at larger perturbation amplitudes. In the present study (data from 12 determinations on 4 ovarian ligaments), increasing the perturbation amplitude from 10 to 70 um decreased the isometric stiffness by 11.9% (±3.2% SD). The phase angle during isometric contraction was essentially unchanged. There was no statistically significant effect due to perturbation amplitude during isometric contraction (in marked contrast to the various results found during stretch; cf. Figs. 4 and 5).

Changes in perturbation amplitude during constant-velocity stretch modified the responses previously described. The amount of change in the phase angle during linear stretch decreased in a curvilinear fashion as the perturbation amplitude increased. The absolute value of the cumulative deviation also decreased (in a nearly linear way) as the perturbation amplitude increased. These effects were more pronounced at low perturbation frequencies (25-35 Hz) than at high (70-100 Hz).

DISCUSSION

While some aspects of the response of smooth muscle to large stretches have been investigated (Gunst, 1983; Johansson, 1983; Meiss, 1982), some important features of the mechanical behavior during stretch remain unclear. If current notions regarding smooth muscle ultrastructure are correct, large stretches should induce crossbridge detachment and cause myofilaments to shear past one another. Possibilities should then exist for transient re-attachment of these broken bridges during the overall muscle lengthening. The present study has used small oscillatory length perturbations to assess mechanical properties during the stretch. The very different responses of the muscle to the sinusoidal perturbations during constant-velocity stretch (as compared to those observed during isometric contraction) indicate fundamental differences in the mechanical properties of the elongating muscle.

Since the muscle force does not continue to rise as steeply during most of the stretch as its initial response would indicate (cf. Figs. 1 and 2, and Meiss, 1982), it is apparent that marked yielding is taking place; in the context of a sliding-filament/crossbridge mechanism of contraction, this yielding must take place at the expense of forced crossbridge detachment. The continued linear rise in force during the stretch (especially at lengths at which parallel elastic contributions are negligible) implies the re-attachment of broken crossbridges. While the force rises during stretch, the smaller increase in stiffness implies the presence of fewer crossbridges than would normally bear that muscle force.

The rate-dependence of the effectiveness of constant-velocity stretch in reducing muscle stiffness may depend upon the rate at which crossbridges can re-attach; at higher rates of myofilament shearing less time would be available for the attachment step, and stiffness should fall accordingly. In active skeletal muscle, applied stretch has been shown to reduce or temporarily abolish the consumption of ATP (Infante, et al., 1964), and in smooth muscle (taenia coli; Butler, et al., 1984), a four-fold reduction in metabolic rate was found in response to forced elongation of isometrically-contracting muscle. These energetic effects are likely due to reduced numbers of effectively-cycling crossbridges during the forced elongation.

The frequency of the perturbation applied to measure the stiffness, however, also has its own effect on the quantities being measured. In the current hypothetical context, the perturbation may affect either the detachment or re-attachment step in the cycle. The lack of a perturbation-frequency dependence in reducing the stiffness of isometrically-contracting muscle suggests that during stretch the sinusoidal perturbation may not be affecting the detachment step, but rather may be interfering with crossbridge re-attachment by further reducing the temporal and/or spatial opportunities for re-attachment. The lack of any relative maxima or minima of the measured responses over the frequency range used in the present study implies that the rate processes involved in crossbridge re-attachment are slower than could be detected with the perturbation frequencies which were employed (cf. Kawai, 1982).

This work was supported by National Institutes of Health grant number R01-AM34385.

REFERENCES

Barden JA (1981). Estimate of rate constants of muscle crossbridge turnover based on dynamic mechanical measurements. Physiol Chem Phys 13:211-219.

Butler TM, Siegman MJ, Mooers SU (1984). Chemical energy usage during stimulation and stretch of mammalian smooth muscle. Pfluegers Arch 410(4):391-395.

Gunst SJ (1983). Contractile force of canine airway smooth muscle during cyclical length changes. J Appl Physiol 55:759-769.

Infante AA, Klaupiks D, Davies RE (1964). Adenosine triphosphate: changes in muscles doing negative work. Science 144:1577-1578.

Johansson B (1983). Responses of the relaxed and contracted portal vein to imposed stretch and shortening at graded rates. Acta Physiol Scand 118(1):41-49.

Kawai, M (1982). Correlation between exponential processes and crossbridge kinetics. In Twarog BM, Levine RJC, Dewey MM (eds): "Basic Biololgy of Muscles: A Comparative Approach," New York: Raven Press, pp 109-130.

Kawai M, Brandt PW (1980). Sinusoidal analysis: a method for correlating biochemical reactions with physiological processes in activated skeletal muscles of rabbit, frog, and crayfish. J Muscle Res Cell Motil 1:279-303.

Meiss RA (1984). Nonlinear force response of active smooth muscle subjected to small stretches. Am J Physiol 246:C114-C124.

Meiss RA (1982). Transient responses and continuous behavior of active smooth muscle during controlled stretches. Am J Physiol 242:C146-C158.

Meiss RA (1978). Dynamic stiffness of rabbit mesotubarium smooth muscle: effect of isometric length. Am J Physiol 234:C14-C26.

Warshaw DM, Fay FS (1983). Cross-bridge activity in single smooth muscle cells. J Gen Physiol 82:157-199.

SMOOTH MUSCLE CONTRACTION: MECHANISMS OF CROSSBRIDGE SLOWING

Newman L. Stephens and C. Y. Seow

Department of Physiology
Faculty of Medicine
University of Manitoba
Winnipeg, Manitoba, Canada R3E 0W3

INTRODUCTION

From a functional point of view, in striated muscle throughout the course of a contraction, whether isotonic or isometric, only one type of actomyosin crossbridge is active. Changes in its cycling rate are brought about by quantitative means. Thus, early in contraction, the rate is submaximal because energy liberating reactions necessary to maximally activate the bridges are submaximal. These reactions rapidly attain and hold a maximum value for the remainder of the contraction. Shortening of the muscle can also result in reduction of cycling rate. This has been well described for skeletal (Taylor and Rudel, 1970), cardiac (Jewell and Blinks, 1968; Brutsaert et al, 1971), and smooth (Stephens et al, 1984) muscles. However it must be pointed out that the range of lengths over which cycling rate is reduced varies: in cardiac muscle this rate, as measured by maximum velocity of unloaded shortening (V_o) starts to fall only after the muscle has shortened by about 0.13 l_o (Brutsaert et al, 1971) where l_o represents optimal length.

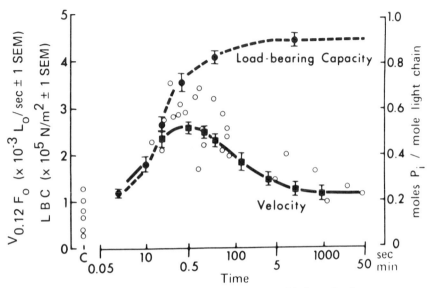

Figure 1. Shortening velocity, myosin light chain phosphorylation and load-bearing capacity of K^+-stimulated swine carotid artery (see text). (Dillon et al, 1981).
©1981 by the American Association for the Advancement of Science

In smooth muscle, importantly and uniquely, the bulk of the diminution in V_o and thus of crossbridge cycling rate has been ascribed to a qualitative change in the crossbridge. Figure 1 is taken from a paper by Dillon et al (1981). It depicts the time course of V_o during an isometric contraction elicited by KCl. The phasic behavior is quite evident. A plot (open circles) of myosin light chain phosphorylation versus time shows correspondence between the two curves. The third curve depicts maximum load bearing capacity. From these sets of data they concluded that velocity of shortening rapidly falls to a very low value and that this course of events is regulated by the state of light chain phosphorylation. As the cycling rate of the bridges falls, their ability to develop force and sustain a load increases. The slowly cycling bridges were termed latch bridges and were postulated to hinder the cycling rate of the remaining bridges resulting in slowing of the overall muscle contraction. The typical properties of the latch bridge are said to be determined by recent dephosphorylation of a normally cycling bridge. However, a considerable controversy exists about the latch bridge and Butler et al, 1986 strongly support the idea that the slowing in muscle contraction is due to progressive slowing of all bridges and not to development of a unique class of latch bridges.

The time course of behavior of myosin light chain phosphorylation has also been the source of controversy. DeLanerolle et al (1980), Butler et al (1982), and Siegman et al (1984) had earlier reported that once light chain phosphorylation developed, it maintained a plateau. However more recently Kamm and Stull (1985) have reported a phasic time course of phosphorylation. We (Kong et al, 1984) have carried out studies of myosin light chain phosphorylation in isolated canine tracheal smooth muscle. Using supramaximal electrical stimulation and fast-freezing techniques we have found that the degree of light chain phosphorylation increases from a resting value of about 20% to a maximum of about 40% upon stimulation; this increase in phosphorylation is almost 75% complete within 500 msec of onset of stimulation. Peak phosphorylation is maintained at a plateau value and does not demonstrate any phasic behavior of the type reported by Dillon et al (1981). Dephosphorylation only occurs after the stimulus is turned off. Why this is so in the canine trachealis is not easy to determine but could be related to differences in tissues, species, or modes of stimulation.

To compound the confusion further, some workers feel that contraction is regulated by an actin-linked non-phosphorylation dependent process. The recent discovery of caldesmon (Sobue et al, 1982; Ngai and Wash, 1984) throws another variable into the melange. Caldesmon is normally actin-linked during the resting state of the muscle. With contraction this linkage is broken and caldesmon binds to the calcium-calmodulin complex. This would act as a secondary control mechanism. However, even here controversy exists. Miyata and Chacko (1986) claim that tropomyosin exerts an important modifying effect on the action of caldesmon; whereas Ngai and Walsh (1984) believes tropomyosin has no effect.

All these controversies aside, our own viewpoint is that there are a number of physiological mechanisms that can reduce the cycling rate of crossbridges in smooth muscle just as they do in skeletal. These effects must first be eliminated before we can identify true latch or slowly cycling crossbridges. We propose now to discuss some of these factors. The discussion is based on the work we have carried out in canine tracheal smooth muscle.

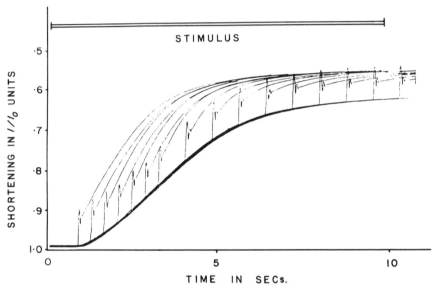

Figure 2. Zero-load clamps applied to an isotonically contracting trachealis. (Stephens et al, 1985).

NORMALLY CYCLING AND LATCH/SLOWLY CYCING CROSS BRIDGES IN CANINE TRACHEAL SMOOTH MUSCLE

We have confirmed the presence of slowly cycling crossbridges in the trachealis. Figure 2 is a reproduction of an oscilloscope recording from a typical experiment. It represents shortening (shown by the ordinate as increasing upwards) versus time. The lowermost heavy line represents isotonic shortening for a strip of trachealis set at 1_o and shortening with the preload alone. At given intervals in subsequent contractions zero load clamps (Brutsaert et al, 1971) were applied. The response consists of two components. The first or rapid transient represents the shortening of the series elastic component. This is followed by a slow transient that represents shortening of the muscle's contractile element. The maximum slope of this transient is the maximum velocity of shortening (V_o) for the zero loaded muscle. This maximum is attained early in the contraction. Uvelius has reported similar data (1979). Though not clear in this figure, examination of records obtained at high speed and gain have shown firstly, that the earliest onset of mechanical activation occurs 600 msec after onset of stimulus and that maximum V_o is attained early in contraction. The records in the figure clearly show that V_o decreases with time.

Figure 3 represents analysis of records obtained from an experiment similar to that shown in Figure 2.

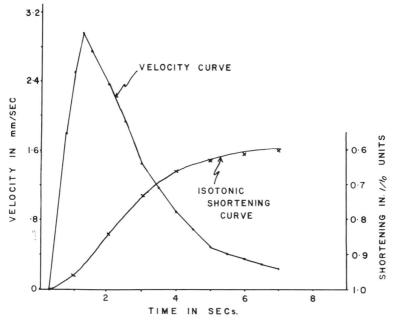

Figure 3. Isotonic shortening and zero-load velocity as functions of time. Stephens et al, 1985).

Maximum V_o develops at about 1.5 sec and then declines. The isotonic shortening curve (during the course of which zero load clamps were applied) is also shown. V_o initially shows an increase because energy liberating reactions are also increasing. The descending limb of the curve demonstrates the progressive decrease in velocity. This decrease, as pointed out before, is due to the development of slowly cycling or latch bridges.

In Figure 4 the results of a conventional compartmental analysis of record (of the type shown in Figure 3) from one experiment are shown. The mean values from six experiments showed that a curve with two exponential terms could be

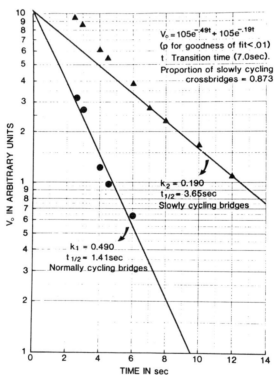

Figure 4. Compartmental analysis of V_o from one experiment. (Stephens et al, 1985).

fitted to the descending limb of the velocity constant. The general form of the equation was $V_o(t) = e^{-0.40t} + e^{-0.10t}$. This confirms that the faster crossbridges cycle almost four times more rapidly than the slower.

Force-velocity curves were obtained 2 seconds after onset of stimulation by applying a series of load clamps of varying magnitude during the course of an isotonic contraction, the load for which was exactly equal to that required to stretch the muscle to l_o. Figure 5 shows the resultant curve. Similar curves were obtained at 8 sec at

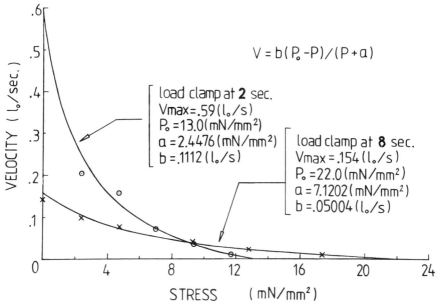

Figure 5. Force-velocity curves obtained at the 2 and 8 sec points in an isotonic contraction.

which time the muscle was shortening very slowly. The difference in Vmax is almost 4 fold. It must be noted that at 8 sec the muscle's contractile element had shortened by about 30% of l_o if one allows for the elastic recoil of the series elastic component. To determine what the effect of this degree of shortening is on V_o (Vmax is being used synonymously with V_o), load clamps were applied during the course of an isometric contraction. Figure 6 shows displacement records from such an experiment.

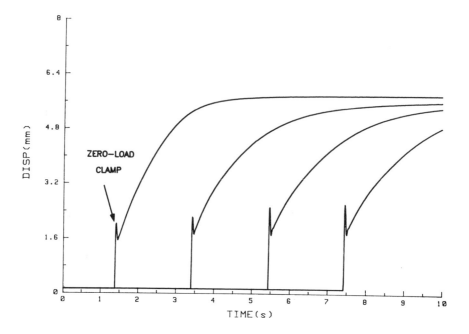

Figure 6. Displacement curves showing zero-load clamps applied to an isometric contraction.

Force-velocity curves were obtained at the 2 sec and 8 sec point, and are shown in Figure 7. The mean Vmax (N = 6) obtained at 2 sec was not significantly different from that obtained in experiments of the type shown in Figure 5. However the Vmax at 8 sec was considerably greater in the isometric experiments. From this we calculated that about 50% of the dimunition in V_o at 8 sec shown in Figure 3 is due to a shortening effect, presumably via a reduced activation effect of the type discussed earlier. Data to support this are presented next.

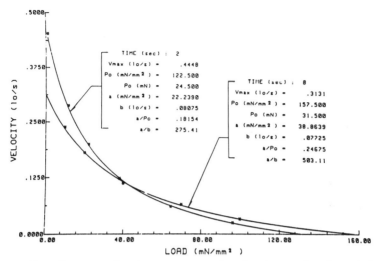

Figure 7. Force-velocity curves obtained at the 2 and 8 sec points in an isometric contraction. (Seow and Stephens, 1986).

REDUCED ACTIVATION AT $1 < 1_o$ IN TRACHEAL SMOOTH MUSCLE

While experiments have not yet been carried out for isotonic shortening, studies of isometric force development have (Kromer and Stephens, 1983). Conventional length-active tension curves were elicited using KCl (100 mM) as an agonist. In one series, the bathing medium contained 2 mM Ca^{2+} and in the other 4.75 mM. P_o at 1_o increased by more than two-fold. This is akin to what has been reported in cardiac muscle (Jewell and Blinks, 1968). In skeletal muscle (Taylor and Rudel, 1970) however no change in P_o is seen. However in both skeletal and cardiac muscle after force is normalized, at lengths below 1_o more isometric tension is seen at the high calcium concentrations. From such experiments it was concluded that at $1 < 1_o$ reduced activation occurs.

In Figure 8 the control length-active tension curve represents the curve elicited from the canine tracheal smooth muscle incubated in normal Krebs-Henseleit solution (2.0 mM Ca^{2+}). For the "high Ca^{2+}" curve the concentration was 4.75 mM. From these curves it is clear that as the muscle shortens under control circumstances it demonstrates reduced activation. From this it is inferred that reduced activation would also develop for shortening ability. However this remains to be experimentally substantiated.

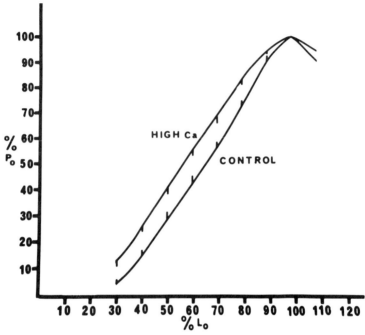

Figure 8. Length-active tension curves of canine trachealis at 2 mM and 4.75 mM Ca^{2+} concentrations in Krebs-Henseleit solution. (Kromer and Stephens, 1983).

We conclude, from what has been written above that about 50% of the slowing of the crossbridge in the later phases of isotonic shortening is due to the shortening effect itself. The data reported by Dillon et al (see Figure 1) indicated that velocity drops considerably also but since this is in the course of an isometric contraction the dimunition cannot be ascribed to shortening inactivation. The trachealis and the hog carotid thus appear to differ quite markedly in some of their mechanical properties. It would appear that the development of latch bridges contributes less to slowing in contraction of trachealis than in the carotid.

Our major concern, with respect to airway smooth muscle contraction, is to determine to what extent latch bridges are operative. This can only be done after allowing for the effects of other physiological factors that may be playing a role. One such factor is the internal resistance to shortening which is discussed next.

INTERNAL RESISTANCE TO SHORTENING

If unstimulated canine trachealis smooth muscle is stretched beyond optimal length and then released its length is automatically restored to optimum. If on the other hand this muscle is electrically stimulated to shorten maximally carrying a light isotonic load and the stimulus is then turned off, the muscle re-elongates to optimal length. Apparently some passive structure/force is responsible for the restoration. In the case of shortening the experiment shows that in the trachealis, as in other muscles, there is an internal resistance to shortening and its recoil results in re-elongation of the muscle. This resistance could load the muscles' contractile element and so reduce its velocity. The reduction would be proportional to shortening.

A simple way to demonstrate these phenomena is with the aid of Voight's model shown in Figure 9. The model choice

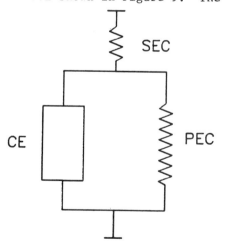

Figure 9. Voigt's model for muscle.

is arbitrary. At optimal length (l_o) the PEC (parallel elastic component) would be almost unstressed in the trachealis since resting tension at l_o is very low (< 0.10 P_o) and the curve at $l < l_o$ shows considerable compliance. Under these conditions, as the muscle shortened the PEC would be rapidly compressed. This compression would retard the shortening ability of the muscle's CE (contractile element). We have developed a method for measuring the mechanical properties of this internal resistance of PEC. It is based on the use of Voigt's model and the further assumption that compression of the PEC loads the CE in the same way that the SEC loads it when an external load is applied. The method uses a combination of the force-balance equation and A. V. Hill's for the force-velocity relation of skeletal muscle (Hill, 1938).

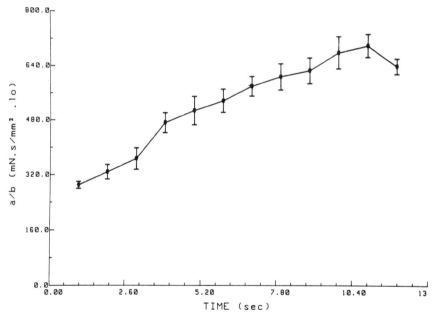

Figure 10. Internal resistance (a/b) as a function of time. (Seow and Stephens, 1986).

In the force-balance equation $F(t) = kX + \alpha V + mA$ where $F(t)$ is force as some function of time, kX, αV and mA represent elastic, viscous and inertial forces. In an isotonic contraction for a given load, the length of the SEC remains constant and kX is then represented by P, and F(t) by $P_o(t)$; mA is generally small enough to be ignored in muscle contraction. The force-balance equation then takes the form $P_o(t) = P + \alpha V$. If time is taken into consideration for the Hill equation, then the latter may be rearranged thus: $\alpha P_o(t) = P + [(P + a)/b]V$ where a and b are constants representing the asymptotes of a rectangular hyperbola, with units of force and velocity respectively. Comparing these two equations since the two left hand terms are identical, we obtain $P + \alpha V = P + [(P + a)/b]V$. From this, α, which we are denoting as an index of internal resistance to shortening may be solved for: $\alpha = (P + a)/b$ when $P = 0$, $\alpha = a/b$.

By applying a range of load clamps during the course of an isometric contraction we were able to delineate force-velocity curves at one sec intervals. From these, a/b could be calculated, and plotted as a function of time. This is depicted in Figure 10 which shows that a/b increases with time from which we concluded that internal resistance to shortening increases with time and reduces velocity of shortening.

It is likely that the site of the internal resistance is the developing latch bridge, however more conventional parallel elastic elements could be contributing. These in general, have been identified as collagen and elastin but the role of cytoskeletal elements cannot be ignored. Recent work on striated muscle (Anderton, 1981) suggests that cytoskeletal elements are highly organized in a sarcomeric (or costameric according to recent terminology) array. Experiments to evaluate the role of the cytoskeleton need to be carried out.

We have recently developed a method for delineating the tension-compression characteristics of the parallel elastic element. This utilizes the assumptions already enumerated, viz that reduction in velocity of shortening of the contractile element at lengths below l_o is due to internal loading by the compressed PEC and this loads the CE in exactly the same way as the SEC loads the CE in transducing an external load.

Zero load clamps were applied at the same instant in time in the course of several isotonic contractions all starting at different lengths determined by different preloads. Figure 11 shows records from a single such experiment.

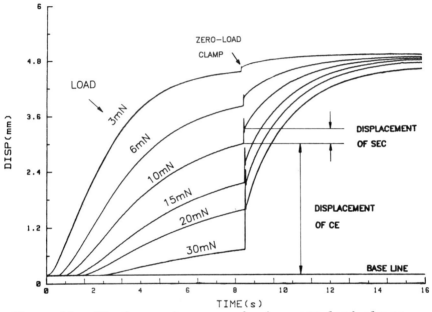

Figure 11. Displacement curves showing zero-load clamps applied to isotonic contractions under different loads.

Figure 12 shows a plot of zero load velocities versus computed contractile element length. A parabolic curve was used to fit the data.

Figure 13 shows the curve of Figure 12 superimposed on a force-velocity curve for the same muscle. Note the abscissa represents load for the force-velocity curve and CE length for the length-velocity phase plane. At isovelocity points the corresponding length and load parameters must represent tension-compression points for the PEC. This is shown in Figure 14. It could also be regarded as the tension compression curve of the structure responsible for internal resistance to shortening.

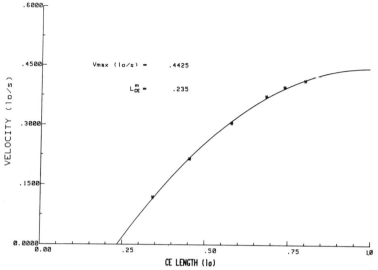

Figure 12. V_o vs. CE length data fitted by a parabolic function: $V_o = V_{max}\ (1-[(1-l_{CE})/(1-l_{CE}^m)]^2)$, where Vmax is the maximum zero-load velocity which occurs at l_o, l_{CE} is the CE length, and l_{CE}^m is the maximally contracted CE length under zero-load.

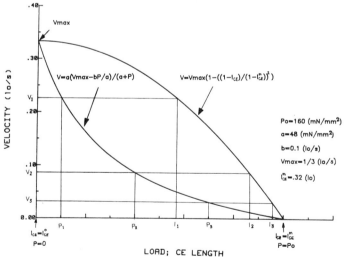

Figure 13. Illustration of the method for obtaining the PEC's tension-compression curve.

Figure 15 shows PEC curves for tracheal smooth muscle obtained from ragweed pollen sensitized dogs and litter mate controls. The increased compliance of the sensitized muscles' PEC is evident. This could account for the greater ability to shorten that we have reported in the past (Antonissen et al, 1979) for the sensitized muscle.

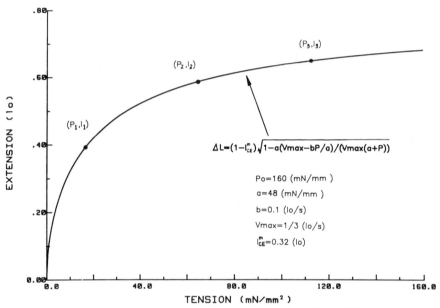

Figure 14. Tension-compression curve for the PEC obtained from Figure 13.

MISCELLANEOUS FACTORS SLOWING CROSSBRIDGE CYCLING RATE

1. Fatigue. In skeletal muscle, as fatigue develops the rate of tension production (dP/dt) and the magnitude of the tension, drop in the course of a series of twitches. This has been ascribed to the development of fatigue; it can develop within 25 sec. No one has as yet reported as to what occurs in isotonic contractions. We have conducted experiments (unpublished) on strips of mouse diaphragm. We observed that fatigue develops much more rapidly in isotonic experiments than in isometric and that both magnitude and

velocity of shortening are reduced. Fatigue could therefore lead to reduced shortening velocity. Whether fatigue develops in the course of a ten second tetanus in smooth muscle remains to be determined.

2. Development of intracellular acidosis. Edman and Mattiazzi (1981) have shown that fatigue in isolated skeletal muscle may be prevented by incubating the muscle in medium with a pH of 7.7. The development of intracellular acidosis could therefore be a mechanism for slowing crossbridges in smooth muscle also. Since tracheal smooth muscle possesses only 10% of the oxidative phosphorylation capacity that skeletal muscle does (Stephens and Wrogemann, 1970) it must rely more heavily on aerobic glycolysis as an energy source than skeletal muscle. This suggests that intracellular acidosis could develop more rapidly in this muscle since lactic acid is a major product of the glycolytic pathway. We plan to determine whether this mechanism could be responsible for the strength of

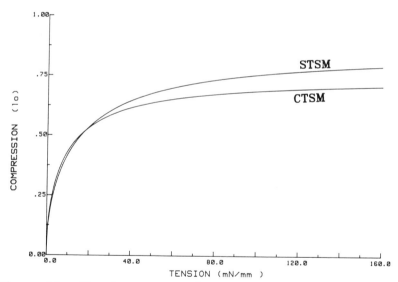

Figure 15. Tension-compression curves for sensitized and control canine trachealis.

crossbridges late in contraction. This will be achieved by measuring zero load velocities at the 8 sec compared to that in a bath of normal pH. If V_o at 8 sec increases at pH 7.7, it will support the idea that acidosis was partly responsible for the reduced velocity.

CONCLUSION

We have shown that reduced muscle length operating via the production of reduced activation, or by recruitment of an internal resistance to shortening could lead to reduced velocity of shortening later on in contraction. We have suggested that fatigue and intracellular acidosis could also contribute to reduced velocity of contraction. It is only after the effects of all these factors have been accounted for can one conclude latchbridges or slowly cycling crossbridges exist.

REFERENCES

Anderton BH (1981). Intermediate filaments: a family of homologous structures. J Muscle Res Cell Mot 2:141-166.

Antonissen LA, Mitchel RW, Kroeger EA, Kepron W, Tse KS and Stephens NL (1979). Mechanical alterations of airway smooth muscle in a canine asthmatic model. J Appl Physiol 46:681-687.

Brutsaert DL, Claes VA and Sonnenblick EH (1971). Effects of abrupt load alterations on force-velocity-length and time relations during isotonic contractions of heart muscle: Load clamping. J Physiol 216:319-330.

Butler TM and Siegman MJ (1982). Chemical energetics of contraction in mammalian smooth muscle. Fed Proc 41(2): 204-208.

Butler TM, Siegman MJ and Mooers SU (1986). Slowing of cross-bridge cycling in smooth muscle without evidence of an internal load. Am J Physiol 251:C945-C950.

DeLanerolle P and Stull JT (1980). Myosin phosphorylation during contraction and relaxation of tracheal smooth muscle. J Biol Chem 255:9993-1000.

Dillon PF, Aksoy MO, Driska SP and Murphy RA (1981). Myosin phosphorylation and the cross-bridge cycle in arterial smooth muscle. Science 211:495-497.

Edman KAP and Mattiazzi AR (1981). Effects of fatigue and altered pH on isometric force and velocity of shortening at zero load in frog muscle fibres. J Muscle Res Cell Mot 2:321-334.

Hill AV (1938). The heat of shrotening and the dynamic constants of muscle. Proc Roy Soc B 126:136-195.

Jewell BR and Blinks JR (1968). Drugs and the mechanical properties of heart muscle. Ann Rev Pharmacol 8:113-130.

Kamm KE and Stull JT (1985). The function of myosin and light chain kinase phosphorylation in smooth muscle. Ann Rev Pharmacol Toxicol 25:593-620.

Kong SK, Shiu RPC and Stephens NL (1984). Role of myosin light chain (MLC) phosphorylation in canine tracheal smooth muscle (TSM) contraction. Fed Proc 43:427.

Kromer U and Stephens NL (1983). Airway smooth muscle mechanics: Reduced activation and relaxation. J Appl Physiol 54(2):345-348.

Miyata H and Chacko S (1986). Role of tropomyosin in smooth muscle contraction: Effect of tropomyosin binding to actin on actin activation of myosin ATPase. Biochemistry 25(9):2725-2729.

Ngai PK and Walsh MP (1984). Inhibition of smooth muscle actin-activated myosin Mg^{2+}-ATPase activity by caldesmon. J Biol Chem 259:13656-13659.

Siegman MJ, Butler TM, Mooers SU, and Michalek A (1984). Ca^{2+} can affect Vmax without changes in myosin light chain phosphorylation in smooth muscle. Pflugers Archiv 401:395-390.

Sobue K, Morimoto K, Inui M, Kanda K and Kakiuchi S (1982). Control of actin-myosin interaction of gizzard smooth muscle by calmodulin- and caldesmon-linked flip-flop mechanism. Biomedical Res 3(2):188-196.

Stephens NL, Mitchell RW and Brutsaert DL (1984). Shortening inactivation, maximum force potential, relaxation, contractility. In: Smooth Muscle Contraction. Ed. N.L. Stephens, Publ. Marcel Dekker, Inc. 91-112

Stephens NL and Wrogemann K (1970). Oxidative phosphorylation in smooth muscle. Am J Physiol 219(6): 1796-1801

Taylor SR and Rudel R (1970). Striated muscle fibers: Inactivation of contraction induced by shortening. Science 167:882-884

ARE CONTRACTION KINETICS AFFECTED BY THE ACTIVATION MODE?

Ulrich Peiper, Brigitte M. Lobnig, and Bruno Zobel
Institute of Physiology
University of Hamburg
Hamburg 20, F.R. Germany

INTRODUCTION

The aim of this study was to find out the answers to the following questions:
(1) are the contraction kinetics of smooth muscle dependent on the mode of activation (e.g., activation of different receptor types, membrane depolarization, barium activation in the place of calcium)?
(2) are the contraction kinetics dependent on the extent of developed force (i.e., are they varied by changes in the resting tension, the addition of calmodulin antagonists, the blockade of calcium channels or the blockade of cholinergic receptors, the variation of the extracellular calcium concentration, or by the additional activation of B-adrenoreceptors)?
(3) is it possible to re-accelerate the down-regulated contraction kinetics by different drugs during sustained activation?

Contraction kinetics have been measured by describing the time constants of tension recovery after a sinusoidal length vibration had produced an immediate fall in the tension of an active muscle preparation (Klemt et al, 1981).

METHODS

The experiments were performed using rat tracheal smooth muscle. One tracheal ring was opened in the middle and mounted between a vibrator (Type 101, Ling Inc.) and a force transducer (Type 300, Cambridge Technology Corp.). The incubation medium employed was Tyrode solution (in mM, NaCl 132.2; KCl 4.8; $CaCl_2$ 2.5; $MgCl_2$ 0.49; $NaHCO_3$ 11.9; NaH_2PO_4 0.36; glucose 5.05; Na-pyruvate 2.0; bubbled with 95% O_2 and 5% CO_2 at 37 °C; pH 7.3). A resting tension of 2 mN was maintained throughout the experiment.

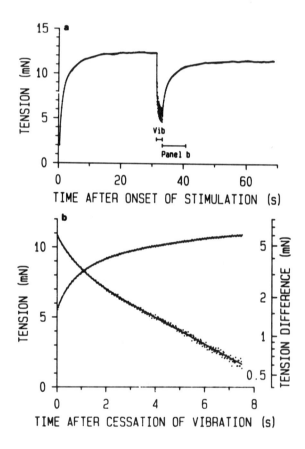

Figure 1. Upper panel: A smooth muscle preparation from the rat trachea was subjected to electrical field stimulation starting at time = 0 s. A 2 s length vibration was applied after 30 s which produced an immediate fall in tension. After cessation of vibration, tension recovered nearly

completely. Digitized values are shown together with the corresponding biexponential functions. Lower panel:The time course of post-vibration tension recovery as well as the semilogarithmic plot of the differences between steady state tension and actual tension are shown in order to demonstrate the two components of tension change.

The modes of activation or inhibition of the muscle preparations are detailed in the following section. At various times after the onset of stimulation, a sinusoidal length vibration (amplitude about 5% of the muscle length; frequency 100 Hz) was applied for 2 s to the activated muscle preparation. The resultant effect is demonstrated in Fig. 1a. Biexponential functions were fitted to the digitized data obtained during (a) the initial tension development after the onset of stimulation, (b) the fall in tension after onset of the sinusoidal length vibration, and (c) the tension recovery after cessation of vibration with continuing activation. The last period is also shown in Fig. 1 b. The semilogarithmic plot of tension changes reveals the existence of two components occurring during post-vibration tension recovery. The time course of tension increase is well described by the biexponential function

$$T = T_{ss} - A_1 \exp(-t/t_1) - A_2 \exp(-t/t_2)$$

where T = actual tension, T_{ss} = steady-state tension after complete tension recovery from inhibiting vibration, A_n = amplitudes and t_n = time constants of post-vibration tension recovery with an initial fast (n=1) and a subsequent slow component (n=2). The fast component of tension recovery reflects the re-attachment of crossbridges detached during vibration, and the following slow component can be related to the kinetics of actin-myosin interaction (Peiper, 1983; Peiper et al., 1984). As the term t_2 is a most useful parameter for measuring contraction kinetics, the presentation of our results has focussed on this time constant. All data are given as mean ± S.E.M., and were tested for their significance by the Student t-test.

RESULTS AND DISCUSSION

I. Different Modes of Stimulation.

   A. <u>Electrical field stimulation</u>. Electrical field stimulation was performed by mounting the tissue between two parallel platinum electrodes and by applying square wave pulses of 0.15 ms duration and 40 volts at a frequency of 30 Hz. This type of activation produces maximum tension and works entirely through the neurogenic release of acetylcholine as the contractile response could be completely abolished by the addition of 0.1 µM atropine. Vibration was performed 30 s after the onset of stimulation, and the time constant t_2 of post-vibration tension recovery was duly calculated. This experimental paradigm served to evaluate the properties of the muscle preparation with respect to the rate and extent of force development. Mean values averaged 5.90 ± 0.14 s and 9.02 ± 0.13 mN for t_2 and T_{ss}, respectively (Fig. 2). On extending the stimulation period prior to the onset of vibration to 5 min, t_2 increases to about 15 s with only a marginal decrease occurring in T_{ss} (Peiper et al, 1984), which represents a down-regulation of the contraction kinetics. The peak value of contraction kinetics was found after 30 s electrical field stimulation. This perid is similar to that of the transient peak of both the myosin phosphorylation and the contraction kinetics described by Gerthoffer and Murphy (1983) and by Kamm and Stull (1985) for the tracheal smooth muscle.

   B. <u>Receptor-mediated activation</u>. Different types of receptors were activated by adding maximally effective concentrations of acetylcholine (100 µM), carbachol (10 µM) or serotonin (10 µM) for 45 min. The first vibration was performed 8 min after the onset of stimulation in order to allow the kinetics to reach steady-state. About 10 vibrations were applied to each muscle preparation. The time constant t_2 and the steady-state tension T_{ss} were (Fig. 2): SER = 10.99 ± 0.2 s, 10.30 ± 0.21 mN (n=85); Ach = 12.59 ± 0.20 s, 9.01 ± 0.13 mN (n = 265); CAR = 17.28 ± 0.50 s, 8.02 ± 0.17 mN (n=71). These t_2 values are of the same order of magnitude but two to three-fold greater than those obtained 30 s after the onset of electrical stimulation (= control experiments).

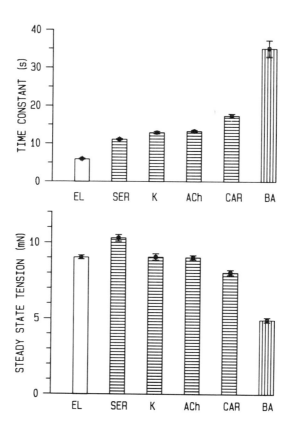

Figure 2: Steady state tension and the time constant of the slow component of tension recovery after vibration. The preparation was activated by different activation modes: EL = electrical field stimulation; SER = serotonin (10 µM); K = high K (137 mM) solution; Ach = acetylcholine (100 µM); CAR = carbachol (10 µM); BA = calcium-free solution + 10 mM barium. t_2 was measured 30 s (EL), 8-45 min (SER, Ach, CAR), or 45-90 min (BA) after the onset of stimulation. Data are means ± SEM.

C. <u>Membrane depolarization</u>. The cell membrane of the smooth muscle was depolarized by a potassium-rich modified Tyrode solution (equimolar substitution of Na by K; 137 mM K). 8-45 min after exposure to high potassium, about 10 vibrations were performed. t_2 and T_{ss} were 12.88 ± 0.32 s and 9.04 ± 0.21 mN, respectively (Fig. 2). These values lie within the same range as those observed during receptor-mediated activation.

D. <u>Contraction induced by barium</u>. Barium is able to replace calcium in the generation of force in smooth muscle (Kreye et al, 1986). Muscle preparations, incubated in calcium-free solution containing 0.1 mM EGTA and 0.1 µM atropine methylnitrate for the purpose of cholinergic receptor blockade, were activated by 10 mM barium chloride. Tension developed slowly and reached about 6 mN. Steady-state conditions occurred for t_2 only after 45 min of barium activation. t_2 and T_{ss} averaged 34.95 ± 2.21 s and 4.92 ± 0.15 mN, respectively, as measured 45-100 min after the addition of barium (Fig. 2).

A striking result from this experimental series was the extraordinary high t_2 values which occurred during steady-state contraction kinetics under barium activation. These time constants were more than twice as high as those obtained from down-regulated preparations during long-term drug-induced activation. One speculative explanation might be that the above-mentioned activation modes work predominantly via the increase of myoplasmic calcium in contrast to barium activation.

II. Different Modes of Inhibition of Contraction

A. <u>Inhibition of the response to electrical field stimulation</u>. The aim of this experimental series was to reduce the active tension developed during electrical field stimulation to about 50%. Three different methods were used for this purpose: the lowering of resting tension (RT) from 2.0 to 0.8 mN; the addition of 7 µM verapamil (Ver); and the addition of 0.4 nM atropine (Atr). The time constants of post-vibration tension recovery, however, were not affected by these changes in active tension and averaged 607 s (Fig. 3). Furthermore, preparations were pretreated with 1 µM trifluoperazine (TFP) at 4°C for 12 h. Control specimens were stored at 4°C for 12 h. Both groups were then activated by electrical field stimulation at 37°C with or without 1 μM TFP. Tension was only slightly decreased by this low TFP concentration. The time constant t_2, however, was significanly increased to 13.64± 1.4 s (Fig. 3) reflecting the retarding effect the calmodulin antagonist exercises on the contraction kinetics.

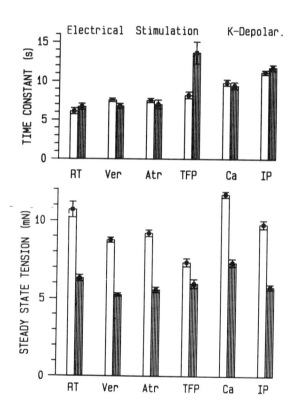

Figure 3: The time constant t_2 and the steady state tension are shown after a partial inhibition of tension produced by either electrical field stimulation or by potassium depolarization. RT = decrease in resting tension from 2 mN to 0.8 mN; Ver = addition of 7 μM verapamil; Atr = addition of 0.4 nM atropine; TFP = addition of 1 μM trifluoperazine after pretreatment at 4° C for 12 h; Ca = decrease in extracellular calcium from 2.5 mM to 0.7 mM; IP = addition of 1.5 μM isoproterenol. Data are means ± SEM.

B. Inhibition of the response to potassium depolarization. In a first experimental series, the preparation was incubated in a calcium-free Tyrode solution for 1 h at 37°C. Subsequently, the medium was replaced by a K-rich and Ca-free solution in order to depolarize the muscle membrane without affected tension. The cumulative addition of calcium chloride resulted in a stepwise tension increase. During the equilibrium tension produced by each concentration, 3 vibrations were applied to measure contraction kinetics. t_2-values were calculatd at the ED_{50} and ED_{100} level, which corresponds to 0.7 mM and 2.5 mM calcium. The time constants obtained at both levels did not significantly differ from each other (Fig. 3). Thus, changes in force which reflect corresponding changes in myoplasmic calcium do not affect contraction kinetics.

In the second experimental series, the muscle was depolarized in a K-rich solution containing 2.5 mM calcium. After an 8 min equilibrium, isoproterenol was added cumulatively and resulted in a stepwise tension decrease. t_2 was calculated at the IC_{50} level and remained within the down-regulated kinetics range. These experiments reveal how marked the stability of contraction kinetics exists even after pronounced variations in force development. The question arises as to whether these stabilized down-regulated kinetics can be re-accelerated during sustained activation.

III. Re-acceleration of Down-Regulated Contraction Kinetics

During sustained activation by 10 μM SER, 100 μM Ach (both at maximally effective concentrations) or by 2 μM ACh (a concentration approximating the ED_{70} level), t_2-values were measured repeatedly 8-45 min after the addition of the drug. These values averaged 11 to 12 s, which is within the normal range of down-regulated kinetics (Table 1). In order to re-accelerate the contraction kinetics, electrical field stimulation was applied to the muscle preparations in addition to continuous drug activation. In the preparation activated by 100 μM Ach, only a marginal increase in tension was observed and t_2 remained at the 11-12 s level. In the preparations activated by 2 μM Ach or 10 μM SER, however, the additional stimulus produced a tension increase of about 3-5 mN. t_2-values measured in the tension recovery period after 30 s of electrical stimulation averaged 6.16 s and 6.45 s, respectively (Table 1). This decrease in the time constants reflects a re-acceleration of the contraction kinetics which could be reproduced several times for each muscle preparation.

These results show that the contractile proteins and their related enzymes are not in a refractory state and are able to react to accelerating stimuli. The nature of the accelerating stimulus is obviously related to increases in the myoplasmic calcium level, as indicated by a corresponding tension increase.

Table 1. Activation of the tracheal smooth muscle preparations by serotonin (SER) or by acetylcholine (ACh).

Mode of Activation	n	T_{ss} (mN)		t_2 (s)	
10 M SER	85	11.19	0.26	10.99	0.20
10 M SER + EL	38	14.53	0.58	6.45	0.23
100 M Ach	118	13.39	0.16	11.45	0.22
100 M Ach + EL	47	13.13	0.27	11.11	0.38
2 M Ach	23	9.18	0.56	11.12	0.51
2 M Ach + EL	26	14.08	0.52	6.16	0.25

(T_{ss} = steady-state tension, t_2 = time constant of the slow component of post-vibration tension recovery. Values are means ± SEM. 8-45 min after onset os stimulation, the preparation was frequently vibrated in order to measure t_2. An additional electrical stimulus (EL) was applied to each muscle preparation, and t_2 was measured 30 s after the onset of electrical stimulation.

REFERENCES

Gerthoffer WT, Murphy RA (1983). Myosin phosphorylation and regulation of cross-bridge cycle in tracheal smooth muscle. Am J Physiol 244:C182-C187.

Kamm KE, Stull JT (1985). Myosin phosphorylation, force and maximal shortening velocity in neurally stimulated tracheal smooth muscle. Am J Physiol 249:C238-C247.

Klemt P, Peiper U, Speden RN, Zilker F (1981). The kinetics of post-vibration tension recovery of the isolated rat portal vein. J Physiol (Lond) 312:218-296.

Peiper U (1983). Alterations in smooth muscle contraction kinetics during tonic activation. Pflugers Arch 399: 203-207.

Peiper U, Vahl CF, Donker E (1084). The time course of changes in contraction kinetics during the tonic activation of the rat tracheal smooth muscle. Pflugers Arch 402:83-87.

Regulation and Contraction of Smooth Muscle, pages 387-398
© 1987 Alan R. Liss, Inc.

HIGH MYOSIN LIGHT CHAIN PHOSPHATASE ACTIVITY IN ARTERIAL SMOOTH MUSCLE: CAN IT EXPLAIN THE LATCH PHENOMENON?

Steven P. Driska

Physiology Department, Medical College of Virginia, Virginia Commonwealth University, Richmond, VA 23298

INTRODUCTION

    The "latch" phenomenon, as used here, refers to a state in which smooth muscles develop high forces with only slight increases in the level of phosphorylation of the 20,000 dalton light chain of myosin (LC20) over those in resting muscle. In addition, the shortening velocities under those conditions are very low. This phenomenon is observed when swine carotid artery smooth muscle strips are stimulated with supramaximal concentrations of various agonists, which causes active force development that can be maintained for for long periods of time (30 minutes). However, if the muscle strips are allowed to shorten isotonically after different periods of isometric stimulation, the shortening velocity against a given load depends on the duration of stimulation, with the highest shortening velocities observed after only brief isometric stimulation, i.e., not long enough for active force to reach its peak value. When the muscles have contracted isometrically for longer times, isotonic shortening velocities are lower. Levels of myosin light chain phosphorylation in this tissue increase from resting values to a peak value of about 0.6 mol P /mol LC20 before peak force is reached. During longer isometric contractions, phosphorylation declines steadily, approaching the levels observed in resting muscle, even though force is essentially unchanged (Dillon et al., 1981). Plots of shortening velocity and phosphorylation against duration of stimulation show a good correlation of shortening velocity with the level of phosphorylation. These observations appeared to be inconsistent with a simple system where the mechanical output of the muscle was regulated only by the

extent of myosin light chain phosphorylation. To explain
these results it was proposed that myosin crossbridges with
phosphorylated light chains cycled and produced shortening,
while dephosphorylation of attached crossbridges resulted in
attached, non-cycling crossbridges ("latchbridges") which
could maintain force. The term "latch state" refers to a
situation where many of the crossbridges are latchbridges.
It was proposed that latchbridges constituted an internal
load on phosphorylated crossbridges thereby slowing the
shortening velocity (Dillon et al., 1981). This hypothesis
was modified in later work which suggested that latchbridges
were slowly-cycling, rather than non-cycling (Gerthoffer and
Murphy, 1983). The decline in the level of phosphorylation
was attributed to the decline in $[Ca^{2+}]_i$ after an initial
transient peak (Morgan and Morgan, 1982). When
K^+-depolarized tissues were used to study the dependence of
force and phosphorylation on the $[Ca^{2+}]$ in the bathing
solution, the steady-state active force response required
less $[Ca^{2+}]$ than the light chain phosphorylation response
(Aksoy et al., 1983). These experiments suggested that LC20
phosphorylation was not the only regulatory mechanism and
were interpreted to mean that force maintenance by
latchbridges was regulated by an unspecified Ca^{2+}-dependent
process which had a lower requirement for Ca^{2+} than did
LC20 phosphorylation.

Our studies of histamine-induced rhythmic contraction
have provided some important insights into regulation by
LC20 phosphorylation. When stimulated by 10 uM histamine,
most swine carotid artery strips develop rhythmicity,
contracting spontaneously at regular intervals, typically
every 1-5 minutes (Stein and Driska, 1984). Near-maximum
forces are developed, and the muscles relax substantially
between the contractions (to as little as 15% of the peak
force). Rhythmic contractions can continue for hours, and
because histamine is present throughout this time, this
preparation allowed us to study events related to
contraction or relaxation without the diffusion delays that
complicate responses to the addition of agonists. The
rhythmicity implies that the individual cells in the tissue
are contracting synchronously. Preliminary studies of
changes in myosin light chain phosphorylation during the
contraction-relaxation cycle indicated that the myosin light
chain phosphatase activity in the intact strip at 37 C was
unexpectedly high, with an estimated phosphatase rate
constant of 0.13 s^{-1} (Driska, 1986). In the absence of

myosin light chain kinase activity, this much phosphatase activity would reduce the level of phosphorylation ten-fold in 17 seconds. Myosin light chain phosphatase appears to be unregulated, and the presence of a highly active, unregulated phosphatase has important consequences for various processes in smooth muscle. The high phosphatase activity implies that there would be substantial turnover of phosphorylated and non-phosphorylated light chains under conditions where myosin light chain kinase (MLCK) activity and net phosphorylation are high. It also means that the ATP usage by the combined kinase and phosphatase system (the "pseudo-ATPase") would be quite high. More importantly, the high phosphatase activity must be considered when interpreting previous and current experiments examining phosphorylation as a regulatory system. The purpose of this article will be to examine some of these consequences.

METHODS

The mathematical model used in these studies is explained in the RESULTS section. To determine steady-state solutions of the simultaneous reactions, an electronic circuit analysis package (SPICE) was used. In this approach, concentrations of reactants are represented as voltages at nodes of a circuit, and reaction rates or fluxes are represented as currents flowing through controlled current sources. The SPICE package allowed steady state solutions to be quickly obtained as individual rate constants were varied over a 1,000-fold range, and allowed the model to be modified and manipulated very easily. Results of these simulations were displayed and plotted using SAS/GRAPH software.

RESULTS

The level of phosphorylation of the light chain represents the balance of the phosphorylation and dephosphorylation reactions, and is therefore related to ratio of the rate constants for these steps. Most previous work on light chain phosphorylation in smooth muscle has only considered the level of phosphorylation or the ratio of the kinase and phosphatase activities. Our estimates of the phosphatase rate constant allow the phosphorylation rate to be calculated in a steady state, and this in turn allows consideration of phosphorylation and dephosphorylation kinetics in relation to the kinetics of crossbridge attachment, crossbridge detachment and latchbridge detachment. While these latter parameters have not been rigorously measured in smooth muscle, various experimental data allow estimates, and a wide variety of values can be investigated in simulations.

The Model

This report examines the effects of changing various first order rate constants in a simple mathematical model of smooth muscle contraction. The model is depicted schematically in Figure 1, and considers four states for the myosin crossbridge as follows: 1) detached, unphosphorylated myosin (M); 2) detached, phosphorylated myosin (PM); 3) phosphorylated myosin attached to actin in a high-force state (APM); and 4) myosin dephosphorylated while attached to actin, a high-force state defined as a latchbridge (LAM). In the model, myosin light chain kinase (MLCK) is equally active on both the attached and detached dephosphorylated species (M and LAM), and myosin light chain phosphatase (MLCP) is equally active on both the attached and detached phosphorylated species (PM and APM). The level of phosphorylation is the sum of PM and APM, and cooperativity is ignored both in the phosphorylation process and in the interaction with actin. For simplicity the model assumes that both LC20 light chains of a single myosin molecule will be phosphorylated simultaneously. The phosphatase rate constant (kp) used in most of the simulations was $0.13\ s^{-1}$, based on our measurements of the dephosphorylation rate in rhythmic contractions (not shown). The MLCK activity (kk) is varied in simulations and is the only Ca^{2+}-dependent step in the model. Since the highest level of phosphorylation we have ever measured in this tissue is 85%, when the tissue is maximally activated the MLCK rate constant is probably about 5 times that of the phosphatase, or $0.65\ s^{-1}$. The model allows only phosphorylated myosin to attach to actin, and attachment is assumed to be rapid, with a rate constant (ka) of $1.3\ s^{-1}$ in most simulations (10 times the phosphatase rate constant). This value is close to the rate constant for the increase in stiffness ($1.1\ s^{-1}$) when tracheal smooth muscle was stimulated electrically, which was suggested to represent the crossbridge attachment rate (Kamm and Stull, 1986). The detachment of phosphorylated crossbridges in this model (ks) represents all the steps of a normal crossbridge cycle, except reattachment, and has been chosen to be 0.43 s^{-1}, or 1/3 of the attachment rate, so that when fully activated, a maximum of 75% of crossbridges will be attached. The crossbridge detachment rate used here is in reasonable agreement with the steady-state actin-activated ATPase activity of porcine arterial myosin (Srivastava et al., 1987). Furthermore, the energetics measurements on intact swine carotid suggest myosin ATPase rates of about $0.2 - 0.5$ s^{-1} (Krisanda and Paul, 1984).

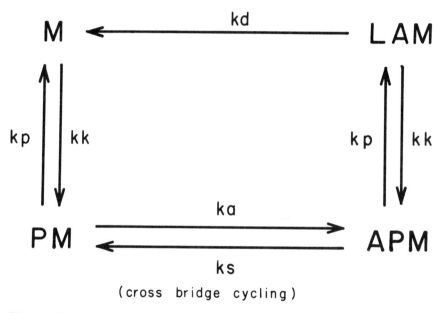

Figure 1. Model of the regulation of smooth muscle contraction by myosin light chain phosphorylation. Explained in text.

Steady-state force is represented in this model as the sum of the attached crossbridges (APM) and latchbridges (LAM), normalized to total myosin, and under most conditions this will not exceed 0.75 because of the choice of the attachment and detachment rate constants. The final, and most interesting step is the latchbridge dissociation step (kd). This step is not reversible in the model. Variation of this rate constant has important consequences for studies on phosphorylation as a regulatory mechanism.

One can see intuitively that if a latchbridge dissociates quickly, it will not contribute much to force development, but if it dissociates slowly, it will continue to exert force for some time and will therefore make an important contribution to force production. If the phosphatase activity is very high, and latchbridges dissociate very slowly, then one can envision a situation where attached crossbridges are dephosphorylated almost as soon as they attach, becoming latchbridges. If this happened there would be a large number of latchbridges, but very few phosphorylated attached crossbridges, and the level of phosphorylation would be low while force was high.

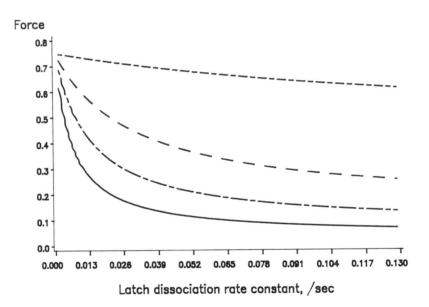

Figure 2. Dependence of force on kd, the latchbridge dissociation rate constant, when kp = 0.13 s^{-1}. Curves show this relationship at the following levels of phosphorylation: ———, 5%; —·—·—, 10%; — — —, 20%; - - - - -, 60%.

The model has been used to examine this possibility quantitatively. Figure 2 shows plots of force, at specified levels of phosphorylation, against the latchbridge dissociation rate. With 60% phosphorylation, the force is insensitive to the latchbridge dissociation rate (kd). However, at lower levels of phosphorylation (20%, 10%, and 5%) the force becomes more critically dependent on kd. Of particular interest is the region where kd is less than 0.013 (i.e. 1/10 of the phosphatase rate). Here, substantial forces are maintained in spite of very low levels of phosphorylation.

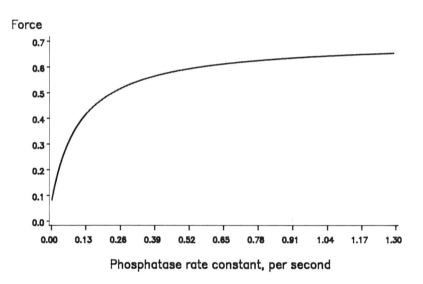

Figure 3. Dependence of force on kp, the phosphatase rate constant, at 10% phosphorylation. Phosphorylation was held at 10% by keeping the kinase and phosphatase rate constants in a 1 : 9 ratio while both were varied over a 1,000-fold range. A latchbridge dissociation rate constant (kd) of 0.013 s^{-1} was used.

Figure 2 showed the importance of the latchbridge dissociation rate (kd) in determining the active force at selected levels of phosphorylation, with the phosphatase rate constant set to 0.13 s^{-1}. However, the figure does not show the importance of the phosphatase rate constant chosen. Figure 3 explicitly shows how force development depends on the phosphatase rate at one chosen level of phosphorylation (10%). A latchbridge dissociation rate (kd) of 0.013 s^{-1} was used because at lower values of kd, force is high at all levels of phosphorylation, making it difficult to explain relaxation, and if higher values of kd are used, latchbridges are not important. It can be seen from the figure that when phosphatase activity is low, a phosphorylation level of 10% only leads to attachment of about 10% of the crossbridges, i.e. force is 0.1. However, as the phosphatase is increased to 1.3 s^{-1}, over 60% of the crossbridges are attached at 10% phosphorylation. When the phosphatase rate is 0.13 s^{-1}, about 40% of the crossbridges in the model are attached. This is likely to be similar to the situation in carotid artery strips after long stimulation, when phosphorylation has declined but force development is still high, i.e. the "latch state".

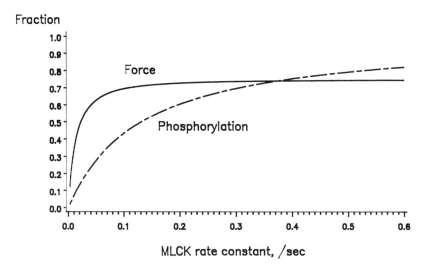

Figure 4. Dependence of force and phosphorylation on the activation level. Ordinate shows the fraction of the total myosin heads that are phosphorylated or attached. Increasing values of the MLCK rate constant represent increased activation of the muscle by Ca^{2+}. The value used for kp was 0.13 s^{-1}, and the value for kd was 0.013 s^{-1}.

Figure 4 shows a surprising result not predicted in the original latchbridge hypothesis, the non-identical dependence of phosphorylation and force on the steady-state level of activation. Figure 4 shows plots of steady-state force and phosphorylation as MLCK activity is increased from very low levels to levels sufficient to achieve about 80% phosphorylation. It is clear that maximum force can be attained at a lower level of activation than maximum phosphorylation, and that the force response is much steeper. Because MLCK is Ca^{2+}-dependent, this also means that force development in the model requires less $[Ca^{2+}]$ than phosphorylation does. It is surprising that this behavior is obtained from a model with only one site for Ca^{2+} action, the activation of MLCK. The conclusion is that it is unnecessary to hypothesize the existence of an additional Ca^{2+}-dependent process besides phosphorylation to explain the lower Ca^{2+} requirement for force development than for phosphorylation.

DISCUSSION

Figure 2 demonstrated quantitatively the behavior proposed in the original latchbridge model (Dillon et al., 1981), showing how force development in the model varied when the latchbridge dissociation rate (kd) was varied. This simulation clearly shows that when the latchbridge dissociation rate is low, substantial forces can be maintained with low levels of phosphorylation.

Figure 3 showed the dependence of the active force, at a low level (10%) of phosphorylation, on the absolute values of kinase and phosphatase activities used to produce that level of phosphorylation. With higher phosphatase (and kinase) activities, force development was greater than would be expected from the level of phosphorylation. This illustrates that force maintenance by latchbridges is quantitatively much more important when the phosphatase activity is higher, a possible explanation for why "latch" behavior is not observed in some other smooth muscles. This work emphasizes the importance of knowing the phosphatase rate in absolute terms, rather than just the kinase : phosphatase ratio. In other words, simply knowing the level of phosphorylation is not enough; information on the light chain phosphate turnover is also needed to understand regulatory mechanisms.

In early studies the turnover of phosphorylated light chains (i.e. the dephosphorylation rate in the steady state) was usually implicitly assumed to be slow relative to the crossbridge cycle. In a slow turnover system the force output would be directly proportional to the level of phosphorylation. The high phosphatase activity we measured means that the swine carotid artery is not a slow turnover system, and therefore regulation by phosphorylation must be re-evaluated in this light. The latchbridge hypothesis was an important first attempt to deal with dephosphorylation during contraction, although the early ideas of latchbridges inaccurately depicted them as a static, rather than dynamic, population. The mathematical model presented here suggests that latchbridges are a more dynamic population, continually being formed by dephosphorylation of attached crossbridges and disappearing by slow dissociation from the thin filament, and, in the presence of calcium, also being converted to rapidly cycling crossbridges by phosphorylation.

The simple model presented here can explain regulation of force development by phosphorylation without a simple proportionality between force and phosphorylation. It can also explain substantial force development with only slight increases in phosphorylation, such as the 0.09 mol P/mol LC20 increase that accompanied force development after stimulation of swine carotid strips with phorbol esters (Chatterjee and Tejada, 1986). It may also explain maximum force generation by functionally skinned chicken gizzard fibers with only 20% phosphorylation (Hoar et al., 1979). Finally, the model can explain the lower $[Ca^{2+}]$ requirement for force than for phosphorylation (Aksoy et al., 1983).

Some comments should be made about the rate constants used. The phosphatase rate constant, k_p, was estimated by measuring levels of phosphorylation during relaxation, assuming instant and total inactivation of MLCK. Because of this, the phosphatase activity may be underestimated. The latchbridge dissociation rate, k_d, may be the major determinant of the isometric force relaxation rate. During relaxation, the rate of force decay may approximate k_d. In the rapid relaxations of rhythmically contracting swine carotid strips, rate constants of force decay are in the range of $0.01 - 0.05$ s^{-1}, a range which includes the value of 0.013 s^{-1} used in Figures 2 and 3.

In conclusion, a mathematical model of phosphorylation as the sole regulatory mechanism, described here, can explain several aspects of smooth muscle contraction that were previously thought to be evidence for the existence of an unknown regulatory mechanism in addition to light chain phosphorylation. It must be realized that such models cannot disprove the existence of an additional regulatory mechanism, but can only demonstrate whether an additional mechanism needs to be postulated. While all the rate constants in the kinetic scheme are important, it seems that the key element is the high phosphatase activity. An appealing possibility is that different smooth muscles may have similar rate constants for crossbridge attachment (ka), crossbridge cycling (ks) and latchbridge detachment (kd), but the muscles may have different amounts of kinase and phosphatase, and therefore different values of kk and kp. In such a situation, the muscles with more kinase and phosphatase would be more likely to demonstrate "latch" behavior. Loss of myosin light chain phosphatase may be the reason why "latch" behavior is not easily demonstrated in skinned preparations. Finally, since the rate constants may have different temperature dependences, "latch" behavior may be observed at some temperatures, but not at others.

REFERENCES

Aksoy MO, Mras S, Kamm KE, Murphy RA (1983). Ca^{2+}, cAMP, and changes in myosin phosphorylation during contraction of smooth muscle. Am J Physiol 245: C255-C270.

Chatterjee M, Tejada M (1986). Phorbol ester-induced contraction in chemically skinned vascular smooth muscle. Am J Physiol 251: C356-C361.

Dillon PF, Aksoy MO, Driska SP, Murphy, RA (1981). Myosin phosphorylation and the cross-bridge cycle in arterial smooth muscle. Science 211: 495-497.

Driska SP (1986). High myosin light chain phosphatase activity in arterial smooth muscle: implications for regulatory mechanisms involving light chain phosphorylation. Biophys J 49:70a (Abstract).

Gerthoffer WT, Murphy RA (1983). Ca^{2+}, myosin phosphorylation, and relaxation of arterial smooth muscle. Am J Physiol 245: C271-C277.

Hoar PE, Kerrick WGL, Cassidy PS (1979). Chicken gizzard: relation between calcium-activated phosphorylation and contraction. Science 204: 503-506.

Kamm KE, Stull JT (1986). Activation of smooth muscle contraction: relation between myosin phosphorylation and stiffness. Science 232: 80-82.

Krisanda JM, RJ Paul (1984). Energetics of isometric contraction in porcine carotid artery. Am J Physiol 246: C510-C519.

Morgan JP, Morgan KG (1982). Vascular smooth muscle: the first recorded Ca^{2+} transients. Pfleugers Arch 395:75-77.

Srivastava S, Sasser G, Driska SP (1987). A simple method of preparing myosin from porcine aorta. Preparative Biochemistry 17(1): 1-8.

Stein PG Driska SP (1984). Histamine-induced rhythmic contraction of hog carotid artery smooth muscle. Circ Res 55:480-485, 1984.

Supported by NIH HL24881 and NIH RCDA HL01198.

DEPENDENCE OF STRESS AND VELOCITY ON Ca^{2+} AND MYOSIN PHOSPHORYLATION IN THE SKINNED SWINE CAROTID MEDIA

Meeta Chatterjee, Chi-Ming Hai, and R. A. Murphy

Department of Physiology, School of Medicine, University of Virginia, Charlottesville, VA 22908

INTRODUCTION

A consistent observation in skinned smooth muscle is that stress development is proportional to Ca^{2+}-induced myosin phosphorylation. In contrast, intact tissues can (very slowly) develop near maximal stress without exceeding 25% phosphorylation (Murphy, et al., this volume). Unfortunately many skinned preparations exhibit lower peak levels of phosphorylation and stress when compared to the values which can be elicited prior to skinning, and the preparations characteristically deteriorate with time and repeated contractions. In an effort to minimize these problems we employed a brief skinning procedure with Triton X-100 in the swine carotid media (Gordon, 1978; Chatterjee and Murphy, 1983). These preparations typically exhibit Ca^{2+}-induced stress development equal to 70% of the preskinning depolarization-induced stress with 70% maximal phosphorylation. The skinned tissues also exhibit stress maintenance when the Ca^{2+} concentration is reduced and phosphorylation falls (Fig.1). This behavior, which we interpreted as analogous to the 'latch' phenomenon in intact tissues, has not been reported in other skinned smooth muscle preparations.

A prediction of the hypothesis diagrammed in Fig. 1C is that the shortening velocity at zero load (V_o) should be greater when the Ca2+ is raised to a given concentration than when it is reduced from a high value to the same concentration. That is, V_o should be reduced in a protocol favoring the latch state. Our objective was to test the model by determining V_o when the $[Ca^{2+}]$ was increased or decreased to the same final concentration.

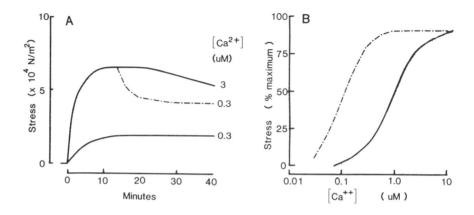

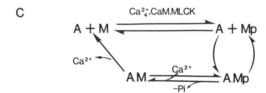

Figure 1. A. Diagrammatic responses of the briefly skinned carotid media to 0.3, and 3.0 uM Ca^{2+}, and 3.0 uM followed by a reduction to 0.3 uM Ca^{2+}. Note that stress is well maintained at low $[Ca^{2+}]$, but deterioration is evident at 3 uM Ca^{2+}. Deterioration can be dramatic at supramaximal $[Ca^{2+}]$. B. Dependence of developed stress (solid curve) when the $[Ca^{2+}]$ is increased and for maintained stress (interrupted curve) when the $[Ca^{2+}]$ is reduced from maximally activating concentrations (from Chatterjee and Murphy, 1983). The solid curve depicts phosphorylation under both conditions. All measurements were made soon after the response to a solution change stabilized to avoid or minimize deterioration. C. The hypothesis (Aksoy, et al., 1983) was that a regulatory mechanism with a greater sensitivity to Ca^{2+} than myosin kinase was involved in control of stress maintenance by dephosphorylated non- or slowly cycling crossbridges. A = thin filament (actin); M = crossbridge (myosin); M_p = phosphorylated, cycling crossbridge; AM = non- or slowly cycling 'latchbridge'; CaM = calmodulin; MLCK = myosin light chain kinase; MLCP = myosin light chain phosphatase.

METHODS

Strips of the carotid media were dissected (Driska, et al., 1981), and attached to the lever apparatus (Cambridge Technology 300-H Dual Mode Servo; Singer, et al., 1986), and equilibrated in physiological salt solution at 22° C for 2 hours. The preparations were then adjusted to the optimum length for stress generation (L_o) and a control response to K+ depolarization was obtained. Tissues were then skinned for 60 min by exposure to a solution containing 0.5% (v/v) Triton X-100 (Chatterjee and Murphy, 1983). Measurements of shortening velocity or phosphorylation were made after a constant stress was achieved when the $[Ca^{2+}]$ was increased from 0.018 uM or decreased from high concentrations. Ca^{2+} concentrations were determined with a CaEGTA/EGTA buffer system using the association constants selected by Fabiato (1981) for the relevant ionic equilibria. Total EGTA was 5 mM in the test solutions, and 0.1 mM in any precontracting solution.

Shortening velocities were obtained after quick-release to a constant load which produced an initial elastic recoil followed by shortening at a rate described by two exponentials. The muscles were allowed to shorten for 1.5 sec. The length vs. time data (collected and stored every 1.6 msec) were fitted with two exponentials and the rate of shortening calculated from the slow exponential term extrapolated to the instant of release. Velocity-stress curves were constructed from releases to loads of 0.062, 0.125, 0.187, 0.250, 0.312, and 0.437 x developed stress in random order during a single contraction over several minutes. V_o was calculated from a transformation of the hyperbolic velocity-stress data by fitting the data with a linear regression in a plot of S_r/S vs $(1 - S_r/S)/V$, where S = steady-state stress, S_r = load after release, and V = the value of dL/dt at $t = 0$ for the slow exponential shortening term.

Phosphorylation of the 20 kilodalton myosin regulatory light chain was determined by two dimensional polyacrylamide gel electrophoresis (Driska, et al., 1981) in tissues quick-frozen in an acetone-dry ice slurry (-80° C). Results are reported as fractional phosphorylation, neglecting satellite light chain species. All values are reported as means $\pm$ 1 SEM. Statistical significance was judged by Student's t-test for unpaired data ($p < 0.05$).

RESULTS

The results of experiments in which the $[Ca^{2+}]$ was increased from about 0.02 uM or decreased from 13 uM are shown in Fig. 2. In these experiments both V_o (Fig. 2A) and stress (Fig. 2B) were proportional to the $[Ca^{2+}_2]$. V_o was proportional to the level of myosin phosphorylation (Fig.3), irrespective of the change in $[Ca^{2+}]$. In these preparations, stress maintenance with decreasing Ca^{2+} and phosphorylation was not observed. These results differ from the experimental data reproduced in Fig. 1B (dashed curve).

The inconsistency between the data shown in Fig. 1B and Fig. 2 prompted an extensive series of prolonged time course studies in the skinned carotid media (Fig. 4). Prolonged exposure of the tissue to high $[Ca^{2+}]$ led to deterioration (not illustrated). Over the course of 3 hours the stress fell progressively to 20-25% of that initially developed. One contributory cause was found to be an irreversible super-contraction of some cells on the dissected edge of the preparation (Moreland and Murphy, in press). Because of these effects we used protocols in which the exposure to high $[Ca^{2+}]$ was limited. After some reduction in the $[Ca^{2+}]$, the stress reached a new lower value within approximately 20 min (Fig. 4). However, this proved to be a pseudo steady-state and was followed by a slow decline to a lower sustained value. While some deterioration cannot be ruled out, the results were very different than at 3 uM Ca^{2+} where stress fell progressively and did not stabilize after 3 hours.

Prolonged exposure to low $[Ca^{2+}]$ also gave unexpected results. After rapid attainment of a modest plateau, stress began to rise (Fig. 4). The steady-state stress at 3 hours was twice the initially developed value. The initial stress developed at low $[Ca^{2+}]$ was also a pseudo steady-state. Such results imply that the kinetics of the responses of the skinned carotid media include some very slow processes which we previously attributed to artifacts. Given sufficient time, there appears to be a unique dependence of stress (and phosphorylation) on the $[Ca^{2+}]$, irrespective of the protocol.

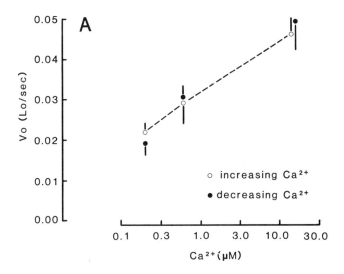

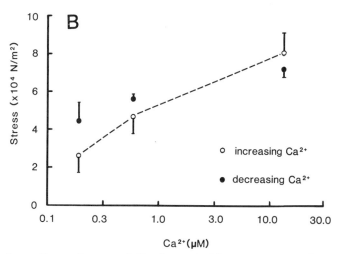

Figure 2. Dependence of V_o (panel A) and stress (panel B) on the [Ca2+] in solutions containing (mM): imidazole (pH 6.7 @ 22° C), 20; K-acetate, 50; $MgCl_2$, 6; ATP, 6; and dithiothreitol, 5. The ionic strength was 0.12, MgATP = 5.0 mM, and Mg^{2+} = 0.8 - 0.9 mM. There were no statistically significant differences in the results at the same [Ca^{2+}] with the two protocols.

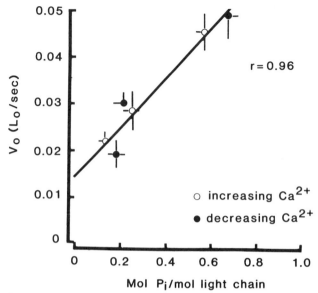

Figure 3. Dependence of V_o on phosphorylation of the myosin regulatory light chain. Line is a linear regression. Experimental protocol as in Methods and legend to Fig. 2.

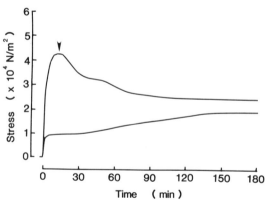

Figure 4. Time course of stress development when the $[Ca^{2+}]$ was increased to 0.2 uM (lower trace) or when Ca^{2+} was reduced from 3 to 0.2 uM (at arrow in upper trace). A large increase in the concentration of the EGTA buffer system would speed the initial relaxation and extend the plateau phase (Moreland and Murphy, 1986).

DISCUSSION

Deterioration and Attainment of a Steady-State

An early observation in our studies of the briefly skinned carotid media preparation was that exposure to maximally activating [Ca^{2+}] produced a marked, progressive fall in stress. Factors which may contribute include diffusional loss of proteins involved in regulation, accumulation of metabolites, activation of a Ca^{2+}-dependent protease (Haeberle, et al., 1985), and irreversible supercontraction of some cells (Moreland and Murphy, in press). We adopted rigorous protocols to minimize this problem including avoidance of supermaximal [Ca^{2+}], brief exposures to high [Ca^{2+}], and limiting experiments to one response of less than 60 min. The results obtained (Chatterjee and Murphy, 1983) were consistent with available information on the intact carotid media where phosphorylation transients inevitably occurred with the agonists employed. That is, if [Ca^{2+}] and phosphorylation fell from high to lower levels, stress maintenance was observed. The inference that the latch state was regulated by an unidentified Ca^{2+}-dependent regulatory system with a high Ca^{2+}-sensitivity was supported by this work.

In contrast to maximal responses, low stresses induced by submicromolar [Ca^{2+}] are very stable. However, Fig. 4 illustrates a new observation in this preparation: a slow component of the response to changes in [Ca^{2+}]. This implies (1) that deterioration may not be a major problem if high stresses are only briefly developed and (2) that initial stress development or stress maintenance is a pseudo steady-state. Over 2 hours were required to attain a true steady-state stress which was approximately proportional to the [Ca^{2+}], irrespective of the protocol (Fig. 4).

Comparisons with Intact Carotid Media

Ca^{2+}-depleted intact tissues contract in the absence of a phosphorylation transient (Murphy, et al., 1987). Small sustained increases in phosphorylation were associated with an initial rapid stress development. This was subsequently followed by a very slow increase in stress to a high value. The stress attained was equal to that rapidly developed during an initial phosphorylation transient induced by the same agonist in tissues which were not Ca^{2+}-depleted. This behavior was qualitatively different from observations in skinned tissues 40 min after the addition of Ca^{2+} (Fig. 1; Moreland and Murphy, in press). However, it is qualitatively the same as the behavior of skinned tissues over three hours (Fig. 4). The differences are quantitative in that (1) attainment of a steady state is considerably faster in the intact tissues and (2) steady-state stresses at low phosphorylation levels are considerably higher in intact tissues. An interpretation is that the kinetics of the response of skinned fibers are extremely slow and that two phases are distinguished in terms of an initial pseudo steady-state and later true steady state; at least under conditions when deterioration is minimized.

Regulation of Stress Maintenance (Latch)

The recent studies of intact tissues showed that stress development or maintenance was always associated with phosphorylation. However, the levels of phosphorylation required were low, with near maximal stress associated with about 25% phosphorylation (Murphy, et al, 1987). The model shown in Fig. 1C was developed to explain early observations of stress maintenance after phosphorylation had fallen to levels not significantly different from the unstimulated controls (~15% - Driska, et al., 1981; Aksoy, et al., 1983). Thus, the model shown in Fig. 1C might be unnecessarily complicated in postulating a second Ca^{2+}-dependent regulatory system for the latch state. A simplified hypothesis (Fig. 5) was modeled. Ca^{2+}-stimulated phosphorylation alone appeared capable of predicting the responses of intact tissues. If the rate of 'latchbridge' detachment is slow (K_7 in Fig. 5) relative to the rates of crossbridge phosphorylation, low levels of Ca^{2+} and phosphorylated crossbridges can theoretically sustain large populations of 'latchbridges.' The model predicts slow stress development to high levels with modest sustained increases in phosphorylation as was observed in the intact tissues.

While the model illustrated in Fig. 1C was the minimum working hypothesis to explain the skinned fiber data during the initial pseudo steady-state, the simplified scheme (Fig. 5) may also explain the long term behavior of skinned tissues. It can also theoretically account for the quantitative differences in the true steady-state behavior

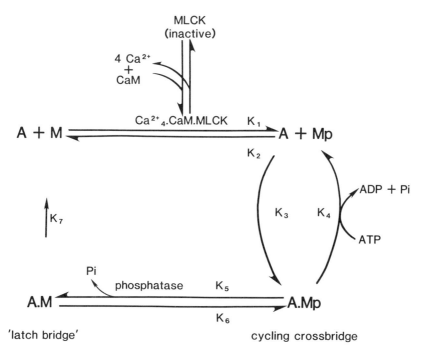

Figure 5. Simplified model which qualitatively predicts changes in crossbridge phosphorylation (= M_p + AM_p) and stress (= AM_p + AM). Symbols as in Fig. 1. Rate constants are: phosphorylation of free (K_1) and attached (K_6) crossbridges, dephosphorylation of free (K_2) and attached (K_5) crossbridges; crossbridge attachment (K_3) and detachment (K_4) from the thin filament; and latchbridge detachment (K_7). The model assumes that latchbridges are produced by dephosphorylation, that latchbridge detachment is the slowest reaction, and that each head of the myosin molecule is phosphorylated and functions independently.

of intact and skinned tissues. The experimentally determined rates of phosphorylation in skinned carotid media preparations are very slow (minutes) compared to intact tissues (seconds). Thus, latchbridge formation would be slowed relative to breakage. The size of the pool of latchbridges which would be maintained at a low constant level of phosphorylation would be reduced. This could explain why stress is low in skinned carotid media at < 25% phosphorylation and increases with higher levels of phosphorylation.

The hypothesis depicted in Fig. 5 may also provide an explanation for the pseudo steady-state observed some 20 min after a reduction in $[Ca^{2+}]$ (Fig. 4). At the initial high $[Ca^{2+}]$, the majority of the crossbridges would be phosphorylated (typically 60 - 70%). Phosphorylation falls when the Ca^{2+} is reduced to 0.2 uM. The model predicts that many of the phosphorylated, cycling crossbridges would be converted to latchbridges. The latter may underlie the initial stress maintenance due to their slow detachment rate. Since the reduced level of phosphorylation cannot regenerate the resulting large pool of latchbridges, stress will slowly decay to the lower steady-state value.

Dependence of V_o on Phosphorylation

Unloaded shortening velocity (Paul, et al., 1983) or V_o (Arner, 1983; Arner and Hellstrand, 1985) increases with the $[Ca^{2+}]$ in skinned smooth muscle. Steady-state V_o is directly proportional to crossbridge phosphorylation in intact smooth muscle. We have interpreted this as reflecting the relative sizes of the pools of phosphorylated, cycling crossbridges and latchbridges. A similar interpretation is not sufficient to explain the data on skinned tissues (Figs. 2 and 3). The postulated ratio of latchbridges to phosphorylated crossbridges should be higher when Ca^{2+} was decreased than when Ca^{2+} was increased to the same concentration at the time period when these measurements were made. It appears that the releases made to obtain the velocity data caused an abrupt loss of the stress maintenance capacity and rapid attainment of a steady-state. Although there was a trend for stress to be higher when the $[Ca^{2+}]$ was decreased (Fig. 2B), the differences proved to be insignificant. Mechanistically, if stress maintenance in the pseudo steady-state is due to a persisting population of latchbridges, a quick-release followed by a return to L_o may lead to detachment. The observed stress would then reflect the level proportional to phosphorylation.

Conclusions

The intact and skinned carotid media differ in some basic characteristics. These include the rates of reaching steady-states and the inability of the skinned preparation to attain stresses approaching those generated by intact tissues at 20% phosphorylation unless the $[Ca^{2+}]$ is high and phosphorylation reaches some 70%. In other respects intact and skinned tissues are similar such as in the direct dependence of V_o on phosphorylation. The hypothesis depicted in Fig. 5 can account for the differences in terms of changes in the rate constants of the depicted reactions after skinning (i.e., reduce K_1 and K_5 to values approaching K_7)1.

While the latch state appears to be consistent with a regulatory system involving only Ca^{2+}-calmodulin activated myosin kinase and a possibly unregulated myosin phosphatase, other control mechanisms may play a role. There is evidence for a number of potential Ca^{2+}-dependent regulatory sites (Marston, 1982). Of particular interest is Hoar, et al.,'s (1985) evidence that Ca^{2+} may slow K_7, the rate of latchbridge detachment (Fig. 5). Our Occam's razor approach in model development to favor the simplest hypothesis does not exclude other regulatory systems. Based on our work with intact tissues and the results described here, it appears plausible that phosphorylation is necessary and sufficient to explain both activation and the latch state in smooth muscle. However, any system affecting the illustrated rate constants (Fig. 5) would modulate the response of the contractile system. Changes in the response could be profound, including the loss of latch behavior. This may provide an explanation for marked differences reported for the properties of various skinned smooth muscle preparations (c.f. Kerrick and Hoar, 1985; Gagelmann, et al., 1985).

REFERENCES

Aksoy MO, Mras S, Kamm KE, Murphy RA (1983). Ca^{2+}, cAMP, and changes in myosin phosphorylation during contraction of smooth muscle. Am J Physiol 245 (Cell Physiol 14): C155-C270.

Arner A (1983). Force-velocity relation in chemically skinned rat portal vein: effects of Ca^{2+} and Mg^{2+}. Pflügers Arch 397:6-12.

Arner A, Hellstrand P (1985). Effects of calcium and substrate on force-velocity relation and energy turnover in skinned smooth muscle of the guinea pig. J Physiol (Lond) 360:347-365.

Chatterjee M, Murphy RA (1983). Calcium-dependent stress maintenance without myosin phosphorylation in skinned smooth muscle. Science 221:464-466.

Driska SP, Aksoy MO, Murphy RA (1981). Myosin light chain phosphorylation associated with contraction in arterial smooth muscle. Am J Physiol 240 (Cell Physiol 9):C222-C233.

Fabiato A (1981). Myoplasmic free calcium reached during the twitch of an intact isolated cardiac cell and during calcium-induced release of calcium from the sarcoplasmic reticulum of a skinned cardiac cell from the adult rat or rabbit ventricle. J Gen Physiol 78:457-497.

Gagelmann M, Arner A, Chacko S (1985). Relation between the level of phosphorylation, force and ATPase during dephosphorylation of the regulatory myosin light chains in chemically skinned taenia coli smooth muscle fibers (1985). Adv Protein Phosphatases II:400 (abstract).

Gordon, AR (1978). Contraction of detergent-treated smooth muscle. Proc Natl Acad Sci USA 75:3527-3530.

Haeberle JR, Collican SA, Evans A, Hathaway DR (1985). The effects of a calcium-dependent protease on the ultrastructure and contractile mechanics of skinned uterine smooth muscle. J Muscle Res Cell Motility 6:347-363.

Hoar PE, Pato MD, Kerrick WG (1985). Myosin light chain phosphatase: effect on the activation and relaxation of gizzard smooth muscle skinned fibers. J Biol Chem 260:8760-8764.

Kerrick WGL, Hoar PE (1985). Regulation of contraction in skinned smooth muscle cells by Ca^{2+} and protein phosphorylation. Adv Protein Phosphatases II:133-152.

Marston SB (1982). The regulation of smooth muscle contractile proteins. Prog Biophys Molec Biol 41:1-41.

Moreland RM, Murphy RA (1986). Determinants of Ca^{2+}-dependent stress maintenance in skinned swine carotid media. Am J Physiol 251:C892-C903.

Murphy RA, Ratz PH, and Hai, C-M (1987). Determinants of the latch state in vascular smooth muscle. IN Siegman MJ (ed):"Smooth Muscle Contraction", New York: Alan R. Liss (Abstract)

Paul RJ, Doerman G, Zeugner C, Rüegg JC (1983). The dependence of unloaded shortening velocity on Ca^{2+}, calmodulin, and duration of contraction in "chemically skinned" smooth muscle. Circulation Res 53:342-351.

Singer HA, Kamm KE, Murphy RA (1986). Estimates of activation in arterial smooth muscle. Am J Physiol 251 (Cell Physiol 20):C465-C473.

Supported by NIH grant 5 P01 HL19242 and fellowships 5 T32 HL07355 from NIH (M. C.) and the American Heart Association, Virginia Affiliate (C.-M. Hai).

DETERMINANTS OF THE LATCH STATE IN VASCULAR SMOOTH MUSCLE

R.A. Murphy, P.H. Ratz, and C.M. Hai.

Department of Physiology
University of Virginia
Charlottesville, Virginia 22908

Tonic stimulation of intact smooth muscle is typically characterized by initial transient increases in crossbridge phosphorylation and high rates of crossbridge cycling followed by declines in both parameters to lower steady-state values. "Latch" was defined to describe tonic stress maintenance with reduced phosphorylation levels and lower cycling rates. Studies of skinned tissues showed that (1) Ca^{2+}-dependent myosin phosphorylation determined stress development and that (2) stress could subsequently be maintained if the [Ca^{2+}] was reduced to values which did not support proportional phosphorylation. These observations suggested that the long-lasting crossbridge attachments characterizing the latch state arise by dephosphorylation of cycling crossbridges and that they were dependent on a second Ca^{2+}-dependent regulatory mechanism with a greater sensitivity to Ca^{2+} than myosin kinase.

Recent studies of intact swine carotid media revealed quantitative differences from skinned preparations. We tested the hypothesis that initial myosin phosphorylation transients were not necessary for the attainment of high steady-state stress. Large transient increases in phosphorylation were largely eliminated when intact strips were activated by Ca^{2+} influx in the presence of histamine or phenylephrine following depletion of Ca^{2+} from the sarcoplasmic reticulum. Elimination of the phosphorylation transient produced a marked decline in the rate of stress development, but did not affect the steady-state levels of stress or phosphorylation. Bay k 8644 was used to bypass receptor-operated activation mechanisms and contract the swine carotid media. This Ca^{2+} channel agonist produced strong, although very slow, contractions and increased myosin phosphorylation only slightly above basal values. These data showed that a large increase in myosin phosphorylation is not obligatory for high stress development in intact tissues. However, agonist-induced phosphorylation transients greatly enhance the rate of contraction and attainment of steady-state stress.

In summary, data from intact tissues demonstrate that (1) steady-state stress exhibits a very steep dependence upon phosphorylation with maximal stress attained at about 25% phosphorylation. (2) Initial transients in myoplasmic Ca^{2+} and phosphorylation are not necessary for the attainment of high values of stress, although they can increase the rate of stress development by 10- to 50-fold. (3) Steady-state shortening velocities with zero external load are directly proportional to the level of crossbridge phosphorylation between 5 and 60%. To explain these observations we have hypothesized a two state model in which the relative sizes of the populations of phosphorylated cycling crossbridges and of non- or slowly-cycling "latchbridges" determine average cycling rates and V_o.

$$
\begin{array}{ccc}
\text{A + M} & \xrightarrow{k_1} & \text{A + M}_p \\
\text{(relaxation)} & \underset{k_2}{} & \\
& & \\
k_7 \updownarrow & & k_4 \updownarrow \; \text{cycling} \; \updownarrow k_3 \\
& k_6 \updownarrow & \\
\text{AM} & & \text{AM}_p \\
\text{(latch)} & & \\
& k_5 &
\end{array}
$$

In this model A = thin filament, M = myosin crossbridge, M_p = phosphorylated crossbridge, AM = latchbridge, k_1 and k_2 are the rate constants for phosphorylation and dephosphorylation of free crossbridges, respectively, k_3 and k_4 are the rate constants for crossbridge detachment and attachment, k_5 and k_6 are the rate constants for dephosphorylation and phosphorylation of attached crossbridges, and k_7 is a slow rate constant for latchbridge detachment.

Computer simulations reveal that this model can quantitatively mimic both transient and steady-state behavior of intact swine carotid media and that fairly small pools of phosphorylated crossbridges can maintain a much larger population of latchbridges and high values of stress. A second regulatory system with a very high sensitivity to Ca^{2+} need not be invoked to explain the latch state. That is, Ca^{2+} stimulated crossbridge phosphorylation coupled with significant rates of dephosphorylation can lead to large AM_p pools. However, this scheme is not inconsistent with a Ca^{2+} dependence for k_7, or latchbridge detachment as postulated by Hoar et al (J Biol Chem 260:8760-8764, 1985).

In conclusion, the behavior of intact swine carotid tissues may be explained by (1) an obligatory crossbridge phosphorylation for crossbridge attachment and cycling, followed by (2) buildup of a population of non- or slowly-cycling latchbridges formed by dephosphorylation. The quantitative differences between the properties of intact and skinned preparations may be explained by differences in the rate constants in the illustrated scheme such that step (2) is greatly reduced in skinned tissues.

Supported by PHS grants 5 P01 HL19242, 1 F32 HL06987, and a fellowship from the Am. Heart Assoc., VA Affiliate.

REGULATION OF CAROTID ARTERY SMOOTH MUSCLE RELAXATION BY MYOSIN DEPHOSPHORYLATION

Joe R. Haeberle, Brett A. Trockman, and Anna A. Depaoli-Roach

Departments of Physiology, (J.R.H. & B.A.T.), Biochemistry (A.A.D.), and the Krannert Institute of Cardiology (J.R.H.), Indiana University School of Medicine, Indianapolis, IN 46223

INTRODUCTION

It is widely accepted that phosphorylation of the 20,000 dalton light chain of smooth muscle myosin (LC_{20}) is both necessary and adequate for the initiation of contraction. The most direct demonstrations of a causal relationship between LC_{20} phosphorylation and contraction have come from work with chemically skinned smooth muscle preparations. Phosphorylation by a calcium-independent form of myosin light chain kinase, in the presence of EGTA, initiated contraction of skinned smooth muscle (Walsh et al, 1982) and the subsequent removal of the kinase or the addition of purified phosphatase resulted in relaxation (Hoar et al, 1985a). Such experiments have shown that phosphorylation and dephosphorylation are adequate to produce contraction and relaxation and that calcium is not directly required.

Experiments with intact smooth muscle, however, suggest that calcium may play a role in the maintenance of isometric force during tonic contractions. Murphy and colleagues originally reported that with tonically contracted muscles, LC_{20} phosphorylation often returned to the same level measured in resting muscles with no loss of isometric force (Dillon et al, 1981). These investigators have suggested that calcium-dependent, slowly-cycling crossbridges are formed by dephosphorylation of contracted muscle in the presence of free calcium in the range of 0.1 - 1 uM (Rembold et al, 1985). Others have previously postulated the existence of attached, non-cycling crossbridges in smooth muscle (Siegman et al, 1980; Meiss, 1971). Murphy and associates have coined the term "latch-bridge" to refer to attached, slowly-cycling crossbridges and have suggested that relaxation or dissociation of latch-bridges requires that intracellular calcium be reduced to less than 0.1 uM (Rembold et al, 1985). Although similar results have been reported for tracheal smooth muscle (Silver et al, 1982; Gerthoffer et al, 1983), we have not seen evidence for force maintenance in the absence of LC_{20} phosphorylation in intact rat uterine smooth muscle (Haeberle et al, 1985a). We found that under stead-state conditions there was a consistent and linear relationship between the stoichiometry of LC_{20} phosphorylation and isometric force maintenance (Haeberle et al, 1985a).

Efforts to demonstrate a latch-state in skinned smooth muscle have been controversial. Others have shown that the addition of a purified smooth muscle phosphatase to a partially contracted skinned gizzard smooth muscle preparation, in the presence of micromolar calcium, produced dephosphorylation but not relaxation (Hoar et al, 1985b). We have reported similar experiments using skinned rat uterine smooth muscle and a phosphatase catalytic subunit from skeletal muscle (Haeberle et al, 1985b). In contrast, we found that addition of the phosphatase to fully contracted muscles, in the presence of saturating calcium and calmodulin, produced rapid and complete dephosphorylation associated with full relaxation.

Others have shown evidence for a latch-state in skinned porcine carotid artery by looking at the calcium sensitivity of both phosphorylation and isometric force maintenance during activation and relaxation (Chatterjee et al, 1983). During relaxation, isometric force was maintained, in spite of LC_{20} dephosphorylation, in the range of 0.1 - 1 uM calcium. Relaxation required that the free calcium concentration be reduced to less than 0.1 uM. Similar experiments, utilizing glycerinated rat uterine smooth muscle, showed only minor differences in the calcium sensitivity of steady-state isometric force maintenance and LC_{20} phosphorylation during contraction and relaxation (Tanner et al, 1986).

These differences between the rat uterine smooth muscle and both the carotid artery and tracheal smooth muscles suggested that the regulation of contraction may differ in

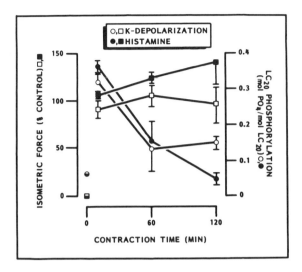

Figure 1. Time-course of isometric force development and LC_{20} phosphorylation for intact porcine carotid artery smooth muscle. Muscles were contracted either by potassium-depolarization (100 mM) or with 10^{-5} M histamine. Points are means ± S.E.M. for 4-8 muscles.

these smooth muscle types. Of particular interest was the possibility that the uterine smooth muscle was not capable of forming latch-bridges due to the lack of a necessary regulatory factor(s). The purpose of the present study was to determine if a latch-state could be demonstrated in carotid artery smooth muscle under the same conditions previously used to study the rat uterus.

STUDIES WITH INTACT PORCINE CAROTID ARTERY SMOOTH MUSCLE

Initially we examined intact porcine carotid artery strips to confirm the results obtained previously by other laboratories showing that LC_{20} dephosphorylation could occur during prolonged contractions of vascular smooth muscle without loss of isometric force. As shown in Fig. 1, we observed a significant decline in LC_{20} phosphorylation during 120 min contractions of porcine carotid artery strips stimulated either with histamine or potassium depolarization. The level of phosphorylation after 120 min of contraction in the presence of histamine was not significantly different from the level measured in the control, unstimulated muscles, and was significantly lower than that measured at 10 min of contraction. Isometric force continued to steadily increase throughout the entire contraction period in spite of the dephosphorylation of LC_{20}.

STUDIES WITH GLYCERINATED PORCINE CAROTID ARTERY

Chemically skinned smooth muscle preparations allow for direct biochemical manipulation of the contractile proteins in a partially intact contractile system while monitoring the effects on contraction and relaxation. The same procedure was used to glycerinate porcine carotid arteries as was previously described for glycerination of rat uterus. This method yielded permeabilized muscle fibers with intact myofilamentous ultrastructure (Haeberle et al, 1985c). These fibers could be contracted repeatedly without loss of peak force and could be phosphorylated up to 1.0 mol PO_4/mol LC20 (Haeberle et al, 1985b). Fibers prepared in this manner could be stored at -70°C for at least 8 months with no apparent loss of function. We have used this same procedure to skin porcine carotid arteries. Segments of whole artery which have been glycerinated and frozen were thawed at the time of an experiment and a thin layer (100-200 um thick) was dissected from the endothelial surface of the tissue. Fig. 2 demonstrates the reproducibility of individual contractions with regard to both peak isometric force development and completeness of relaxation. All experiments reported here involved only the first two contractions for any given muscle.

The major focus of the skinned muscle experiments was to determine if relaxation of vascular smooth muscle was mediated by dephosphorylation or by a combination of dephosphorylation and reduced calcium as has been proposed by others (Murphy et al, 1983). By using a skinned muscle preparation it was possible to manipulate LC_{20} phosphorylation and calcium independently so that the relative contributions of each to the regulation of smooth muscle relaxation could be determined. We first looked at dephosphorylation of contracted muscle in the presence of saturating calcium and calmodulin by adding purified

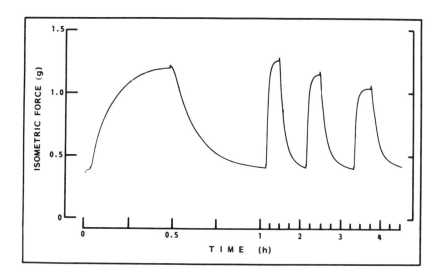

Figure 2. Repeated contractions of glycerinated porcine carotid artery smooth muscle. The contracting solution contained 20 uM free calcium, 10 uM calmodulin, 1 mM free magnesium and 6 mM MgATP. The relaxing solution contained no added calcium and 5 mM EGTA.

phosphatase to the muscle bath. We then examined the calcium-sensitivity of contraction and relaxation to determine if a shift in the calcium sensitivity to lower calcium concentrations occurred during relaxation as described by others (Chatterjee et al, 1983).

The catalytic subunit of a type 2A, skeletal muscle phosphatase (Haeberle et al, 1985b), was used to dephosphorylate the LC_{20} of fully contracted carotid arteries in the presence of saturating calcium (30 uM) and calmodulin (10 uM). Fig. 3 shows the effect of adding 0.28 uM phosphatase catalytic subunit (PCS) to the muscle bath of a fully contracted skinned muscle. The PCS initiated relaxation of the muscle and dephosphorylation of LC_{20} as shown in Table I. The time-course of dephosphorylation and relaxation was significantly slower than for rat uterine smooth muscle under similar conditions. The half-time of relaxation of the skinned rat uterus was 0.9 min compared to 7.0 min for the carotid artery. Fig. 4 demonstrates that

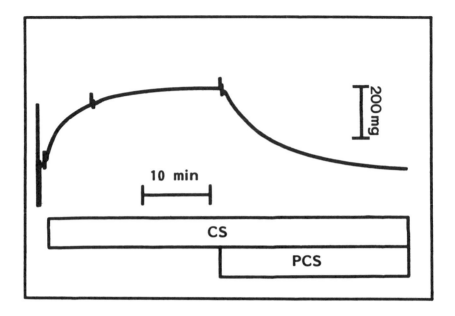

Figure 3. Relaxation of glycerinated porcine carotid artery smooth muscle by exogenous phosphatase catalytic subunit (PCS). PCS was added to a final concentration of 0.23 μM into (CS) contracting solution containing 20 uM Ca^{2+} and 10 uM calmodulin.

Table 1. Dephosphorylation and Relaxation of Skinned Porcine Carotid Artery Smooth Muscle by Phosphatase Catalytic Subunit

% Maximum Isometric Force	mol PO_4/mol LC20
100	0.85 ± 0.028 *
50	0.36 ± 0.043
0	0.25 ± 0.065

*Means ± S.D., n=3

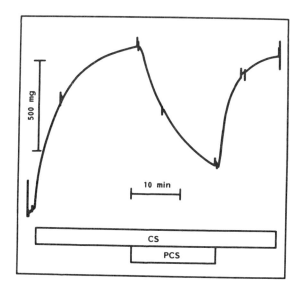

Figure 4. Reversal of PCS induced relaxation. The skinned muscle was in contracting solution (CS) containing 20 uM calcium and 10 uM calmodulin. PCS was washed out of the muscle by two changes of CS.

the effect of the PCS was completely reversed by washing the muscle with fresh contracting solution without PCS.

To control for the possibility that the effect of the PCS on relaxation was not causally related to dephosphorylation, we thiophosphorylated the skinned muscles to prevent dephosphorylation. Others have shown that thiophosphorylated myosin is resistant to dephosphorylation by myosin phosphatases and that thiophosphorylation of skinned smooth muscles prevents both dephosphorylation and relaxation in the presence of EGTA (Hoar et al, 1979). As shown in Fig. 5 the addition of 0.28 uM PCS to the muscle bath following thiophosphorylation of LC_{20} had no effect on force maintenance. In other experiments, in rat uterus, we have further demonstrated that the relaxing effect of the PCS is a direct result of LC_{20} dephosphorylation. The relaxing effect of the PCS could be reversed by subsequently adding either purified MLCK, to increase the level of LC_{20} phosphorylation, or by adding thiophosphorylated LC_{20}, to

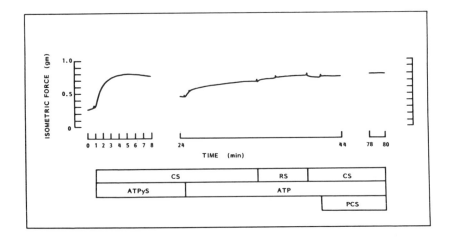

Figure 5. Inhibition of PCS induced relaxation by thiophosphorylation of LC_{20}. CS, contracting solution containing 20 uM Ca^{2+} and 10 uM calmodulin; RS, relaxing solution containing no added calcium; ATPyS, 2 mM thio-ATP + 6 mM ATP; ATP, 6 mM ATP; PCS, 0.28 uM phosphatase catalytic subunit.

competitively inhibit the PCS (Haeberle et al, 1985).

We have also measured the calcium-sensitivity of force maintenance during contraction and relaxation of the glycerinated carotid artery smooth muscle preparation. As shown in Fig. 6, there was no indication that the calcium sensitivity of force maintenance was different for contracting and relaxing muscles. The calcium sensitivity of force maintenance observed in these studies was similar to that reported by Chatterjee for contraction of detergent skinned porcine carotid artery (Chatterjee et al, 1983). The major difference between these two studies is that we did not observe a shift in the calcium sensitivity of force maintenance to lower calcium concentrations during relaxation.

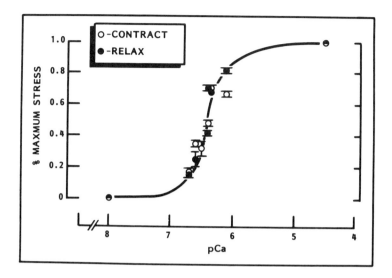

Figure 6. Calcium-sensitivity of isometric foce maintenance during contraction and relaxation of glycerinated porcine carotid artery. Maximum isometric stress averaged 23 ± 6.5 mN/mm^2 (mean ± S.D.).

DISCUSSION

Although it is now clear that myosin phosphorylation plays a central role in the regulation of smooth muscle contraction, a large body of experimental evidence exists which suggests that additional calcium-dependent regulatory processes may be involved in the regulation of smooth muscle contraction (see Kamm et al., 1985 for review). Studies in this laboratory with both intact and skinned uterine smooth muscle have demonstrated that isometric force maintenance and LC_{20} phosphorylation were closely coupled. Steady-state isometric force was linearly correlated with the stoichiometry of LC_{20} phosphorylation in both intact and chemically skinned muscle preparations (Haeberle et al, 1985a; Haeberle et al, 1985). Furthermore, there was no indication that the relationship between phosphorylation and force maintenance was different for contracting and relaxing muscles. Our observations with uterine smooth muscle are consistent with a simple regulatory model in which phosphorylation regulates the number of attached, cycling crossbridges. In this regard, Sellers (1985) has shown that phosphorylation regulates inorganic phosphate release and the formation of the high-affinity actomyosin complex.

It is intriguing that isometric force maintenance in intact carotid artery and tracheal smooth muscle is independent of the extent of phosphorylation under some conditions (Silver et al, 1982; Gerthoffer et al, 1983; Dillon et al, 1981), while in glycerinated muscles, phosphorylation and force are closely coupled under all conditions examined. Although we were able to show isometric force maintenance in the absence of LC_{20} phosphorylation in the intact carotid artery preparation, we were not able to show this with the glycerinated carotid artery. Our observations with glycerinated carotid artery smooth muscle were distinctly different from those reported by others using different skinned muscle preparations. In contrast to previous studies showing either a shift in the calcium-sensitivity of force maintenance during relaxation (Chatterjee et al, 1983), delayed mechanical relaxation relative to dephosphorylation (Chatterjee et al, 1983), or calcium-dependent force maintenance subsequent to dephosphorylation (Hoar et al, 1985b), no such effects were found in the current study. Our observation that dephosphorylation by a phosphatase catalytic subunit in the presence of saturating calcium and calmodulin produced simultaneous relaxation and dephosphorylation of the carotid artery smooth muscle suggests that dephosphorylation per se does not lead to latch-bridge formation. It is possible that soluble factors are lost during the skinning procedure which are necessary for the regulation of latch-bridge formation. On the other hand, an intact sarcolemma may be required for the initiation of latch-bridges. Finally, we must consider the possibility that proteins, in addition to LC_{20}, were dephosphorylated by the phosphatase catalytic subunit used in the present studies. This enzyme has a relatively broad substrate specificity compared to the smooth muscle phosphatase used by Hoar et al. (1985b) which is more specific for myosin. Arguing against this latter explanation is the fact that we did not see evidence for latch bridges when skinned muscles were relaxed by reducing calcium without adding phosphatase.

In conclusion, the results of the current study with glycerinated porcine carotid artery indicate that phosphorylation and dephosphorylation of LC_{20} are adequate to produce contraction and relaxation, respectively, of glycerinated vascular smooth muscle. Although we were able to demonstrate isometric force maintenance in the absence of elevated phosphorylation with intact carotid artery, we were not able to do so with skinned carotid artery. Our inability to initiate a latch-state, as others have reported, either by graded reduction of calcium or by dephosphorylation in the presence of calcium and calmodulin, suggests that latch-bridge formation may require the presence of additional regulatory factors which are either lost or inactivated in the glycerinated muscle preparation. Comparative examination of detergent extracted and glycerinated skinned muscle preparations might provide useful information concerning the regulation of latch-bridge formation in smooth muscle.

ACKNOWLEDGMENTS

This work was supported by grants from the American Heart Association, Indiana Affiliate; the National Institutes of Health (AM 35822); and the Krannert Institute of Cardiology.

REFERENCES

Chatterjee M, Murphy RA (1983). Calcium-dependent stress maintenance without myosin phosphorylation in skinned smooth muscle. Science 221:464-466.

Dillon PF, Aksoy MO, Driska SP, Murphy RA (1981). Myosin phosphorylation and the cross-bridge cycle in arterial smooth muscle. Science 211:495-497.

Gerthoffer WT, Murphy RA (1983). Myosin phosphorylation and regulation of cross-bridge cycle in tracheal smooth muscle. Am J Physiol 244:C182-C187.

Haeberle JR, Hott JW, Hathaway DR (1985a). Regulation of isometric force and isotonic shortening velocity by phosphorylation of the 20,000 dalton myosin light chain of rat uterine smooth muscle. Pflugers Arch 403:215-219.

Haeberle JR, Hathaway DR, DePaoli-Roach AA (1985b). Dephosphorylation of myosin by the catalytic subunit of a type-2 phosphatase produces relaxation of chemically skinned uterine smooth muscle. J Biol Chem 260:9965-9968.

Haeberle JR, Coolican SA, Evan A, Hathaway DR (1985c). The effects of a calcium dependent protease on the ultrastructure and contractile mechanics of skinned uterine smooth muscle. J Muscle Res Cell Motil 6:347-363.

Hathaway DR, Haeberle JR (1985). A radioimmunoblotting method for measuring myosin light chain phosphorylation levels in smooth muscle. Am J Physiol 249:C345-C351.

Hoar PE, Kerrick WG, Cassidy PS (1979). Chicken gizzard: relation between calcium-activated phosphorylation and contraction. Science 204:503-506.

Hoar PE, Pato MD, Kerrick WG (1985a). Myosin light chain phosphatase. Effect on the activation and relaxation of gizzard smooth muscle skinned fibers. J Biol Chem 260:8760-8764.

Hoar PE, Pato MD, Kerrick GL (1985b). Myosin light chain phosphatase: Effect on the activation and relaxation of gizzard smooth muscle skinned fibers. J Biol Chem 260:8760-8764.

Kamm KE, Stull, JT (1985) The function of myosin and myosin light chain kinase phosphorylation in smooth muscle. Annu Rev Pharmacol Toxicol 25:593-620.

Meiss RA (1971). Some mechanical properties of cat intestinal muscle. Am J Physiol 220:2000-2007.

Murphy RA, Aksoy MO, Dillon PF, Gerthoffer WT, Kamm KE (1983). The role of myosin light chain phosphorylation in the regulation of the cross-bridge cycle. Fed Proc 42:51-56.

Rembold MC, Hai C, Murphy RA (1985). Myoplasmic Ca2+ and activation of vascular smooth muscle. Adv Prot Phosphatases II:89-101.

Sellers JR (1985). Mechanism of the phosphorylation-dependent regulation of smooth muscle heavy meromyosin. J Biol Chem 260:15815-15819.

Siegman MJ, Butler TM, Mooers SU, Davies RE (1980). Chemical energetics of force development, force maintenance and relaxation in mammalian smooth muscle. J Gen Physiol 76:609-629.

Silver PJ, Stull JT (1982). Regulation of myosin light chain and phosphorylase phosphorylation in tracheal smooth muscle. J Biol Chem 257:6145-6150.

Tanner JA, Haeberle JR, Meiss RA (1986). The relationship between isometric force and stiffness and the stoichiometry of myosin phosphorylation during steady-state contraction of skinned smooth muscle. Biophys J 49:72a.

Walsh MP, Bridenbaugh R, Hartshorne DJ, Kerrick WG (1982). Phosphorylation-dependent activated tension in skinned gizzard muscle fibers in the absence of Ca2+. J Biol Chem 257:5987-5990.

CALCIUM/CALMODULIN ACTIVATION OF GIZZARD SKINNED FIBERS AT LOW LEVELS OF MYOSIN PHOSPHORYLATION

Jürgen Wagner, Gabriele Pfitzer, J. Caspar Rüegg,

II. Physiologisches Institut, University of Heidelberg, Im Neuenheimer Feld 326, D-6900 Heidelberg, Federal Republic of Germany

INTRODUCTION

Calmodulin occupied by calcium ions activates myosin light chain kinase thereby causing the phosphorylation of smooth muscle myosin and smooth muscle contraction. Here we present evidence that in addition calcium and calmodulin may also activate the contractile mechanism of smooth muscle independent of myosin phosphorylation. This mechanism may be distinct from other activating mechanisms which are said to be independent of both myosin light chain kinase and calmodulin. These include activation by leiotonin (Ebashi, 1980) or other protein factors (Persechini et al., 1981; Cole et al., 1982) or by direct calcium binding to smooth muscle myosin (Kaminski and Chacko, 1984).

MATERIALS AND METHODS

Immediately after slaughtering the chicken the gizzard was removed and dissected and strips from the outer circumferential layer were immersed for 30 min into a solution containing (in mM): EGTA 5, KCl 50, sucrose 150, imidazole 20 (pH 7.4), DTE 2. They were subsequently skinned for 4 hrs by the addition of Triton X-100 (1% vol/vol) at 4°C. The skinned fibers were then stored at -20°C in a solution containing 50% vol/vol glycerol and 50% relaxing solution containing (in mM): EGTA 4, KCl 50, MgCl 5, imidazole 25, ATP 1; pH 7.0 (Sparrow et al., 1981). Preparations stored for less than 1 week were considered to be "fresh". Aged preparations were obtained after storing the fibers for 6 to 8 weeks at -20°C or alternatively by storing the fibers at 35°C for 10 hrs. To measure isometric contractions, thin fiber bundles were glued to an AME 801 force transducer (SensoNor, Horten, Norway) and first suspended in calcium free relaxing solution containing (in mM): EGTA 4, KCl 50, MgCl 5, imidazole 25, ATP 1, creatine phosphate 1, creatine phosphokinase 0.4 mg/ml, as well as calmodulin at the concentrations indicated. The pH was 7.0, T=20°C.

Contractions were induced by increasing the Ca^{2+} concentration to the desired level by adjusting the proportion of EGTA and CaEGTA, and the free calciumion concentration was calculated by using an apparent binding constant of $2\times10^{6} M^{-1}$ (Portzehl et al., 1964). The activating solution contained calmodulin from bovine testis (Gopalakrishna and Anderson, 1982) in varying concentrations. Isometric contractions were induced in fiber bundles of 50-200 μm thickness and LC-20 phosphorylation was determined in parallel in other fiber bundles which were subjected to the same experimental protocol except that the fiber bundles were immersed in 15% ice-cold trichloroacetic acid at the desired time (20 min after onset of contraction). Phosphorylated and non-phosphorylated light chains were separated by 2D-gelelectrophoresis and quantified by densitometric scanning according to Gagelmann et al. (1984) and in some experiments according to Haeberle et al. (1984). Myosin light chain satellites were not observed. Unloaded shortening velocity was determined when force reached peak tension by using the slack test method, as adapted by Paul et al. (1983).

RESULTS AND DISCUSSION

As shown in Fig. 1, in freshly skinned gizzard muscle fibers an increase in calmodulin concentration enhances the rate and extent of force development elicited in activating solution in the presence of 1.6 μM Ca^{2+}. Note that the calmodulin-dependent activation is not associated with a corresponding increase in the extent of myosin light chain phosphorylation. The latter remains almost at basal levels (about 0.1 moles phosphate per mole light chain) as long as the calmodulin concentration does not exceed 0.1 μM (Fig. 2).

FIGURE 1

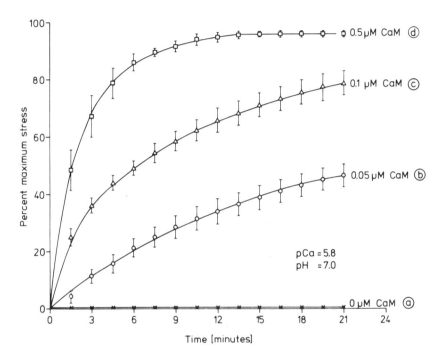

Fig. 1. Time course of increase in force of skinned fibers induced in activating solution at pCa 5.8 at various concentrations of calmodulin.

At higher concentrations of calmodulin, however, the extent of myosin phosphorylation increases, too, and reaches 0.48±0.12 moles phosphate per mole of light chain at 5 μM calmodulin and pCa 5.2. Since nearly 50% of the maximal tension may be developed at very low levels of phosphorylation (0.08-0.11 μmoles phosphate per mole of light chain) it is of interest whether the MgATP substrate may be replaced by MgITP which is not a substrate of the myosin light chain kinase (cf. Cassidy et al., 1979).

FIGURE 2

Calmodulin-dependence of stress and LC-2-phosphorylation in "native" fibres

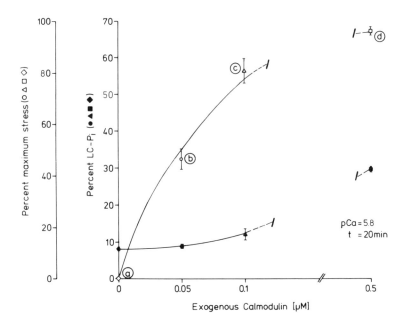

Fig. 2. Influence of exogenous calmodulin on phosphorylation of regulatory light chain and force of skinned fibers at pCa 5.8.

As shown in Fig. 3, skinned chicken gizzard muscle fibers contract in a calmodulin-dependent manner even when MgITP is used as a substrate for the actomyosin ATPase. The contractile force increases at a given calciumion concentration with increasing calmodulin concentration, and at maximal calmodulin concentration (2 µM) the force development attains about 50% of the maximal value in the presence of MgATP. In this respect, skinned chicken gizzard fibers differ from skinned guinea pig taenia coli muscle fibers that do not contract at all with MgITP as a substrate (cf. Arner et al., 1985). Though skinned chicken gizzard muscle fibers contract strongly at basal levels of myosin phosphorylation, their shortening velocity is extremely low under these conditions (~0.01 lengths/sec).

FIGURE 3

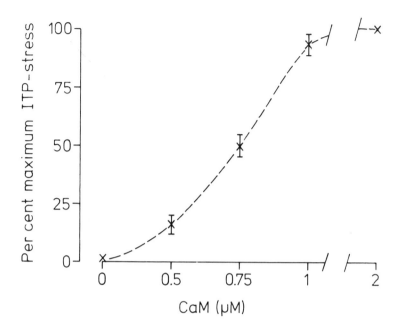

Fig. 3. Contractile force of skinned fibers induced in activating solution containing MgITP in place of MgATP, pCa 5.2; calmodulin concentration as indicated.

Fig. 4a shows the effect of calmodulin concentration on the unloaded shortening velocity, as determined by the slack test method. For comparison, the calmodulin-dependency of isometric force development of the same fibers is also shown. Note that force doubles when the calmodulin concentration increased from 0.05 to 0.1 µM at 1.6 µM Ca^{2+}, whereas the shortening velocity increased by a factor of 5. The shortening at basal levels of phosphorylation is, however, small compared with the maximal velocity of unloaded shortening (about 0.2 fiber length per second) which is obtained at high calmodulin concentration when levels of myosin light chain phosphorylation are high (>0.3 mole phosphate per mole light chain).

FIGURE 4

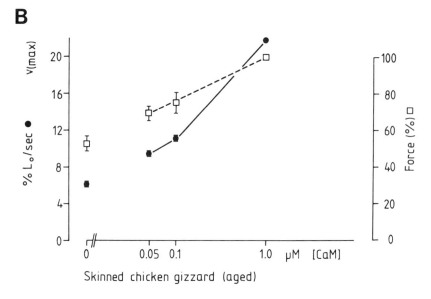

Fig. 4A and B. Effect of calmodulin concentration on force (open symbols) and unloaded shortening velocity (closed symbols) of freshly skinned fibers (above) and aged fibers (below) at pCa 5.8.

In conjunction, the experiments presented so far may be taken to mean that calmodulin exerts an activating effect on chicken gizzard skinned fibers which cannot be accounted for by an increased calmodulin-induced activation of myosin light chain kinase. It is tempting to speculate then that freshly skinned chicken gizzard fibers may contain a factor which slows down or even inhibits the contractile process but which may be overcome by calcium and calmodulin. This hypothetical factor appears to be fairly labile, however, as suggested by experiments with aged skinned fibers.

Fib. 4b shows the calmodulin dependency of force and unloaded shortening velocity in skinned fibers that had been stored in glycerol buffer solution at -20°C after having been exposed to body temperature for several hours. Note that at maximal activation with calcium and calmodulin the unloaded shortening velocity is similar to that in freshly skinned fibers (about 0.2 fiber lengths per second). At 0.1 µM calmodulin and 1.6 µM Ca^{2+}, however, the unloaded shortening velocity is about three times lower in fresh fibers than in aged ones while the relative force development is comparable (about 60% of the maximal force development). At 0.05 µM calmodulin force development is considerably inhibited in fresh but not in aged fibers and the shortening velocities differ by an order of magnitude. In the absence of exogenous calmodulin aged fibers still produce 50% of the maximal tension while fresh fibers do not contract at all. Under these conditions, the extent of myosin light chain phosphorylation amounts to above 0.35 moles phosphate per mole of light chain whereas freshly skinned fibers incorporate only 0.1 mole phosphate per mole of light chain. Yet the ratio of myosin phosphatase activity and myosin light chain kinase activity is similar in aged and fresh fibers (Bialojan, pers. communication). In aged fibers myosin phosphorylation is much higher than in fresh fibers and there is a nearly linear relationship between force development and myosin phosphorylation when the calmodulin or calcium concentration is increased, as shown in Fig. 5.

We propose that a protein factor may be present in fresh but not in aged skinned chicken gizzard fibers which inhibits the shortening velocity and force development as well as the myosin light chain phosphorylation. It is tempting to speculate that this factor resembles caldesmon in that it binds to and may be neutralized by calmodulin in the presence of calcium.

FIGURE 5

Relationship of stress and LC-2-phosphorylation of stored fibres

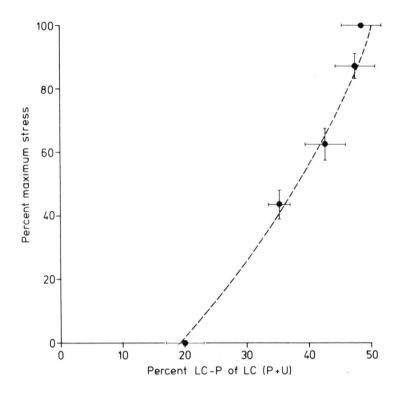

Fig. 5. Relationship of light chain phosphorylation and isometric force in aged skinned fibers.

The question arises, however, as to how it is possible to induce a contracture at low levels of phosphorylation. One possibility has been pointed out by Ebashi who suggested that a factor called leiotonin may activate contraction without additional phosphorylation (Ebashi, 1980). Alternatively (cf. Murphy et al., 1986), a small population (perhaps about 10%) of phosphorylated force producing and cycling crossbridges might be transformed into dephosphorylated slowly cycling crossbridges which maintain the developed force. As this happens, a further portion (again about 10%) of the available crossbridges may be recruited by phosphorylation, and these would then produce an additional force which is superimposed on the force already maintained by the latch-bridges. In this manner, force would develop gradually and slowly as more and more crossbridges are recruited by phosphorylation and transformed into latch-bridges by dephosphorylation. In this process of slow force generation at basal levels of phosphorylation (about 0.1 mole phosphate per mole light chain), caldesmon, an inhibitor of actomyosin ATPase, might play an important role if it had the property of inhibiting ADP release and crossbridge detachment thereby promoting the latch-state (cf. Walsh, 1986).

REFERENCES

Arner A, Wagner J, Rüegg JC (1985) Ca^{2+}-calmodulin-activation of contraction at stabilized levels of myosin phosphorylation in chemically skinned smooth muscle. Acta Physiol Scand 124 (Suppl 542):210

Cassidy P, Hoar PE, Kerrick WGL (1979) Irreversible thiophosphorylation and activation of tension in functionally skinned rabbit ileum strips by ^{35}S ATPγS. J Biol Chem 254:11148-11153

Cole HA, Grand RJA, Perry SV (1982). Non-correlation of phosphorylation of the P-light chain and the actin activation of the ATPase of chicken gizzard myosin. Biochem J 206:319-328

Ebashi S (1980) Regulation of muscle contraction. Proc R Soc (London) B 207:259-286

Gagelmann M, Rüegg JC, DiSalvo J (1984). Phosphorylation of the myosin light chains and satellite proteins in detergent-skinned arterial smooth muscle. Biochem Biophys Res Commun 120:933

Gopalakrishna R, Anderson WB (1982). Ca^{2+}-induced hydrophobic site on calmodulin: application for purification of calmodulin by phenylsepharose affinity chromatography. Biochem Biophys Res Commun 104:830-836

Haeberle JR, Hott JW, Hathaway DR (1984). Pseudophosphorylation of the smooth muscle myosin light chain: an artefact due to protein modification. Biochim Biophys Acta 790:78-86

Kaminski EA, Chacko S (1984). Effects of Ca^{2+} and Mg^{2+} on the actin-activated ATP hydrolysis by phosphorylated heavy meromyosin from arterial smooth muscle. J Biol Chem 259: 9104-9108

Murphy RA, Ratz PH, Hai C-M (1986). Determinants of the latch state in vascular smooth muscle. This volume

Paul R, Doerman G, Zeugner C, Rüegg JC (1983). The dependence of unloaded shortening velocity on Ca^{++}, calmodulin and duration of contraction in "chemically skinned" smooth muscle. Circ Res 53:342-351

Persechini A, Mrwa U, Hartshorne DJ (1981). Effect of phosphorylation on the actin-activated ATPase activity of myosin. Biochem Biophys Res Commun 98:800-805

Portzehl M, Caldwell PC, Rüegg JC (1964). The dependence of contraction and relaxation of muscle fibres from the crab Maia squinado on the internal concentration of free Ca^{2+}-ions. Biochim Biophys Acta 79:581-591

Sparrow MP, Mrwa U, Hofmann F, Rüegg JC (1981). Calmodulin is essential for smooth muscle contraction. FEBS Lett 125: 141-145

Walsh MP (1986). Caldesmon, a major actin- and calmodulin-binding protein of smooth muscle. This volume

ACKNOWLEDGEMENTS

We thank Isolde Berger for the careful preparation of the manuscript, Claudia Zeugner for technical assistance and the Deutsche Forschungsgemeinschaft for supporting this work (Ru 154/14-2).

NON-Ca^{2+}-ACTIVATED CONTRACTION IN SMOOTH MUSCLE

W. Glenn L. Kerrick and Phyllis E. Hoar

Departments of Physiology and Biophysics (W.G.L.K., P.E.H.), and Pharmacology (W.G.L.K.), University of Miami School of Medicine, Miami, Florida 33101

INTRODUCTION

Considerable evidence has accumulated suggesting that smooth muscle is primarily regulated by phosphorylation of smooth muscle light chains (Kamm and Stull, 1985; Hartshorne and Siemankowski, 1981; Adelstein et al., 1982; Kerrick, 1982). Additional findings suggest that other mechanisms directly or indirectly involving Ca^{2+} can also play a role in regulating actomyosin interactions (Dillon et al., 1981; Silver and Stull, 1982). One undefined mechanism is referred to as the latch-bridge mechanism which hypothesizes Ca^{2+} by some unknown mechanism to cause myosin cross-bridges to slowly cycle following activation by myosin light chain phosphorylation (Dillon et al., 1981). Caldesmon has also been shown capable of altering the actomyosin interactions in vitro (Ngai and Walsh, 1984). Other TnC-like thin filament regulators of smooth muscle contraction have also been demonstrated in vitro (Smith and Marston, 1985).

Skinned smooth muscle cells are about as close as one can come to a system which is structurally intact and in which one can control the intracellular environment and measure the physiological response "force" at the same time. In such a system myosin light chain phosphorylation seems to be the primary regulator of contraction (Hoar et al., 1979; Walsh et al., 1982). However recent evidence suggests that there are circumstances where myosin light chain phosphorylation does not seem to be well correlated with force. For example the relationship between myosin light chain phosphorylation and tension shows hysteresis when going from low Ca^{2+} to high Ca^{2+} and back (Chatterjee and Murphy, 1983). Also, when myosin light chains are dephosphorylated in the presence of Ca^{2+} and nucleotide triphosphate smooth muscle cells remain contracted (Hoar et al., 1985). The reasons and possible explanations for such observation have been pursued in this laboratory.

One possible explanation for the hystersis and maintenance of tension in the presence of Ca^{2+} involves the buildup of nucleotide diphosphate in the fibers following activation by Ca^{2+}. An increased ratio of ADP/ATP would have the effect of slowing down the cross-bridge cycling rate when ADP competes with ATP forcing the cross-bridge to remain in the attacted or force generating state for a longer period of time.

METHODS AND PROCEDURES

Small bundles of skinned chicken gizzard cells 100-200 μm in diameter and 3 mm long were prepared and myosin light chain phosphorylation determined as described by Hoar et al. (1985). Standard solutions contained 85 mM K^+ + Na^+, 2 mM $MgATP^{2-}$, 1 mM Mg^{2+}, 7 mM EGTA (ethylene glycol bis(β-aminoethyl ether)-N,N,N´,N´-tetraacetic acid), 10^{-9}-$10^{-3.8}$ M Ca^{2+}, and propionate as the major anion. When ATPγS (adenosine-5´-O-(3-thiotriphosphate)) or CTP were used they were direct substitutions for ATP in the test solutions. Inorganic phosphate (Pi) was included in some solutions. In other solutions $[Mg^{2+}]$ was 0.1 to 15 mM. In some solutions $MgADP^-$ or $MgCDP^-$ was included at 0-5 mM. The Mn^{2+} solutions contained 1 mM Mn^{2+}, 2 mM $MnATP^{2-}$, 85 mM K^+ + Na^+ and no Ca^{2+}. Dithiothreitol (DTT) was included in some Mn^{2+} solutions. Solutions containing the creatine phosphate (CP^{2-}, 15 mM) - creatine phosphokinase (CPK, 15 U/ml) ATP-regenerating system were similar to the standard solutions but had 70 mM K^+ + Na^+. $[MgATP^{2-}]$ was 15 mM in some solutions. A wash solution (pCa 9, 7 mM EDTA, and 85 mM K^+ + Na^+) was used in some cases to quickly remove Mg^{2+} and $MgATP^{2-}$ from the fibers. Imidazole propionate was used to maintain the pH at 7.00 ± 0.02 and to adjust the ionic strength to 0.15 in all solutions described above.

RESULTS AND DISCUSSION

We have been able to show that millimolar concentrations of ADP can cause an activation of smooth muscle in the absence of Ca^{2+} (Kerrick and Hoar, 1985). A fiber which is caused to contract in ADP can be caused to contract further if CTP is substituted for ATP (Fig. 1, top). An explanation for this is that ADP can compete better relative to CTP than to ATP for myosin. The opposite effect on tension is observed when CDP is used to compete for ATP in the fibers (Fig. 1, bottom). Because CDP has such a low affinity for myosin relative to ATP the fiber cannot be activated by the CDP. It has been shown that CDP has a lower affinity for myosin than does ADP (Hiratsuka, 1984). What we believe is happening is that in the relaxed state a certain low number of cross-bridges are cycling at a low rate which does not manifest itself in tension. When ADP is added to solutions containing ATP or CTP the few cycling cross-bridges are made to remain attached in the force generating or strong binding state state for a very long period of time (Eisenberg and Hill, 1985). This has the effect of causing the number of cross-bridges attached to increase and thus manifests itself in tension. This tension development would be and is very slow as can be seen in the tension records (Fig. 1).

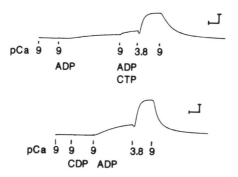

Figure 1. Effect of $MgADP^-$ and $MgCDP^-$ on tension of skinned chicken gizzard fiber bundles in the absence of Ca^{2+}. All solutions contained $MgATP^{2-}$ except when CTP was substituted for ATP as indicated. Five mM $MgADP^-$ or $MgCDP^-$ were included where indicated. Horizontal calibration bars, 5 min.; vertical calibration bars, 12.8 mg.

If ADP induces the fibers to contract in the absence of Ca^{2+}, one would expect that ADP would also cause the fibers to relax slower when Ca^{2+} is removed because the fibers would act as a diffusional barrier to ADP and the ADP released from one myosin would bind to another on its way to the outside of the fibers. One way to test this hypothesis would be to lower the free Mg^{2+} to a level that does not significantly bind to ADP forming $MgADP^-$ which could bind to myosin. Figure 2 shows that in the presence of high Mg^{2+} the fiber relaxes much slower that it does at low concentrations of Mg^{2+}. Thus this data is consistent with the hypothesis that a buildup of ADP in a contracting fiber which is in part attached to the myosin is partly responsible for the slow relaxation of skinned smooth muscle fibers.

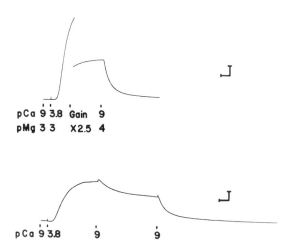

Figure 2. Effect of Mg^{2+} on relaxation of tension. Fibers relaxed faster and more completely in low Mg^{2+} (pMg 4, top) than in high Mg^{2+} (pMg 2, bottom) solutions. Horizontal calibration bars, 5 min.; vertical calibration bars, 25.1 mg. (top at end of trace after change in amplifier gain), 10.7 mg. (bottom).

Previously we have shown that in the presence of high Ca^{2+} contracting skinned smooth muscle cells will not relax although myosin light chains are dephosphorylated (Hoar et al., 1985) in the presence of CTP. One explanation is that, in the presence of high Ca^{2+}, Ca^{2+}-activated ATPases other than actomyosin cause a buildup of CDP in the skinned cells which in turn causes the cross-bridges to remain attached and the muscle contracted. Consistent with this hypothesis is data in Figure 3 which shows that in the presence of CTP more tension is generated by a skinned cells in the presence than in the absence of Ca^{2+} when the cells are activated by increasing concentrations of CDP at low or high Ca^{2+}. Since myosin light chains are not phosphorylated in the presence of CTP (Hoar et al., 1985) or CTP plus CDP (data not shown), the cells are activated by CDP and possibly Ca^{2+}, rather than by myosin light chain phosphorylation. Another intrepretation is that the Ca^{2+} can directly activate the cells in CTP in addition to effects of CDP.

It has also been observed that smooth muscle can be activated by high Mg^{2+} in the absence of Ca^{2+} (Saida and Nonomura, 1978; Iino, 1981). One explanation for this has

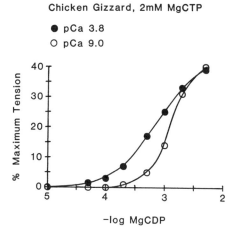

Figure 3. Relationship between percentage of maximum tension and -log [$MgCDP^-$]. High Ca^{2+} (O, pCa 3.8) increases the tension generated in the presence of $MgCTP^{2-}$ plus $MgCDP^-$ in the range 0.2 to 1 mM $MgCDP^-$ over that generated in the absence of Ca^{2+} (O, pCa 9). Data was collected simultaneously at pCa 9 and pCa 3.8 for each of two pairs of fibers.

recently been hypothesized to result from conformational changes in the myosin which are similar to the conformations conferred by myosin light chain phosphorylation (Ikebe and Hartshorne, 1985). It is also possible to explain the results as due in part to a buildup of MgADP$^-$ in the fibers because of the high concentration of Mg^{2+}. If so, then including an ATP regeneration system in the solutions should result in a decrease in force. Figure 4 shows that including an ATP regenerating system in the solutions causes a much decreased activation by Mg^{2+}.

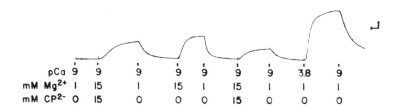

Figure 4. Effect of an ATP regenerating system on Mg^{2+}-activated tension in skinned chicken gizzard cell bundles. The ATP regenerating system consisted of CP (15 mM CP^{2-}) and CPK (15 U/ml). The CPK and MgATP^{2-} were included in all solutions. Horizontal calibration bar, 5 min; vertical calibration bar 21.0 mg.

If the Mg^{2+} contractions are due partly to an increased ratio of ADP to ATP then increasing the concentration of ATP to very high levels should reduce to level of Mg^{2+} contractions. This effect has been reported by Iino (1981) for smooth muscle and the results confirmed by ourselves in Figure 5.

Recently Dillon and coworkers using P^{31} NMR found that, during prolonged contractions of smooth muscle, ADP builds up to levels which would cause contractions to occur in our skinned smooth muscle preparations (personal communication). Thus this could be an explanation for the latch-mechanism which is observed in intact smooth muscle.

However another hydrolysis product of ATP in skinned smooth muscle cells is inorganic phosphate (Pi). Pi has

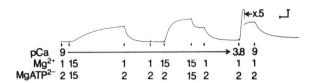

Figure 5. Effect of high [MgATP^{2-}] (15 mM) on Mg^{2+}-activated contractions. Horizontal calibration bar, 10 min; vertical calibration bar 37.9 mg. after the tension response was reduced to 50 %.

been shown by Ruegg and co-workers (Schneider et al., 1981) to increase the relaxation rate in skinned smooth muscle cells. Pi also has the effect of decreasing the maximum tension in skinned smooth muscle cells activated by maximal Ca^{2+} (Fig. 6) or thiophosphorylation of the myosin light chains (Fig. 7). The effects seem to be greater in Ca^{2+}-activated cells than in thiophosphorylated activated cells. This effect of Pi may be related to the fact that Pi binding to the myosin·ADP complex forces some of the force generating cross-bridges into the weak binding state (non-force generating). Pi thus has the opposite effect of ADP and would have a tendency to offset effects of ADP.

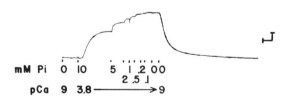

Figure 6. Effect of Pi on maximal Ca^{2+}-activated tension in chicken gizzard skinned fibers. Horizontal calibration bar, 5 min; vertical calibration bar 18.9 mg.

In our search for Ca^{2+}-activated mechanisms other than myosin light chain phosphorylation which might play a role

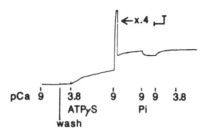

Figure 7. Effect of Pi (10 mM) on maximum tension of fully thiophosphorylated chicken gizzard fibers in the absence of Ca^{2+}. Horizontal calibration bar, 5 min; vertical calibration bar 33.9 mg. after the tension response was reduced to 40 %.

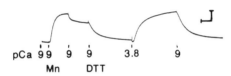

Figure 8. Effect of DTT (50 mM) on relaxation of Mn^{2+}-activated contraction of skinned chicken gizzard fibers. Horizontal calibration bar, 5 min; vertical calibration bar 12.6 mg.

in the activation of smooth muscle contraction it was discovered that Mn^{2+} could activate skinned smooth muscle cells. Once the cells have been activated by Mn^{2+} they do not relax again in a normal relaxing solution containing no Ca^{2+} and Mg^{2+}, but relax if DTT is added (Fig. 8). DTT was used here at 50 mM but also was effective at substantially lower concentrations. The cells again act like control fibers which can be reversibly activated by Ca^{2+} (Fig. 8). Thus it appears that cells activated in the presence of Mn^{2+} are activated in part by protein oxidation. From these experiments it is not possible to determine what is being oxidized, but it is interesting to note that the same solutions will activate skeletal muscle skinned fibers and in contrast they do not need to be treated with DTT to relax again in the absence of Mn^{2+}. This DTT-dependent relaxation following Mn^{2+} activation thus is unique to smooth muscle. Walsh and Ngai (1986) have reported oxidative changes altering smooth muscle myosin conformation which are reversed by DTT. The oxidized myosin actin-activated ATPase is independent of myosin phosphorylation.

Mn^{2+} activation is probably not entirely due to oxidation since the cells contract rapidly (in fact more rapidly than when stimulated with Ca^{2+}) in the presence of Mn^{2+} and DTT and subsequently relax rapidly (more rapidly than after a Ca^{2+} stimulated contraction) in a normal relaxing solution without DTT (Fig. 9). Mn^{2+} activation is not due to activation of myosin light chain kinase as judged by the lack of ^{32}P incorporation into the light chains (Fig. 10). Therefore it seems likely that Mn^{2+} causes skinned smooth muscle cells to contract by acting at sites other than those involved in myosin light chain phosphorylation. Since Mn^{2+} does not act through myosin light chain kinase, possible sites would be either the thin filament which contains divalent Mn^{2+} binding sites (Brauer and Sykes, 1982) or myosin light chains which also bind Mn^{2+} (Bagshaw and Kendrick-Jones, 1980).

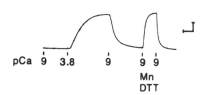

Figure 9. Mn^{2+}-activated contraction in the presence of DTT (50mM). Horizontal calibration bar, 5 min; vertical calibration bar 18.9 mg.

CONCLUSIONS

In conclusion it is possible to cause skinned smooth muscle cells to contract in the absence of Ca^{2+} and myosin

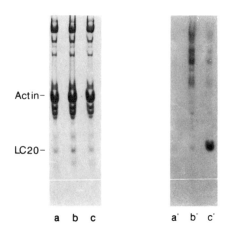

Figure 10. Absence of phosphorylation of the 20,000 dalton myosin light chain (LC20) in the presence of Mn^{2+} in chicken gizzard skinned fibers (a,a'). Coomassie Blue stained SDS polyacrylamide gel (a,b,c) and autoradiogram (a',b',c') of skinned chicken gizzard fibers labeled with γ-[^{32}P]-ATP. Lanes a,a': 1 mM Mn^{2+}, 2 mM $MnATP^{2-}$, no Ca^{2+}; lanes b,b': pCa 9, 1 mM Mg^{2+}, 2 mM $MgATP^{2-}$; lanes c,c': pCa 3.8, 1 mM Mg^{2+}, 2 mM $MgATP^{2-}$. Phosphorylation of LC20 in lane c' represents 20 % phosphorylation.

light chain phosphorylation. We show here that Mg-nucleoside diphosphate will cause skinned smooth muscles to contract in the presence of ATP or CTP. This activation by nucleoside diphosphate is determined by the ratio of nucleoside diphosphate/nucleotide triphosphate. In the activation of skinned cells by high Mg^{2+} concentrations it is probably a build-up of $MgADP^-$ in the cell bundles which is in part responsible for the activation. Likewise the slow relaxation of skinned cell bundles at high Mg^{2+} is also probably due to slow diffusion of $MgADP^-$ from the cells since in experimental conditions which prevent the formation of $MgADP^-$ (low Mg^{2+} concentration) the cell bundles relax faster. It is also shown that oxidation of some contractile proteins will also cause the cells to contract in a reversible manner. However this oxidation activation is not likely to play any physiological role. It is also possible to activate the skinned cell bundles with Mn^{2+} alone in the absence of any myosin light chain phosphorylation under conditions which prevent oxidation by Mn^{2+}. Further investigation of this mechanism may provide insight into Ca^{2+} regulation of smooth muscle contraction.

ACKNOWLEDGMENTS

We gratefully acknowledge the outstanding technical assistance of Deborah Montague. This research was supported by the Muscular Dystrophy Association, the American Heart Association (National and Florida Affiliate) and NIH BRSG S07 RR-05363.

REFERENCES

Adelstein RS, Sellers, JR, Conti, MA, Pato MD, deLanerolle P (1982). Regulation of smooth muscle contractile proteins by calmodulin and cyclic AMP. Fed Proc 41:2873-2878.
Bagshaw CR, Kendrick-Jones J (1980). Identification of the divalent metal ion binding domain of myosin regulatory light chains using spin-labelling techniques. J Mol Biol 140:411-433.
Brauer M, Sykes BD (1982). Effects of manganous ion on the phosphorus-31 nuclear magnetic resonance spectrum of adenosine triphosphate bound to nitrated G-actin: proximity of divalent metal ion and nucleotide binding sites. Biochemistry 21:5934-5939.
Chatterjee M, Murphy RA (1983). Calcium-dependent stress maintenance without myosin phosphorylation in skinned smooth muscle. Science 221:464-466.
Dillon PF, Aksoy MO, Driska PS, Murphy RA (1981). Myosin phosphorylation and the cross-bridge cycle in arterial smooth muscle. Science 211:495-497.
Eisenberg E, Hill TL (1985). Muscle contraction and free energy transduction in biological systems. Science 227:999-1006.
Hartshorne DJ, Siemankowski RF (1981). Regulation of smooth muscle actomyosin. Ann Rev Physiol 43: 519-530.
Hiratsuka T (1984). Affinity labeling of the myosin ATPase with Ribose-Modified fluorescent nucleotides and vanadate. J Biochem 98:147-154.

Hoar PE, Kerrick WGL, Cassidy PS (1979). Chicken gizzard: Relationship between Ca^{2+}-activated phosphorylation and contraction. Science 204:503-506.

Hoar PE, Pato MD, Kerrick WGL (1985). Myosin light chain phosphatase: Effect on the activation and relaxation of gizzard smooth muscle skinned fibers. J Biol Chem 260:8760-8764.

Ikebe M, Hartshorne DJ (1985). Effects of Ca^{2+} on the conformation and enzymatic activity of smooth muscle myosin. J Biol Chem 260:13146-13153.

Iino M (1981). Tension responses of chemically skinned fibre bundles of the guinea-pig taenia caeci under varied ionic environments. J Physiol 320:449-467.

Kamm KE, Stull JT (1985). The function of myosin and myosin light chain kinase phosphorylation in smooth muscle. Ann Rev Pharmacol Toxicol 25:593-620.

Kerrick WGL (1982). Myosin light chain kinase in skinned fibers. In Cheung, WY (ed): "Calcium and Cell Function, Vol II," New York: Academic Press, pp 279-295.

Kerrick WGL, Hoar PE (1985). Regulation of contraction in skinned smooth muscle cells by Ca^{2+} and protein phosphorylation. Adv in Protein Phosphatases II:133-152.

Ngai PK, Walsh MP (1984). Inhibition of smooth muscle actin-activated myosin Mg-ATPase activity by caldesmon. J Biol Chem 259:13656-13659.

Saida K, Nonomura Y (1978). Characteristics of Ca^{2+}- and Mg^{2+}-induced tension devcelopment in chemically skinned smooth muscle fibers. J Gen Physiol 72:1-14.

Schneider M, Sparrow M, Ruegg JC (1981). Inorganic phosphate promotes relaxation of chemically skinned smooth muscle of guinea-pig taenia coli. Experientia 37:980-982.

Silver PJ, Stull JT (1982). Regulation of myosin light chain and phosphorylase phosphorylation in tracheal smooth muscle. J Biol Chem 257:6145-6150.

Smith CWJ and Marston SB (1985). Disassembly and reconstitution of the Ca^{2+}-sensitive thin filaments of vascular smooth muscle. FEBS Letters 185:115-119.

Walsh MP, Bridenbaugh R, Hartshorne DJ, Kerrick WGL (1982). Phosphorylation-dependent activated tension in skinned gizzard muscle fibers in the absence of Ca^{2+}. J Biol Chem 257:5987-5990.

Walsh MP, Ngai PK (1986). Phosphorylation-independent activation of smooth muscle actomyosin. Biophys J 49:184a.

SYNTHESIS OF INOSITOL PHOSPHOLIPIDS IN CARBACHOL-STIMULATED CANINE TRACHEALIS MUSCLE.

Carl B. Baron and Ronald F. Coburn

Department of Physiology
University of Pennsylvania
Philadelphia, PA 19104

Specific radioactivities of [^{3}H]-inositol labeled lipids were determined in carbachol-stimulated and control smooth muscle strips by radioactive scanning of TLC separated lipids followed by charring and scanning densitometrty.

During a 30 min incubation with radiolabel only small amounts of the lipids were labeled. After removal of exogenous label, contraction was initiated by addition of carbachol, 5.5 µM. Between 2 and 15 min, there were parallel rates of increase in specific radioactivities of all three inositol phospholipids.

Rate of increase of individual inositol phospholipids (2-15 min) (± S.E.M.) dpm/(nmole)(min)

	Control	Stimulated
PI	10.7 ± 1.3	521 ± 31
PIP	26.1 ± 2.3	434 ± 19
PIP_2	22.7 ± 2.7	495 ± 34

At 0 min, control levels of PI, PIP and PIP_2 were respectively 980, 2087 and 679 dpm/mole. Myo-inositol pool size, 18.1 ± 0.4 nmoles/100 nmoles total lipid Pi (± S.E.M.; n = 18) was constant ± carbachol and independent of time, as was its specific radioactivity. The pool sizes of PIP and PIP_2 also remained constant and were respectively 0.14 and

0.34 nmoles/100 nmoles total lipid Pi while the pool sizes of PI and PA in stimulated tissue decreased and increased between 0 and 3 min and thereafter remained constant. The apparent rate of stimulated PI synthesis was 0.071 nmoles myo-inositol incorporated/(100 nmoles total lipid Pi)(min). Rates of metabolic flux in phosphoinositides and inositol phosphates was estimated to be 3X greater during development of tension than during tension maintenance.

Appearance of radioactivity in the inositol phosphates (IP, IP_2 and IP_3) (Berridge et al, Biochem. J. (1983) 212: 473), after a delay of about 2 min, increased in relative linear fashion until 15 min.

Rate of increase of label in total pool
(2-15 min) (± S.E.M.)

dpm/(100 nmoles lipid Pi)(min)

	Control	Stimulated
PI	58.5 ± 5.2	1570 ± 50
PIP	1.8 ± 0.2	69 ± 4
PIP_2	16.4 ± 3.2	217 ± 28
IP	-7.0 ± 1.6	282 ± 41
IP_2	-2.4 ± 0.5	99 ± 9
IP_3	-1.1 ± 0.2	51 ± 4

During maintenance of contraction, we calculated that ATP required to fuel phosphoinositide transduction accounted for approximately 0.7% of total J_{ATP} and 5.3% of the incremental increase over resting muscle.

We conclude that: (1) synthesis rates of inositol phospholipids are 20 - 50X greater during maintenance of contraction than at rest; (2) metabolic flux rates are greater during development of tension than during maintenance of tension and; (3) a high proportion of PI may be linked to PIP and PIP_2 turnover, since the rates and levels of increases in specific radioactivities are similar. (HL 19737)

MYOSIN HEAVY CHAIN ISOFORMS IN HUMAN SMOOTH MUSCLE

N. DeMarzo, S. Sartore, L. Saggin, L. Fabbri, and S. Schiaffino

Institute of General Pathology (S.S., L.S., S.Sc.) and Institute of Occupation Medicine (N.D., L.F.) University of Padova, Padova, Italy

To determine whether different isoforms of myosin are present in human smooth muscle from different organs, we examined surgical specimens from human popliteal artery and vein, non-pregnant uterus, taenia coli, bronchi, and lung parenchymal strips. Specimens were homogenized and extracted with 40 mM NaPPi, 1 mM $MgCl_2$, 1 mM EGTA, pH 9.5, for 30 minutes at 4° C, and then immediately electrophoresed on SDS-5% polyacrylamide gel: all the extracts consistently yielded three electrophoretic bands corresponding to myosin heavy chains (MHC-1, MHC-2, and MHC-3, in terms of increasing mobility). The evidence for multiple myosin heavy chains was also confirmed by Western blotting, using a monoclonal antibody (BF-48) specific for bovine smooth muscle myosin heavy chains. The three electrophoretic bands were selectively labelled by this antibody, suggesting the presence of a common epitope in all the three myosin heavy chain isoforms. In addition, a different distribution of the three myosin isoforms in smooth muscles from different organs was observed. In particular, MHC-2 was more abundant in arteries compared to veins, and MHC-3 was more abundant in pulmonary parenchymal strips compared to bronchi and other tissues. In conclusion, we observed that 1) different myosin heavy chain isoforms are present in human smooth muscles and 2) they are variably distributed among and within different organs, suggesting that these differences may be related to distinct physiological properties (e.g., small vs. large airways). (Supported in part by National Research Council, Grant no. 85.00681.04).

CONTRACTION OF HOG CAROTID ARTERIAL SMOOTH MUSCLE CELLS
PREPARED BY DIGESTION WITH PAPAIN

S.P. Driska, M. Desilets, and C.M. Baumgarten

Department of Physiology and Biophysics
Medical College of Virginia
Richmond, Virginia 23298

The purpose of this study was to prepare and characterize isolated smooth muscle cells from the swine carotid artery. The use of cells isolated from this artery has three advantages: 1) it allows cell responses to be studied free of interactions with other types of cells that are present in muscle strips, such as endothelial cells and nerve terminals; 2) it allows electrophysiological measurements to be made; and 3) it will allow biochemical measurements (e.g. myosin light chain phosphorylation) to be made in stirred cell suspensions, eliminating the agonist diffusion delays that have complicated such measurements in multicellular preparations. Pressurized arteries were digested with papain in calcium-free solution. The cells were released when fine strips were dissected from the enzyme-treated arteries. Only about 20% of the cells produced in this way are viable (as judged by exclusion of trypan blue). These cells are quite long for arterial smooth muscle cells; the mean length was 240 µm in calcium-free solution. Most cells shorten somewhat (5-20%) when exposed to 1.6 mM calcium, but still remain fairly elongated; mean cell length in 1.6 mM calcium was 194 µm. To rule out the possibility that the cells were stretched to abnormal lengths during the isolation procedure, and to compare the isolated cell lengths with cell lengths in muscle strips, tissues were fixed with glutaraldehyde at L_o, the optimal length for force generation. The fixed strips were treated with nitric acid to release fixed cells. The mean length of the fixed cells was 211 µm. Thus the lengths of the papain-dissociated cells were close to the optimum cell length for tissue force development. When stimulated, the viable cells can contract to half their initial length or less. Most cells in 1.6 mM calcium contracted in response to histamine, 5-hydroxytryptamine, acetylcholine, angiotensin II, and A23187. Only a few cells contracted in response to 100 mM KCl, and none contracted when exposed to norepinephrine. Phorbol dibutyrate caused the cells to contract extremely slowly, demonstrating that this contractile agonist has a direct effect at the cellular level. Membrane potential of the isolated cells was about -45 mM and unstable. The noise consisted of hyperpolarizing spikes which disappeared when the membrane was hyperpolarized beyond about -50 mM. The amplitude of the spikes increased with depolarization, and they were blocked by tetraethylammonium ion. Caffeine and ryanodine both eliminated the membrane noise. The hyperpolarizing spikes appear to be caused by increases in calcium-dependent K conductance, triggered by release of calcium from the sarcoplasmic reticulum. We are currently attempting to increase the yield of cells and the percentage of viable cells so that myosin light chain phosphorylation measurements can be made in stirred cell suspensions. We have achieved purification to 90-95% viable cells by density gradient centrifugation, and have increased the yield of cells to the point that measurements of light chain phosphorylation by electrophoretic techniques now appear feasible. Supported by NIH grant HL24881 and RCDA HL01198.

GUANYLATE-CYCLASE-DEPENDENT GATING OF RECEPTOR-OPERATED CALCIUM CHANNELS IN VASCULAR SMOOTH MUSCLE

T. Godfraind

Laboratoire de Pharmacodynamie Generale et de
Pharmacologie
Universite Catholique de Louvain
Bruxelles, Belgique

Several vasorelaxing agents stimulate guanylate-cyclase (G.C.) activity and increase cyclic GMP content in smooth muscle cells. The purpose of this study was to examine how activation and inhibition of G.C. could influence Ca entry and contraction in response to vasoconstrictors. Therefore, the effects of EDRF (endothelium derived factors), of methylene blue (MB), an inhibitor of soluble guanylate cyclase and of the lipid soluble analogue of cyclic GMP, 8-Bromo-cyclic GMP were investigated on Ca influx and on contraction of rat aorta.

Removal of endothelium enhanced the contractile response to norepinephrine 10^{-8} M. The contractile force developed by aorta rings without endothelium was six-fold higher than the force developed by preparations with endothelium present ($P < 0.01$) (Godfraind, 1986). A similar amplification of the response was observed in preparations pretreated with methylene blue, 3×10^{-6} M during 10 minutes ($P < 0.01$). Treatment of aorta without endothelium by 8-Bromo-cyclic GMP (10^{-4} M) reduced the contractile response to norepinephrine ($P < 0.01$). The contractile response to K-depolarization was not significantly influenced by the presence of endothelium.

Removal of endothelium evoked a small but significant increase of ^{45}Ca entry in unstimulated preparations. Norepinephrine-dependent ^{45}Ca entry was two-fold higher in preparations without endothelium than in preparations with endothelium ($P < 0.01$). Pretreatment during 30 min of the preparations containing endothelium by MB, 3×10^{-6} M, also enhanced norepinephrine-dependent ^{45}Ca entry ($P < 0.01$). Pretreatment of preparations without endothelium by 8-Bromo-cyclic GMP, 0.1 mM, during 30 min, resulted in a reduction of norephinephrine-dependent ^{45}Ca entry close to the level found in preparations with endothelium preserved ($P < 0.01$). ^{45}Ca entry evoked by a 2 minute exposure in K-depolarizing solution was not changed by these manipulations.

Calcium entry responsible for smooth muscle tonic contraction is thought to occur through receptor-operated and potential-operated calcium channels. Results reported here indicate that the gating of receptor-operated but not of potential-operated Ca channels is sensitive in G.C. activity.

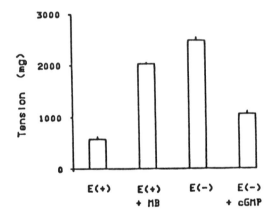

Figure 1. Tonic contractile force (expressed in mg) developed by rings of rat aorta stimulated by norephinephrine, 10^{-8} M, each value is the mean of 6 determinations, S.E. is indicated on the graph: E(+), preparations with endothelium intact; E(+) + MB, preparations with endothelium intact pretreated with methylene blue (3 µM) during 10 minutes before epinephrine; E(-), preparations where endothelium was removed; E(-) + cGMP, preparations where endothelium was removed and pretreated with cyclic GMP, 0.1 mM, during 30 minutes before norepinephrine.

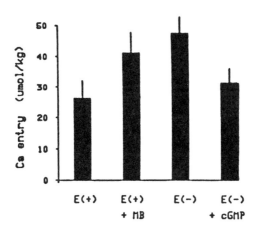

Figure 2. ^{45}Ca influx (expressed in mol/kg wet weight) in preparations treated with norepinephrine, 10^{-8} M, for 2 minutes. (Mean ± S.E., N = 12 for each value). The abbreviations designating treatments are the same as in Figure 1.

This work was supported by F.R.S.M. (grant No. 3.9001.83) and by F.M.R.E.

REFERENCES

Godfraind, T (1986). Acta Pharmacol Toxicol 58 (suppl. II): 5-30.

BIREFRINGENCE OF RAT ANOCOCCYGEUS: CORRELATION WITH THE
DENSITY OF MYOSIN FILAMENTS

A. Godfraind-DeBecker, M.L. Cao and J.M. Gillis

Physiologie des Muscles
Universite Catholique de Louvain
Brussels, Belgium

At rest, the mean birefringence ($\pm$ s.e.m.) of the anococcygeus amounts to 0.88 ($\pm$ 0.05) x 10^{-3} (n = 27). During maximal isometric contraction (norepinephrine 10^{-5} M), it increases by 0.42 ($\pm$ 0.05) x 10^{-3} (n = 40). This effect is totally reversible at relaxation (Godfraind-DeBecker, 1982).

We have now examined the density of myosin filaments in pairs of muscles fixed in three steps (glutaraldehyde; tannic acid; OsO_4), one muscle being relaxed, the other maximally contracted. Ten micrographs were taken for each muscle in 4 pairs. In each picture (magnification 33825), 10 squares of 1 cm^2 cytoplasm were selected at random, wherein the myosin filaments were counted (400 squares for each condition).

The frequency histogram of the contracted muscles was shifted to the right, the mean ($\pm$ S.D.) of the counts being 7.6 $\pm$ 4.2 and 14.7 $\pm$ 5.6 filaments/cm^2, for relaxed and contracted muscles respectively. This corresponded to a density of 86 $\pm$ 47 and 168 $\pm$ 66 filaments/μm^2 of cytoplasm. The difference was highly significant (test of Wilcoxon, P < 0.001).

In a morphometric study on the same muscles, the profiles of 120 relaxed and 151 contracted cells were measured. In the latter, the mean cell area decreased by 16%. The increase of filament density observed can thus be explained for a minor part by a concentration effect, leaving an excess of about 60-70 filaments/μm^2.

The observations strengthen the hypothesis that the increase of birefringence during contraction may reflect an aggregation of myosin molecules. At present, we cannot discriminate between an elongation of pre-existing filaments or a net increase of their number.

REFERENCES

Godfraind-DeBecker A, JM Gillis (1982). Arch Int Physiol Biochim 90:6-7.

MECHANISM OF SHORTENING-INDUCED DEPRESSION OF CONTRACTILITY IN CANINE TRACHEAL SMOOTH MUSCLE

S.J. Gunst and J.Q. Stropp

Mayo Clinic and Foundation
Rochester, Minnesota 55905

In a previous study of canine tracheal smooth muscle, a decrease in muscle length imposed during active contraction was shown to cause a depression of both the rate and magnitude of force redevelopment immediately subsequent to the shortening (Gunst, Am. J. Physiol., 1986). The degree of force depression was proportional to the imposed length change in muscles shortened from different initial lengths to the same final length. In the present study, the effect of length history on the velocity of isotonic shortening was investigated.

Muscle strips were mounted in a tissue bath in 37° physiologic saline and bubbled with 95% O_2-5% CO_2. They were tightly tied to a stationary hook at one end and to a servo-controlled electromagnetic lever at the other (Cambridge Tech, Model 300H). They were then equilibrated for 90 min during which time the length of maximal active force, L_o, was determined using a maximal electrical stimulus (12V, 60Hz). In initial experiments, each of 5 muscles was contracted isometrically using electrical stimulation at an initial length of either 100, 85, or 70% L_o. After 10 s, the muscles were rapidly released to 70% L_o where they were allowed to shorten isotonically under a series of afterloads ranging from 5 to 30% of the maximum isometric force at 70% L_o. For each afterload, velocity-length phase-planes were generated for releases from each initial length and for no release. Shortening velocities at each load were significantly higher in muscles not subjected to a length step than in muscles initially contracted at either 85% L_o or L_o. Values of maximal shortening velocity (Vmax) were estimated by linearization of force-velocity data for both sized length steps and for no length step. Vmax was depressed by 19 ± 3% after release from 85% of L_o, and by 34 ± 3% after release from L_o (n = 5). The proportional depression of Vmax and isometric force redevelopment following release was not significantly different.

To investigate the mechanism for this effect, a modified slack test was used in which muscles were subjected to successive releases from initial lengths of L_o, 95, 90, 85, 80, 75, 70, and 65% L_o to a final length of 50% L_o. At 50% L_o they were allowed to contract freely until they reached a force of 3% of their maximal isometric force at L_o, at which point they began to shorten isotonically. The "slack time" for each length step was measured as the delay between release and the onset of isotonic shortening. In plots of release distance versus slack time, the slope progressively decreased at longer release distances, indicating a decline in shortening velocity. Concurrently, the isotonic shortening velocity subsequent to slack period increased from 3.0 ± 0.5 L_o/s after release from L_o to 50% L_o to 11.1 ± 2.0 L_o/s after release from 65 to 50% L_o (n = 7). The results suggest that the shortening velocity decays progressively during the period of intrinsic shortening when the muscle is slack. As the size of the length step is increased, the time period for which the muscle remains slack increases, and a greater decay in the rate of shortening occurs prior to the measurement of the isotonic shortening velocity. This decay in shortening velocity may also result in a lower level of isometric force development.

Supported by HL 29289 from the National Institutes of Health, and the Parker B. Francis Foundation.

DOES A LIMITATION OF ENERGY SUPPLY TO THE CONTRACTILE
APPARATUS UNDERLIE THE RELAXATION INDUCED BY HYPOXIA IN
SMOOTH MUSCLE?

Y. Ishida, M. Hashimoto, R.J. Paul

Department of Physiology and Biophysics
University of Cincinnati
Cincinnati, Ohio 45267

In a variety of vascular and other smooth muscles, the rate of oxygen consumption is strongly correlated with the level of isometric force (Paul, 1980). Most (70 ∼ 90%) of the ATP production of smooth muscles can be attributed to oxidative phosphorylation in normal aerobic conditions. Hypoxic conditions induce relaxation in many smooth muscles, though not all. Energy limitations have been proposed as a mechanism; however, evidence in support of this hypothesis is not unambiguous. The present experiments were undertaken to investigate the role of phosphagens and rate of metabolic ATP synthesis in limiting contractility of guinea pig taenia caeci and hog coronary arteries.

High-K (45.4 mM for the taenia; 85.4 mM for the coronary) elicited a sustained isometric contraction under normoxia (95% O_2; 5% CO_2); isometric force (mN/mm^2) was 200 for the taenia and 180 for the coronary. Treatment with hypoxia (95% N_2; 5% CO_2) in the presence of high-K elicited a decrease in tension to approximately 10% of control in the taenia and to approximately 70% of control in the coronary. Increasing (glucose) from 5.5 to 55 mM in the presence of high-K during hypoxia restored tension to 40% of control in the taenia, whereas this did not alter the level of tension in the coronary artery.

Under normoxia with high-K, phosphagen contents (in μ mol/g) were 1.4 ATP and 2.3 PCr in the taenia, and 0.26 ATP and 0.22 PCr in the coronary. The level of phosphagens was not significantly different between stimulated and unstimulated muscles. In the taenia treated with high-K, hypoxia elicited a rapid loss of PCr to 10% of control and a gradual decrease in ATP to one-half of control. Increasing (glucose) in the presence of high-K during hypoxia increased ATP to at least 70% of control; PCr was slightly increased. On the other hand, neither hypoxia nor increasing (glucose) under hypoxia significantly changed the level of phosphagens in the coronary artery.

In the presence of high-K, hypoxia increased a lactate release from the muscle: from 80 to 720 nmol/min·g in the taenia and from 230 to 500 nmol/min·g in the coronary. Increasing glucose under hypoxia further increased the lactate release to 1200 nmol/min·g in the taenia, whereas no change was seen in the coronary. This suggests that the glycolytic activity of the coronary artery but not the taenia is saturated at the low concentration (5.5 mM) of glucose. Under normoxia, high-K elicited an increase in oxygen consumption: from 260 to 450 nmol/min·g in the taenia and from 160 to 220 nmol/min·g in the coronary. Thus, rates of ATP synthesis were estimated as follows (μmol/min·g):

	Taenia	Coronary
Unstimulated + normoxia	1.7	1.2
High-K + normoxia	3.0	1.7
High-K + hypoxia	0.9	0.6
High-K + hypoxia + high glucose	1.5	0.6

From the results the rate of ATP utilization per unit isometric force (tension cost) was estimated: 6.7 for the taenia and 2.8 nmol ATP/min·g/mN/mm^2 for the coronary. These results suggest that hypoxia limits the energy required for contraction both in the taenia caeci and the coronary arteries. The hypoxia-induced inhibition of tension was larger in the taenia than the coronary. This may be attributed to the larger tension cost of the taenia.

CULTURED CIRCULAR SMOOTH MUSCLE FROM THE RABBIT COLON

H.W. Kao, S.E. Finn, A. Gown, J. Lechago,
N. Lachant, W.J. Snape, Jr.

Departments of Medicine and Pathology
Harbor-UCLA Medical Center, Torrance, CA
University of Washington, Seattle, WA

While cultured vascular smooth muscle cells (VSMC) have been extensively characterized and investigated, there are very few studies of cultured intestinal smooth muscle cells. The aim of this study was to culture colonic smooth muscle cells (CSMC) from the rabbit colon. Freshly isolated CSMC from the circular muscle layer of the distal colon were prepared by mild mechanical dispersion following collagenase digestion. Cultures were plated in Dulbecco's Modified Eagle's Medium (DMEM) with high glucose and 10% fetal calf serum. In primary culture, CSMC attached to the culture vessels by 48 to 72 h, proliferated (modulated) by 3 to 7 days and reached confluency by 14 to 17 days with a "hill-and-valley" pattern. Spontaneous contractions were not observed at any time at 21° C or 37° C. Confluent primary cultures were 90% CSMC, as identified by intensely positive immunofluorescent staining to smooth muscle-specific CGA-7 and HHF-35 monoclonal antibodies. Absence of staining for factor VIII excluded endothelial cells. Transmission electron microscopy revealed ultrastructural features consistent with smooth muscle cells. Confluent primary cultured CSMC were compared to human skin fibroblasts (HSF) with respect to glucose 6-phosphate dehydrogenase (G6PD), 6-phosphogluconate dehydrogenase (6PGD), hexokinase (HK), and pyruvate kinase (PK) activities ($IU/10^7$).

	G6PD	6PGD	HK	PK
CSMC (n=5)	0.23±0.02	0.16±0.04	0.22±0.06	4.60±1.53
HSF (n=2)	0.76	0.22	0.28	14.14

In conclusion, we have successfully cultured CSMC of the rabbit from freshly isolated cells and validated these CSMC by immunohistochemical staining, electron microscopy and biochemical assay of glycolytic enzymes. These highly pure primary cultures may be used to investigate the membrane-associated events leading to second messenger generation in CSMC.

A GEOMETRIC 3-DIMENSIONAL THERMODYNAMIC MODEL FOR
CROSSBRIDGE (XB) ATTACHMENT IN SMOOTH MUSCLE (SM)

M. Li and D.M. Warshaw

Med. Biostat. and Physiology & Biophysics
University of Vermont
Burlington, Vermont 05405

Smooth muscles generate active stress comparable to that in skeletal muscle but with less myosin. An increased probability of XB attachment followed by a long-lived force producing XB state may in part account for the enhanced stress generating capabilities in SM. We therefore developed a 3-dimensional thermodynamic model for XB attachment based upon the models of Schoenberg (Biophys J 30:51, 1980) and Eisenberg et al (Biophys J 29:195, 1980). The model assumes a single myosin XB capable of interacting with multiple actin sites in a plane normal to the filament's longitudinal axis. Assuming an S1 length of 16 nm and a S2 of 40 nm, the model predicts that XB attachment to actin is restricted to azimuthal angles between ± 60° due to geometric limitations. Since in SM 12-15 actin filaments surround each myosin filament as compared to 6 in skeletal muscle, the increased probability of attachment in SM was assessed by the model and predicted to be 1.5-2.0 times greater in SM. Therefore an increased number of attached force generating XBs per myosin filament may exist in SM resulting from more numerous actin filaments surrounding each myosin filament.

Supported by NIH AM34872 and HL35684.

CYTOCHALASIN-LIKE ACTIVITY IN RAT AORTA SMOOTH MUSCLE CELLS

W. Magargal

University of Alabama at Birmingham
Birmingham, Alabama 35294

Actin and proteins which bind to actin are highly conserved and make up a dynamic filamentous network in all eucaryotic cells. Actin binding proteins can cap, crosslink, bundle, sever, or anchor actin filaments or bind actin monomers. The purpose of this study was to determine if proteins which have cytochalasin-like activity, i.e. cap the barbed end of actin filaments, were present in vascular smooth muscle cells. Three assays were used to detect cytochalasin-like activity; reduction of actin gelatin, inhibition of actin nuclei induced polymerization, and inhibition of [^{3}H]-cytochalasin B binding to actin nuclei. 6×10^6 g·min supernates of rat aorta smooth muscle cells homogenized in low ionic strength buffer, reduced the extent of purified actin gelatin. Approximately 10 μg of supernate reduced the falling ball viscosity of 125 μg of actin by 50%. The activity did not appear to be Ca sensitive since supernates made in Ca free buffer, dialyzed against Ca free buffer, and assayed in the absence of Ca using Mg-actin gels still had activity ($ED_{50} \sim 12$ μg). The activity appeared to be additive with cytochalasin B. Ostwald viscometry measurements showed that smooth muscle cell supernates had no effect on the polymerization of actin in 2 mM $MgCl_2$. However the supernates did inhibit nuclei induced polymerization in 0.4 mM $MgCl_2$ ($ED_{50} \sim 20$ μg). [^{3}H] cytochalasin B binding to actin nuclei was also inhibited by the supernates. Partial purification of the cytochalasin-like activity using DEAE cellulose and S-200 chromatography has yielded a fraction containing proteins with molecular weights between 30 and 75 k-daltons as measured on SDS acrylamide gels. The fraction is about 500-fold more active.

The results of this study indicate that rat aorta smooth muscle cells contain a protein which appears to cap the barbed end of actin filaments in a manner similar to cytochalasin B. This protein may be related to that which is found in chicken embryo fibroblasts supernates and which is increased in supernates of Rous sarcoma virus transformed fibroblasts. The role of this protein in contraction and/or growth of smooth muscle cells is under investigation.

EFFECTS OF CYCLIC AMP-DEPENDENT PROTEIN KINASE IN SKINNED CORONARY ARTERY

J.R. Miller and J.N. Wells

Department of Pharmacology
Vanderbilt University School of Medicine
Nashville, Tennessee 37232

It has been proposed that beta-adrenergic-induced relaxation of smooth muscle results from phosphorylation of myosin light chain kinase (MLCK) by cyclic AMP-dependent protein kinase (cAMP-PK), leading to a reduction in the sensitivity of MLCK to calcium-calmodulin. We have examined a corollary of this hypothesis, that addition of cAMP-PK activity to a skinned smooth muscle preparation should reduce the sensitivity of the preparation to calcium. Strips of porcine coronary artery media were skinned for one hour in 0.5 percent Triton X-100. Incubation of the strips with the catalytic subunit of cAMP-PK (100 mM) decreased neither the intial rate nor the extent of contraction elicited by 0.5 μM free calcium. MLCK activity ratio analysis (Miller et al, Mol. Pharmacol. 24:235, 1983) indicated that incubation of skinned strips with 100 nM cAMP-PK did not lead to a change in sensitivity of the endogenous MLCK to activation by calcium-calmodulin. Skinned muscle strips were also incubated with 100 nM cAMP-PK catalytic subunit in the presence of [^{32}P]-ATP. An immunoprecipitate obtained from an extract of the skinned strips using tracheal smooth muscle anti-MLCK was subjected to SDS-polyacrylamide gel electrophoresis. cAMP-PK treatment increased by 82 percent the amount of ^{32}P in a band corresponding to M_r 145,000, suggesting that cAMP-PK was active and able to permeate the skinned tissue. Our results do not support the hypothesis that phosphorylation of MLCK mediates relaxation of smooth muscle following activation of cAMP-PK. Supported by NIH grant HL19325 and grant from AHA, Middle Tennessee Chapter.

LATCHBRIDGES AND PHOSPHORYLATED CROSSBRIDGES ARE ACTIVATED INDEPENDENTLY IN ARTERIAL SMOOTH MUSCLE

Robert S. Moreland and Suzanne Moreland

Bockus Research Institue, Graduate Hospital
Philadelphia, PA (RSM)
Squibb Institute for Medical Research
Dept. of Pharmacology, Princeton, NJ (SM)

Ca^{2+}-dependent phosphorylation of the 20,000 M_r myosin light chain (MLC) is believed to be the primary activation step in the development of stress by smooth muscle. Recent evidence suggests that the developed stress is maintained by conversion of the phosphorylated crossbridges to Ca^{2+}-dependent, non-phosphorylated, slowly cycling latchbridges. The purpose of this series of experiments was to investigate the possibility that these two populations of crossbridges are activated independently. Strips of swine carotid media were dissected and mounted for isometric force recordings or estimation of isotonic shortening velocity. Experiments were performed on intact and Triton X-100 skinned fibers. MLC phosphorylation levels were measured by 2-D gel electrophoresis. The following results were obtained in INTACT FIBERS. 1) Stimulation with 110 mM KCl resulted in maintained stress and a transient in both MLC phosphorylation and maximal shortening velocity (V_o). 2) Stimulation with 110 mM KCl plus nifedipine resulted in a transient contraction in which the peak level of developed stress was unchanged compared to control, but stress maintenance was abolished. The peak level of V_o was unaffected, but steady state levels were significantly depressed. MLC phosphorylation levels were not different from control at any time. 3) Stimulation with Bay k 8644 resulted in slow stress development to levels similar to those with 110 mM KCl, however, neither V_o nor MLC phosphorylation were increased above basal. These results with intact fibers suggest that stress can develop in the absence of significant increases in MLC phosphorylation and that stress maintenance and V_o can be affected independent of changes in MLC phosphorylation. The following results were obtained in SKINNED FIBERS. 1) Increasing Ca^{2+} resulted in stress development that correlated with the level of MLC phosphorylation, however, once contracted, decreasing Ca^{2+} resulted in stress maintenance without proportional levels of MLC phosphorylation. 2) Increasing the free Mg^{2+} (6-20 mM) in the absence of Ca^{2+} resulted in development of 75% of the maximal stress developed by Ca^{2+}, but there were no measurable changes in MLC phosphorylation. 3) Stress maintained by Ca^{2+} and CTP (a substrate for myosin ATPase but not for MLC kinase) with basal levels of MLC phosphorylation was relaxed by addition of calmodulin antagonists. These results with skinned fibers suggest that stress can be developed without a measurable increase in MLC phosphorylation and that the Ca^{2+} dependence of the MLC phosphorylation independent stress (latch) involves a calmodulin-like protein. In summary, the results from both intact and skinned fibers studies indicate that latchbridges and phosphorylated crossbridges may be activated independently and that the site of activation of latchbridges may be a calmodulin-like protein.

Supported in part by funds from NIH HL 34409 (RSM) and AHA, Mass. Affiliate 13508845 (RSM).

INHIBITION OF CYCLING AND NONCYCLING CROSSBRIDGES IN CHEMICALLY SKINNED SMOOTH MUSCLE BY VANADATE

R.A. Nayler and M.P. Sparrow

Department of Physiology
University of Western Australia
Western Australia, Australia

This study investigated the action of vanadate (Vi) on cycling and noncycling crossbridges in skinned smooth muscle. Guinea pig taenia coli and tracheal smooth muscle was chemically skinned with Triton X-100 (1%) to increase membrane permeability. Using CaEGTA/EGTA buffers, skinned bundles of smooth muscle were maximally contracted by 20 μM calcium (Ca) at 30° C (0.1 μM calmodulin) within 5 min and relaxed in EGTA (< 10 nM Ca). The characteristics of tension development in the presence of Vi depended on whether fibres were exposed to Vi during the previous contraction or during the relaxed state prior to the second contraction. After incubation with Vi (30-300 μM) in the relaxed state, tension to Ca developed with a peak phase after 1-3 min followed by a lower plateau phase. When Vi was incubated during the initial contraction, tension was inhibited (ED: = 23 μM, taenia coli) and the peak phase during the second contraction (in the presence of Vi) was absent. Plateau tension of the second contraction was the same in both cases (r = 0.998). Similar responses were seen with the trachea. The 'peak' response was lost at 20° C but occurred if calmodulin was increased from 0.1 to 1 μM. Thus, the 'peak' response was a kinetic effect related to a rapid initial rate of contraction and was dependent on [Vi]. The effect of Vi on tension development is consistent with Vi binding to some site exposed during crossbridge cycling but absent in the relaxed state. In view of Vi's ability to bind to isolated skeletal myosin it appears that Vi binds to the 'active-site' on smooth muscle myosin, exposed during the crossbridge cycle, forming an actomyosin.ADP.Vi complex that prevents further crossbridge cycling and tension generation. A weak antagonism between phosphate (Pi) and Vi on cyclinig crossbridges was demonstrated during contractions activated by myosin thiophosphorylation. High concentrations of Pi (6-12 mM) were needed to produce a small inhibition (10%) of maximum Ca-activated tension.

Following a contraction skinned fibres relaxed slowly on Ca removal, with an initial faster phase followed by a slower phase which approximated a single exponential process (half-time = 24 min). This slow phase of tension loss corresponded with the loss of an active state (determined by the rate of tension recovery after a quick release) suggesting that tension was now being maintained by noncycling crossbridges. Both Vi (0.1 and 1 mM) and Pi (1 and 6 mM) increased the rate of tension loss (up to 10 fold) but Vi was 5-10 times more potent than Pi. It seems likely that Vi exerts its inhibitory effect at the same site as on cycling crossbridges. It is suggested, therefore, that Vi and Pi both act on the active-site but that Pi has a more efficacious action on slowly cycling than rapidly cycling crossbridges. Thus, it is proposed that Vi binds to the A.M.ADP complex of smooth muscle (in a manner similar to that seen in skeletal muscle) dissociating actin and myosin and inhibiting further crossbridge cycling. An analogous complex is suggested to arise in the noncycling crossbridges found in membrane skinned smooth muscle. Characterization of the inhibitory Vi complex may prove useful in the study of crossbridge cycle intermediates in smooth muscle.

THE EFFECT OF 2,3-BUTANEDIONE MONOXIME (BDM) ON SMOOTH MUSCLE MECHANICAL PROPERTIES.

C.S. Packer, M.L. Kagan, S.A. Robertson, and N.L. Stephens

Department of Physiology
University of Manitoba
Winnipeg, MB, Canada R3E OW3

2,3, butanedione monoxime (BDM) has been reported to selectively block crossbridge interaction in skeletal and cardiac muscle and BDM has been shown to cause a dose-dependent decrease in smooth muscle maximum tension development (P_o). With the relatively recent descriptions of two types of crossbridges operating in smooth muscle, the normally-cycling crossbridges recruited early in contraction and then fairly rapidly -- within 30% of the muscles' contraction time -- replaced by very slowly-cycling or "latch" crossbridges, it became important to know whether BDM is a specific inhibitor of one type of crossbridge or the other. In this study it was shown that 7.5 mM BDM decreased the ability of canine tracheal smooth muscle (TSM) to develop P_o by 29%. BDM has an even greater effect on maximum shortening ability (ΔL_{max}) having decreased the TSM ΔL_{max} by 47%. BDM treatment did not alter "transition time (t_T)" i.e. the time in contraction at which most of the normally-cycling bridges have been replaced by "latch" bridges in TSM. Velocity of shortening early in contraction i.e. prior to t_T, was decreased by 48%, while velocity late in contraction i.e. post t_T, was not decreased with BDM treatment. BDM caused a decrease in maximum load bearing capacity or maximum force potential (MFP) at all times in contracting TSM. This investigation supports the suggestion that BDM inhibits crossbridge cycling rate in smooth muscle. In particular BDM appears to specifically inhibit normally-cycling crossbridges as it has no effect on cycling rate of very slowly-cycling or "latch" crossbridges in TSM. Supported by the Canadian Heart Foundation and Medical Research Council of Canada.

EFFECTS OF ANTIBODIES TO TURKEY GIZZARD (TG) MYOSIN LIGHT CHAIN KINASE (MLCK) ON CONTRACTION AND MYOSIN PHOSPHORYLATION ($MLC-P_i$) IN SKINNED GUINEA PIG TAENIA COLI (TC)

Richard J. Paul, John D. Strauss, and Primal de Lanerolle

Departments of Physiology and Biophysics
University of Cincinnati, Cincinnati, Ohio
University of Chicago, Chicago, Illinois

We have used an immunological approach to investigate the relation of $MLC-P_i$ to other putative Ca^{2+}-regulatory mechanisms in the control of contractility in smooth muscle. Our aim is to selectively inhibit MLCK so that the dependence of other Ca^{2+}-regulatory systems may be unmasked. Anti-MLCK antibodies were produced by immunizing goats with MLCK purified from turkey gizzards. An IgG fraction was prepared and Fab fragments were generated by digesting the IgG fraction with papain. The anti-MLCK Fab were purified by applying the digest to an affinity resin made by coupling purified turkey gizzard MLCK to Sepharose 4B. The anti-MLCK Fab were eluted from the column, concentrated and dialyzed in the appropriate buffers. Anti-MLCK Fab inhibit the activity of purified TG MLCK and interact monospecifically with MLCK in various mammalian smooth muscles as demonstrated by a Western Blot analysis. Guinea pig taenia coli were made permeable to proteins (skinned) as described by Rüegg and Paul (Circ. Res. 50:394-399, 1982). Small fibers approximately 100 microns in diameter and 4 mm long were connected to transducers and maintained at $25°$ C. The fibers were relaxed ($pCa^{2+} > 8$) or contracted ($pCa^{2+} = 5.18$) by immersion in calcium/EGTA buffers. The fibers were also incubated with 6×10^{-8} M anti-MLCK Fab dialyzed in relaxing or contracting solution. Myosin phosphate content was quantitated by first separating the muscle proteins by IEF PAGE. The proteins were then transferred to nitrocellulose paper and incubated with anti-LC_{20} antibodies followed by peroxidase labeled and ^{125}I-labeled second antibodies. The immunoreactive material was visualized by developing the peroxidase reaction. Two types of experiments were performed. First, the fibers were incubated by anti-MLCK Fab in relaxing solution and then contracted. The post-Fab contraction, after 75 mins, developed about 25% of the force of a test contraction performed prior to Fab. Fibers incubated with Fab for shorter or longer times and then stimulated with Ca^{2+} generated proportionately more or less tension. Second, contracted fibers incubated with Fab relaxed completely in about 90 mins despite the presence of Ca^{2+}. No significant effect on isometric force was seen in the presence of affinity-purified mouse Fab against the Fc region of human IgG. $MLC-P_i$, which was increased from basal values of 0.13 ± 0.03 mol P_i/mol LC_{20} to 0.62 ± 0.01 in pCa 5.18 solutions, was lowered to 0.36 ± 0.01 after 90 min in the presence of Fab. Moreover, Fab relaxed ATPγS-induced contraction without any apparent decrease in $MLC-P_i$. These data suggest that anti-MLCK Fab may relax TC by both inhibiting MLCK activity and by another undefined mechanism.

Supported by U of I BRSG98425, NIH HL35808, NIH 22619, and AHA Established Investigatorship (RJP).

EFFECTS OF REDUCED EXTRACELLULAR CALCIUM ON CALCIUM METABOLISM IN VASCULAR SMOOTH MUSCLE

L.N. Russek and R.D. Phair
Department of Biomedical Engineering
The Johns Hopkins University School of Medicine
Baltimore, Maryland 21205

In order to determine whether exposure to low extracellular calcium alters intracellular as well as extracellular calcium metabolism, long (8 hour and 16 hour) effluxes of ^{45}Ca from 2 cm long segments of rabbit thoracic aorta were carried out in normal (1.5 mM) and reduced (0.15 mM) $[Ca]_o$. The results were analyzed using the SAAM computer simulation package and a recently developed model of calcium metabolism in vascular muscle. (Phair and Hai, Circ. Res., in press, 1986).

Effluxes observed during steady state exposure to 0.15 mM $[Ca]_o$ could not be modelled assuming either 1) that mass action is the only determinant of calcium metabolism, or 2) that only plasma membrane rate constants change, in particular, that they change to maintain intracellular calcium constant.

A step decrease from 1.5 mM to 0.1 mM $[Ca]_o$ in the perfusing solution caused a step increase in the efflux of ^{45}Ca. The existence of a very large, very slowly turning over compartment was required to fit this data. This slow compartment may comprise as much as half of the total tissue calcium.

The results suggest that exposure to low extracellular calcium does more than remove extracellular bound and free calcium. Low $[Ca]_o$ appears to cause complex changes in aortic calcium fluxes, including release from a large calcium store.

STIFFNESS OF TRACHEAL SMOOTH MUSCLE DURING ACTIVE SHORTENING: STIFFNESS AND LENGTH RELATIONSHIP

C.Y. Seow and N.L. Stephens

Department of Physiology
University of Manitoba
Winnipeg, Manitoba, Canada R3E 0W3

Stiffness ($\Delta P/\Delta L$) of canine tracheal smooth muscle during active shortening was estimated by applying small force perturbations and measuring the amplitude of the resulting length perturbations. The force perturbation the muscle subjected to was a train of rectangular force waves varying from 0 to about 5 mN (10% of P_o), with a frequency of 10 Hz. The oscillation was critically damped. Experimental results indicated that stiffness increased as shortening proceeded. Plot of stiffness vs. muscle length revealed an almost linear relationship during contraction. However during relaxation there was an initial sharp drop in stiffness without much lengthening of the msucle followed by a slower decrease in stiffness as the muscle lengthened. $\Delta P/\Delta L$ at resting state was 515.1 (mN/mm$^2 \cdot l_o$) 71.8 (SE) and at the peak of contraction it was 4576.5 (mN/mm$^2 \cdot l_o$) 632.1 (SE). If the series elasticity of the muscle resides in the non-overlap zones in sarcomeres (or at least part of it does), then the increase in stiffness during shortening can be explained by the decrease of the total length (in series) of the non-overlap due to shortening.

Supported by grants from the Medical Research Council of Canada.

THE ASSOCIATION OF INTRACELLULAR CA^{2+} RELEASE WITH CONTRACTION IN COLONIC MUSCLE

N. Sevy, H.W. Kao, W.J. Snape, Jr.

Harbor-UCLA Medical Center
Torrance, California 90510

The release of calcium from intracellular stores is thought to initiate a contractile response in a variety of smooth muscles. The aim of this study was to compare the role of intracellular Ca^{2+} stores with extracellular $[Ca^{2+}]$ in bethanechol stimulation of two different muscles of the rabbit colon. The optimal length (L_o) for isometric contraction was found in thin strips (200-300 μm width) of distal circular smooth muscle (CM) and proximal taenia coli (TC) of the rabbit colon. Caffeine, an agent which is believed to release calcium from the sarcoplasmic reticulum in a number of muscle types, elicited a significantly greater contraction in the TC (11.7 ± 2.8% of max bethanechol response, n = 7) than in the CM (4.4 ± 1.2% of max bethanechol response, n = 7), $P < 0.01$. Both TC and CM were superfused in a "zero calcium" solution (chelexed solution with 0.2 mM EGTA, free $[Ca^{2+}] = 10^{-7}$ measured by a calcium sensitive electrode and stimulated with 10^{-5} betehanechol. The time required for a dropoff to 50% of the maximal bethanechol contraction (T 1/2) was significantly different for TC (2.39 ± 0.11 minutes, n = 7) compared with CM (1.05 ± 0.08 minutes, n = 7), $P < 0.001$. Next studies were performed to determine the dependence of contraction on the intracellular concentration of calcium by permeabilizing the tissues with a concentration of saponin (200-300 μg/ml) that caused disappearance of the caffeine induced response. A half maximal contraction (EC_{50}) occurred at an identical pCa, 6.1 ± 0.04 in TC (n = 7) and 6.1 ± 0.1 in CM (n = 7). These studies suggest that 1) both TC and CM require the same intracellular concentration of calcium for maximal contraction, 2) the intracellular calcium stores of TC are significantly greater than those of CM, and 3) therefore CM may rely more on influx of extracellular calcium than the TC for contraction.

CALCIUM, MAGNESIUM AND MgATP^{2+} DEPENDENCE OF SHORTENING IN SKINNED SINGLE SMOOTH MUSCLE CELLS

D.M. Warshaw and M.S. Hubbard

Department of Physiology and Biophysics
University of Vermont
Burlington, Vermont 05405

Most studies of skinned smooth muscle have been performed in whole tissue preparations. In this study, we report the development of a chemically skinned single smooth muscle cell preparation from the toad, Bufo marinus, stomach. Isolated smooth muscle cells were skinned using 30 μg/ml of saponin for 12 minutes. The effect of various ionic environments (i.e. changing free Ca^{2+}, $MgATP^{2+}$ and Mg^{2+}) on skinned cell contractile response was assessed by measuring cell lengths from populations of cells using a computer assisted length measuring system. Comparison of cell length histograms were used to determine the extent of cell shortening in response to a given ionic perturbation. Once skinned, the single cells shortened with a sensitivity to free calcium (ED50 = 1.5 M Ca^{2+}) that was 3 orders of magnitude lower than potassium depolarized cells (ED50 = 1.5 mM Ca^{2+}). In addition to the calcium sensitivity, the effect of free $MgATP^{2+}$ and Mg^{2+} on the extent of cell shortening was investigated. The extent of cell shortening was dependent upon free $MgATP^{2+}$ with a Km of 50 μM $MgATP^{2+}$ at pCa 4 which may reflect the dependence of the myosin light chain kinase for $MgATP^{2+}$. In addition, lowering free Mg^{2+} to 0.05 mM had a profound effect upon the extent of cell shortening suggesting that Mg^{2+} plays a key role in excitation-contraction coupling other than the need for its being complexed to ATP.

Supported by Am34872 and HL35684.

Index

A-3, and inhibition of smooth muscle phosphorylation, 252–260
Acetylcholine, and receptor-mediated activation of smooth muscle, 380, 381
Acidosis, intracellular, and crossbridge slowing mechanisms in smooth muscle, 373–374
Actin
-activated myosin ATPase activity
effect of tropomyosin binding on, 144–149
-actomyosin S1-nucleotide complex, and kinetic model for striated muscle actomyosin S1, 61–62
binding
to caldesmon, 124–126
effect on rate of conformation change in smooth muscle, 64–65
in model of regulation of smooth muscle contractions, 119–121
-myosin interactions
and calcium ions binding to myosin, 132–134
and caldesmon, 127–130
dissociation of myosin light chain kinase from activation of, and calcium regulation in smooth muscle, 109–116
effects of aortic polycation-modulable protein phosphatases, 204–205
and protein kinases, 251–252
tropomyosin binding to
effect of ionic strength on, 147–149
and smooth muscle actomyosin magnesium-ATPase stimulation, 176–177
see also Actin filaments in smooth muscle cells; F-actin
Actin filaments in smooth muscle cells, 2–4
and crossbridge structure of smooth muscle, 29–32; *see also* Crossbridge(s)
isolated, skinned saponin-exposed, 8–17
insertion into dense bodies, association with myosin filaments with crossbridges, 27
rat aorta, cytochalasin-like activity, 469–470
α-Actinin
fluorescently-labeled antibody to, and organization of contractile elements of smooth muscle cells, 6–8
in fusiform dense bodies of smooth muscle cells, 16
Actomyosin ATPase activity
assays, 162–163
effects of caldesmon on, 129–131, 149–154
modulation by thin filament-associated proteins, 143–155
in molluscan muscle, head-tail junction and myosin-linked regulation of, 91–92
regulation of, 62–65
relationship between superprecipitation and, 110
of smooth muscle and striated muscle protein, 27
thin filament-linked regulation of, 121–122
and tropomyosin binding to actin
effects of calcium and magnesium ions on, 144–147
effects of ionic strength on, 147–149
Actomyosin S1
binding to actin in presence of ATP, calcium and, 59
of striated and smooth muscle, kinetic mechanism for non-regulated, 59–62
see also Heavy meromyosin
Actomyosin, smooth muscle

489

and calcium regulation in smooth muscle, 110–116
effect of actin binding on rate of conformation change, 64–65
effect of phosphorylation and ionic strength on kinetic mechanisms of heavy meromyosin and myosin, 63–64
and kinetic mechanism of regulation distinguished from structural mechanisms, 59–65
mechanism of regulation, compared with striated muscle actomyosin, 59–65
and regulation of smooth muscle actomyosin ATPase, 62–65
and smooth and striated muscle heavy meromysin, 62–63, 65
see also Actomyosin ATPase activity; Actomyosin S1
ADP
and activation of smooth muscle in absence of calcium ions, 437–447
effects on skinned smooth muscle
and ATP-induced relaxation from rigor, 52–54
and calcium-activated contraction, 44–46
and force-velocity relationship, 46–51
and kinetic mechanism for striated muscle actomyosin S1, 61
α-Adrenoreceptors, and norepinephrine stimulation of spontaneously hypertensive rat vascular smooth muscle, 277–285
β-Adrenoreceptors
agonists, effects of muscarinic tone on response in airway smooth muscle to, 264–267
and norepinephrine stimulation of vascular smooth muscle of spontaneously hypertensive rats, 277–285
Airway smooth muscle
regulation of cyclic AMP-dependent protein kinase activity in, 263–274
tone
and functional antagonism concept, 264
integration of contractile and relaxant pathways, 263–264

Amino acid composition, of chicken gizzard and bovine aorta caldesmons, 124, 125
Amphibian smooth muscle cells, compared with other vertebrate and mammalian smooth muscle cells, 2–4
AMPPNP, effects on rate of ATP-turnover in skinned smooth muscle, 51–52
Antibodies
and immunofluorescent studies of organization of contractile elements of smooth muscle cells, 6–8
to skeletal muscle myosin, and subunit exchange studies, 87–88, 89
to turkey gizzard myosin light chain kinase, effects on contraction and myosin phosphorylation in skinned guinea pig taenia coli, 479–480
see also Monoclonal antibodies, cross-reactions with smooth muscle myosin light chain kinases; Polyclonal antibodies
Aorta. See Bovine aorta; Aortic entries
Aortic polycation-modulable protein phosphatases structure and function, 195–205
and actin-myosin interactions, 204–205
gel filtration studies, 199, 200–201
and polycation-modulable phosphatases from other tissues, 196
and polypeptide composition of polycation-modulable phosphatases, 196–201
tryptic digestion studies, 201–204
Aortic smooth muscle from hypertensive rats, sensitivity to contractions in, 237–238
Aortic strips, and inhibition of smooth muscle phosphorylation, 253
Arterial smooth muscle
activation of latchbridges and phosphorylated crossbridges, 473–474
carotid cells prepared by digestion with papain, contraction of, 453–454
ATP
and A-3 inhibition of myosin light chain kinase reaction, 255–260
and activation of striated muscle heavy meromysin, 62–65

Index / 491

β-v-imido-. See AMPPNP, effects on rate of ATP-turnover in skinned smooth muscle
and binding of smooth muscle tropomyosin to smooth muscle actin, 171
and effect of calcium on binding of actomyosin S1 to actin, 59
and effect of smooth muscle phosphatases on glycogen metabolism, 214–215
effect in skinned smooth muscle
and calcium-activated contraction, 44–46
and force-velocity relation, 47, 51
and induced relaxation from rigor, 52–54
hydrolysis rate. See Vmax effect
in isolated, saponin-exposed single smooth muscle cells, 8–17
and kinetic mechanisms for striated muscle actomyosin S1, 61
measurement of turnover of, 44
in model of regulation of smooth muscle contraction, 119–121
nonhydrolyzable analogue of. See AMPPNP, effects on rate of ATP-turnover in skinned smooth muscle
turnover during smooth muscle contraction, effects of ATP and AMPPNP on, 51–52
and turnover rate of phosphorylated myosin in rabbit taenia coli, 291–299
utilization rates
and effects of calcium on smooth muscle energetics, 324–328
and hypoxia, 464
see also Caged ATP
Atropine, and inhibition of smooth muscle response to electrical field stimulation, 382–383

Barium-induced smooth muscle contractions, 382
Bay K 8644
and inhibition of myosin light chain kinase, 243
and latch state in vascular smooth muscle, 411–413

Birefringence of rate anococcygeus, and density of myosin filaments, 459–460
Bovine aorta
caldesmon, amino acid composition, 124, 125
and inhibition of smooth muscle phosphorylation, 253–254
and stomach, calcium regulation of, 110–116
Bovine tracheal strips, effect of muscarinic tone on response to β-adrenoreceptor agonists, 264–266
8-Bromo-cyclic GMP, and calcium influx and contraction of rat aorta, 455–457
Bronchoconstriction. See Cyclic AMP-dependent protein kinase
Bronchorelaxants. See Cyclic AMP-dependent protein kinase
Bufo marinus isolated smooth muscle cells, 2
stomach muscularis, force:velocity relationship and helical shortening in, 303–315
2,3-Butanedione monoxime, effect on smooth muscle mechanical properties, 477

Caffeine, and contraction of isolated vascular smooth muscle preparations, 281
Caged ATP
initiation of crossbridge transients in smooth muscle by, 32–38
photolysis of, and detection of crossbridge transients in smooth muscle, 27
Calcium, calcium ions
-activated contraction in skinned smooth muscle, effects of ADP on, 44–46
and activation of striated muscle heavy meromyosin, 62–65
binding to calmodulin, inhibition of, and myosin light chain phosphorylation, 239
binding to myosin, and actin-myosin interactions, 132–134
binding to troponin, conformational changes in skeletal muscle and, 159
and calmodulin activation of smooth muscle skinned fibers at low myosin phosphorylation, 427–435

channels, receptor-operated in smooth muscle, guanylate-cyclase dependent gating of, 455–457
dependence of shortening in skinned single smooth muscle cells, 487
effect on tropomyosin binding to actin, 144–149
effects on smooth muscle mechanics and energetics, 319–329
extracellular, effects on calcium metabolism in vascular smooth muscle, 481
fluxes
 and α-adrenoreceptor mediated response in vascular smooth muscle of spontaneously hypertensive rats, 280–285
 and norepinephrine stimulation of spontaneously hypertensive rat vascular smooth muscle, 277
and hypertension, 279–280
and myosin phosphorylation in skinned smooth muscle, stress development and velocity and, 399–409
inhibition, and inhibition of myosin light chain kinase, 241–243
intracellular, and contraction of rabbit smooth muscle, 485–486
in isolated, saponin-exposed single smooth muscle cells, 8–17
and kinetic model for striated muscle actomyosin S1, 61
and latch state
 in glycerinated porcine carotid artery, 418–422
 in vascular smooth muscle, 411–413
metabolism in vascular smooth muscle, effects of reduced extracellular calcium on, 481
in model of regulation of smooth muscle contractions, 119–121
modulation of actin-activated ATP hydrolysis, 143
reattachment of detached crossbridges in absence of, 36
regulation of activity of striated muscle actomyosin, 59

regulation in smooth muscle, leiotonin and, 109–116
 and myosin light chain kinase activity and leiotonin activity of bovine stomach myosin light chain kinase, 109–112
 and removal of leiotonin activity without affecting myosin light chain kinase activity, 112–113
 and suppression of myosin light chain kinase activity without affecting leiotonin activity, 113–115
role in smooth muscle contractions, 415–416
and shortening velocity of skinned smooth muscle, 322–324
smooth muscle contraction without, 437–447
and tension development from rigor state in smooth muscle and photolysis of caged ATP, 32–38
Caldesmon, smooth muscle, 122–136
and actin-activated ATPase activity of smooth muscle myosin, 143–144
amino acid composition of, 124, 125
associated with "contractile unit," 127–128
calmodulin and actin binding, 124–126
discovery of, 122
effects on actin-myosin interaction, 127–130
effects on actomyosin ATPase, 149–154
and enhancement of viscosity of F-actin, 130
and inhibition of actomyosin ATPase, 129–131
molecular weight and native structure, 123–124
phosphorylation of, 130–131
possible role in latchbridge formation, 131–136
purification of, 122–123
subcellular localization, 127–128
and thin filament-linked regulation of actomyosin ATPase, 122
tissue and species distribution, 126–127
Calmodulin, smooth muscle
binding to caldesmon, 124–126

and calcium activation of smooth muscle skinned fibers at low myosin phosphorylation, 427–435
and discovery of caldesmon, 122
inhibition of activation of myosin light chain kinase, 239–244
inhibition by W-7 and its derivatives, 252–256
in model of regulation of smooth muscle contractions, 119–121
and reversal of caldesmon inhibition of actomyosin ATPase, 151–152
-Sepharose affinity chromatography, and caldesmon binding to calmodulin, 124–125
and vascular smooth muscle contractions, 251
Canine tracheal smooth muscle
carbachol-stimulated, inositol phospholipids synthesis in, 449–450
mechanism of shortening-induced dependence of contractility in, 461–462
phosphodiesterase isozymes in, 269–273
see also Smooth muscle contractions
Carbachol
and receptor-mediated activation of smooth muscle, 380, 381
-stimulated canine trachealis muscle, inositol phospholipids synthesis in, 449–450
Carotid artery smooth muscle relaxation by myosin dephosphorylation. See Smooth muscle contractions
Casein kinase I, and actin-myosin interaction, 251
Casein kinase II, and actin-myosin interaction, 251
Cat duodenal muscle. See Crossbridge(s)
Chicken gizzard
ADP and contractions in, 44
calcium regulation, 110–116
caldesmon
amino acid composition, 124, 125
isolation of, 122
heavy meromyosin, binding of substrate and product intermediates, 62
myosin and heavy meromyosin, conformation changes associated with phosphorylation, 91–104

skinned fibers, calcium/calmodulin activation at low levels of myosin phosphorylation, 427–435
Colonic smooth muscle, rabbit
association of intracellular calcium ions with contraction, 485–486
culture of, 465–466
Competition experiments, and kinetic mechanism for striated muscle actomyosin S1, 61
Conformational changes
and kinetic model for striated muscle actomyosin S1 activation, 62
in model of regulation of smooth muscle contractions, 119–121
in myosin and heavy meromyosin from chicken gizzard, phosphorylation and, 91–104
and ATPase activity, 93–103
and degradation of 20 kDa light chain by papain of dephosphorylated heavy meromyosin, 98
and location of reactive thiols of smooth muscle myosin, 100–101
and N-ethyl maleimide (MalNEt) stimulation of ATPase activity of myosin, 101–103
and proteolytic digestion of smooth muscle myosin and heavy meromyosin, 93–98
and relationship of flexing of heavy meromyosin at head-tail junction to structure and function, 98–99
and role of head-tail junction, 93–103
and SDS polyacrylamide gel electrophoresis of papain digested heavy meromyosin, 96
in skeletal muscle, and tropomyosin response of calcium binding to troponin, 159
Contractile apparatus of smooth muscle cells. See Smooth muscle cell filaments, molecular structure and organization
Contracting relaxation phosphodiesterase. See Phosphodiesterase isoenzymes in canine trachealis

Contractions in smooth muscle. *See* Smooth muscle contractility; Smooth muscle contractions
Cooperativity, and relaxation in smooth muscle, 36–37
Copolymer formation, by smooth muscle myosin filaments, 84–90
Coronary artery, skinned, effects of cyclic AMP-dependent protein kinase in, 471
Crossbridge(s)
  attachment in smooth muscle, geometric three-dimensional thermodynamic model for, 467
  compliance, and stiffness and energetics of active shortening in skinned smooth muscle, 333–343
  cycle kinetics, and force:velocity relations in single smooth muscle cells, 313–314
  cycling
    in model of regulation of smooth muscle contractions, 120
    and noncycling, inhibition by vanadate in skinned smooth muscle, 475–476
    rate in mammalian smooth muscle, 289–300
  and effects of ADP on shortening velocity in smooth muscle, 50–51
  and influence of ATP, ADP, and AMPPNP on contractions in skinned smooth muscle, 43–55
  interaction, 2,3-butanedione monoxime and, 477
  phosphorylated, activation in arterial smooth muscle, compared with latchbridges, 473–474
  phosphorylation, and latch state in vascular smooth muscle, 413
  projections, in studies of molecular structure and organization of smooth muscle cell filaments, 11–15
  properties of smooth muscle during stretch, 347–355
  and isometric stiffness, 351
  slowing mechanisms and smooth muscle contraction, 289–302, 357–374

structure of smooth muscle, 29–32
transients. *See* Smooth muscle
turnover, and active shortening in skinned smooth muscle, 341–343
see also Latchbridge state
Cyclic AMP
  and β-adrenoreceptor-mediated response to norepinephrine stimulation in vascular smooth muscle of spontaneously hypertensive rats, 278–280
  content of airway smooth muscle, regulation of, 263–274
  see also Cyclic AMP-dependent protein kinase
Cyclic AMP-dependent protein kinase
  and actin-myosin interaction, 251
  activity in airway smooth muscle, regulation, 263–274
  and altered vascular reactivity in hypertension, 233–244
  and effects of muscarinic tone on response to β-adrenoreceptor agonists, 264–267
  effects in skinned coronary artery, 471
  inhibitors, 256–258
  and norepinephrine stimulation of adenylate cyclase activity in vascular smooth muscle of spontaneously hypertensive rats, 278–279
  and phosphodiesterase isozymes in canine trachealis, 269–273
  and specificity of functional antagonism, 266–267
Cyclic GMP
  and guanylate-cyclase-dependent gating of calcium channels in smooth muscle, 455–457
  and phosphodiesterase isozymes in canine trachealis, 269–273
Cyclic nucleotide phosphodiesterase. *See* Phosphodiesterase isozymes in canine trachealis, identity and role of
Cyclic nucleotides. *See* Cyclic AMP; Cyclic GMP
Cytochalasin-like activity, in rat aorta smooth muscle cells, 469–470

DEAE chromatography, in study of calcium regulation in smooth muscle, 110–111

Dense bodies of smooth muscle cells. *See* Fusiform dense bodies of smooth muscle cells
Dephosphorylation of proteins
  and contractile activity in smooth muscle cells, 207
  and polycation-modulable protein phosphatases, 195–205
Diltiazem, and inhibition of myosin light chain kinase activity, 242
Dissociation rates of substrate and products, and kinetic mechanisms for striated muscle actomyosin S1, 61

Efficiency and crossbridge cycling rate, 289–302, 328–329
Electrical field stimulation studies
  of myosin phosphorylation in tracheal smooth muscle cells, 188–192
  of smooth muscle contraction kinetics, 380, 381–383
Electron microscopy
  of contractile apparatus of vertebrate smooth muscle cells, 4
  of filaments and dense bodies from isolated, skinned single smooth muscle cells, 12, 13
  and subunit exchange between smooth muscle myosin filaments, 87–88
  see also Optical microscopy of smooth muscle cells
Electrophoresis, of myosin heavy chain isoforms from human smooth muscle from different tissues, 451; *see also* Pyrophosphate gel electrophoresis, and detection of myosin isoforms in smooth muscle; SDS polyacrylamide gel electrophoresis
Endothelium-derived factors, and calcium influx and contraction of rat aorta, 455–457
Energetics of smooth muscle contractions
  active shortening in skinned smooth muscle, stiffness and, 333, 343
  effects of calcium on mechanics and, 319–329
  energy supply to contractile apparatus in hypoxia-induced relaxation, 463–464
  energy usage
    length dependence of, 27
    and slowing of crossbridge cycling rate, 289–300
  see also Latchbridge state
Enzymes
  of cultured rabbit colonic smooth muscle cells, 465–466
  and interactions between contractile and cytoskeletal proteins in smooth muscle, 1
  kinetic analysis of tropomyosin stimulation of actin-myosin interaction, 176
  see also entries *under* specific enzymes
Epidermal growth factor-stimulated kinase, and actin-myosin interaction, 252
Equilibrium constants, and kinetic mechanism for non-regulated actomyosin S1, 59–62
N-Ethylmaleimide, and suppression of myosin light chain activity without affecting leiotonin activity in gizzard extract, 113–115

F-actin
  caldesmon and enhancement of viscosity of, 130
  cofactor activity, and tropomyosin, 159
  removal of caldesmon from, 133–135
Fatigue, and crossbridge slowing mechanisms in smooth muscle, 372–373
Felodipine, and inhibition of myosin light chain kinase activity, 241
Filamin, effects on myosin isoforms of smooth muscle, 67–68, 74–77
Force development, in skinned chicken gizzard fibers, calmodulin and, 428–433
Force:velocity relationship in smooth muscle
  and crossbridge slowing mechanisms, 364
  effects of ADP on, 46–51
  and helical shortening, 303–315
Functional antagonism
  defined, 263
  specificity in airway smooth muscle, 266–267
Fusiform dense bodies of smooth muscle cells, 4
  antibodies to α-actinin of, in study of contractile elements, 6–8

immunochemical labeling, 16–17
interactions with actin and myosin filaments, 4, 5–6
in isolated skinned smooth muscle cells and contraction, 20–22
and saponin exposure, 8–17

Gel filtration, of aortic polycation-modulable protein phosphatases, 199, 200–201
Gizzard muscle. *See* Chicken gizzard
Glycerination of porcine carotid artery, and latch state, 418–422
Glycogen metabolism, and smooth muscle phosphatases, 214–215
Gold binding
and indirect labeling of fusiform dense bodies in isolated smooth muscle cells, 16
in subunit exchange studies of smooth muscle myosin filaments, 87–88, 89
GTP
and norepinephrine stimulation of adenylate cyclase activity in vascular smooth muscle of spontaneously hypertensive rats, 278
-utilizing enzymes, and A-3 inhibition of myosin light chain kinase reaction, 255–260
Guanylate-cyclase-dependent gating, of receptor-operated calcium channels in vascular smooth muscle, 455–457
Guba-Straub extracts of native smooth muscle myosin, myosin isoforms and filamin in, 74–77
Guinea pig
portal vein saponin skinned strips, in study of crossbridge transients in smooth muscle, 28–38
skinned taenia coli
contraction and myosin phosphorylation, antibodies to turkey gizzard myosin light chain kinase and, 479–480
stiffness measurements and energetics of active shortening in, 334–343
vanadate inhibition of crossbridges in, 475–476

skinned tracheal smooth muscle, vanadate inhibition of crossbridges in, 475–476
smooth muscle cells, 2, 3
uterus, myosin isoforms in, 70
see also Smooth muscle contractions

Head-tail junction, and conformational changes in myosin and heavy meromyosin, 91–104
Heavy meromyosin
from chicken gizzard, conformational changes associated with phosphorylation, 91–104
of smooth and striated muscle, comparison of kinetic mechanisms of regulation, 65
unphosphorylated
effect on binding of tropomyosin to F-actin in absence of ATP, 172, 173
flexing at head-tail junction, and conformational changes, 98–99
Helical shortening and force:velocity relationship in smooth muscle cells, 303–315
Histamine-induced contractions in smooth muscle, and effect of high myosin light chain phosphatase on latch phenomenon, 388–397
Hog carotid artery. *See* Swine carotid artery
Human
skin fibroblasts, compared with cultured rabbit colonic smooth muscle cells, 465–466
smooth muscle, myosin heavy chain isoforms in different tissues, 451
Hypertension, and cyclic AMP-dependent kinase and myosin light chain kinase, 233–244; *see also* Spontaneously hypertensive rat vascular smooth muscle, molecular mechanisms of increased contractility to norepinephrine stimulation of
Hypoxia-induced relaxation in smooth muscle, and energy supply to contractile apparatus, 463–464

Immunofluorescent studies

and characterization of fusiform dense bodies in single skinned smooth muscle cells, 16–17
indirect immunofluoresence microscopy, and subcellular localization of caldesmon, 127–128
of organization of contractile elements of smooth muscle cells, 6–8
Immunogold. *See* Gold binding
Inositol phospholipids
1,4,5-inositol-trisphosphate, accumulation in vascular smooth muscle, and norepinephrine stimulation, 282
synthesis in carbachol-stimulated canine trachealis muscle, 449–450
Internal load and crossbridge cycling rate, 289–302, 328–329
Ionic strength
effect on tropomyosin binding to actin, and actin-activated myosin ATPase, 147–149
and kinetic mechanism of heavy meromyosin and myosin in absence of actin, 63–64
and sedimentation velocity and ATPase activity of heavy meromyosin, 96, 97–98
and structural change in smooth muscle myosin and heavy meromyosin, 62–63
Isoforms of myosin. *See* Myosin, smooth muscle
Isometric force
and effects of calcium on smooth muscle mechanics, 320–322
and effects of calcium on smooth muscle energetics, 324–329
Isometric stiffness. *See* Stiffness
Isoproterenol
effect of muscarinic tone in tracheal strips on response to, 264, 265
and role of phosphodiesterase isozymes in canine trachealis, 272–273
Isotonic releases, and force:velocity relationships in smooth muscle cells, 307–308, 311–313

Kinetic mechanism of regulation of smooth muscle and striated muscle actomyosin, compared with structural mechanism, 59–65
Km effect
and effect of tropomyosin in activation of myosin ATPase by actin, 169–171
and Km values for synthetic peptide substrates of myosin light chain kinases, 187–188

Lactate release from smooth muscle, and hypoxia, 463–464
Laser flash photolysis. *See* Caged ATP
Latchbridge state
high myosin light chain phosphatase activity in arterial smooth muscle and, 387–397
mathematical model for, 390–395
and myosin phosphorylation
and calcium ions, 406–408, 409
and dephosphorylation, 121
in skinned smooth muscle, and differences between types of smooth muscle, 416–417
in vascular smooth muscle, determinants of, 411–413
see also Latchbridges
Latchbridges
activation in arterial smooth muscle, compared with phosphorylated crossbridges, 473–474
formation, possible role of caldesmon in, 131–136
model, and slowing of crossbridge cycling rate in mammalian smooth muscle, 290–300
see also Latchbridge state
Leiotonin activity, and calcium regulation in smooth muscle
and myosin light chain kinase activity and leiotonin activity of bovine stomach myosin light chain kinase, 109–112
and procedures to remove leiotonin activity without affecting myosin light chain kinase activity, 112–113
and procedures to suppress myosin light chain kinase activity without affecting leiotonin activity, 113–115
Light chain kinase activity. *See* Myosin light chain kinase activity

Magnesium ADP, and skinned chick gizzard fiber tension in absence of calcium ions, 439–447; *see also* ADP

Magnesium-ATPase activity of smooth muscle myosin
  actin-activation of, and tropomyosin binding to actin, 144–149
  conversion of myosin to 10S form, and thiol reagents, 101–104
  dependence of shortening in skinned single smooth muscle cells, 487
  and force development in skinned chicken gizzard fibers, 430
  myosin phosphorylation and actin-activation of, 143
  and proteolytic digestion of smooth muscle myosin and heavy meromyosin, 93–98
  and stability of copolymers composed of dephosphorylated and phosphorylated smooth muscle myosin, 88, 90
  and stability of synthetic smooth muscle myosin filaments and phosphorylation of myosin light chain, 82–83

Magnesium-CDP, and skinned chicken gizzard fiber tension in absence of calcium ions, 439–447

Magnesium ions
  dependence of shortening in skinned single smooth muscle cells, 487
  effect on tropomyosin binding to actin, 144–149
  and skinned chicken gizzard fiber tension in absence of calcium ions, 444–445
  see also Magnesium ADP, and skinned chicken gizzard fiber tension in absence of calcium ions; Magnesium-ATPase activity of smooth muscle myosin, actin-activation of, and tropomyosin binding to actin

MalNEt, stimulation of ATPase activity of smooth muscle myosin, 101–104

Mammalian smooth muscle
  actin, myosin and fusiform dense bodies in cells of, 4
  slowing of crossbridge cycling rate in, 289–300
  see also entries under other mammals, e.g. Humans; Rats

Maximum force potential, and 2,3-butanedione monoxime, 477

Maximum shortening velocity in smooth muscle cells, 303

Mechanics of smooth muscle contractions
  and 2,3-butanedione monoxime, 477
  and crossbridge properties during stretch, 347–355
  effects of calcium on energetics and, 319–329
    and isometric force, 320–322
    and shortening velocity, 322–324
    and shortening-induced dependence of contractility in canine tracheal smooth muscle, 461–462
  stiffness and energetics of active shortening
    in skinned smooth muscle, 333–343
    in tracheal smooth muscle, 483
  stress and velocity in skinned smooth muscle, and calcium ions and myosin phosphorylation, 399–409
  see also Force:velocity relationship in smooth muscle

Membrane depolarization of smooth muscle, 382

Methacholine, effect of muscarinic tone of tracheal strips on response to, 264–266

Methylene blue, and calcium influx and contraction of rat aorta, 455–457

Monoclonal antibodies, crossreactions with smooth muscle myosin light chain kinases, 185–186

Monomer pool of subunits. *See* Myosin filaments, subunit exchange between

Muscarinic tone, effect on response to β-adrenoceptor agonists in airway smooth muscle, 264–266

Myosin filaments, smooth muscle
  with crossbridges, association with actin filaments inserting into dense bodies, 27
  in smooth muscle cells, 2–4

and crossbridge structure of smooth
muscle, 29-32
density of, and birefringence of rat
anococcygeus, 459-460
and isolated, skinned saponin-exposed
single cells, 8-17
tilting experiments and, 13, 15-16
subunit exchange between, 81-90
and ability of myosin species to copolymerize, 84-90
compared with skeletal muscle myosin
filaments, 90
electron microscopy studies of, 87-88
and exchange of biotinylated smooth
muscle myosin into, 88, 89
and sedimentation patterns, 84-85
and stability of copolymers composed
of dephosphorylated and phosphorylated smooth muscle
myosin, 88, 90
and state of phosphorylation of regulatory light chain, 82-83
see also Synthetic myosin filaments, subunit exchange studies with
Myosin light chain kinase activity
and activation of actin-myosin interaction, 109
and altered vascular reactivity in hypertension, 233-244
antibodies to, effects on contraction and
myosin phosphorylation, 479-480
calmodulin and calcium activation of, and
myosin phosphorylation, 427
development of modulators of, and altered vascular reactivity in spontaneously hypertensive rats,
233-244
and discovery of caldesmon, 122
and effect of phosphatase catalytic subunit
on latch state, 421-422
inhibition of calcium-calmodulin activation of, 239-244
inhibitors, 252-256
cyclic AMP-dependent protein kinase,
280
in model of regulation of smooth muscle
contractions, 119-121
and myosin phosphorylation in smooth
muscle cells, 183-192, 207

in chicken gizzard, conformation
changes associated with,
91-104
and monoclonal antibody studies,
185-186
synthetic peptide substrates and,
186-188
and Western blotting studies, 184-188
procedure for removal from bovine stomach extract without affecting leiotonin activity, 112-115
and protein kinase C, 258-259
purification, 160-161
and vascular smooth muscle contraction,
251
Myosin light chain phosphatase
in arterial smooth muscle, latch phenomenon and, 387-397
purification, 160-161
Myosin phosphorylation in smooth muscle,
47
and actin-activation of magnesium-
ATPase activity of smooth muscle
myosin, 143
assays, 162-163, 164-165
and conformational state, and modification of tropomyosin stimulation of
vertebrate smooth muscle actomyosin ATPase activity, 159-177
inhibitors, 235
low levels of, and calcium/calmodulin activation of smooth muscle skinned
fiber contractions, 427-435
and myosin light chain kinases, 183-192,
207
and monoclonal antibody studies,
185-186
synthetic peptide substrates studies,
186-188
in tracheal smooth muscle cells,
188-192
and Western blotting studies, 184-188
pharmacological modulators of, 239-244
and protein kinase C activation in skinned
vascular smooth muscle, 223-229
as rate limiting step in force development
in smooth muscle, 33-34, 38
and regulation of contractions in smooth
muscle, 119-121, 415

in skinned guinea pig taenia coli, effects of antibodies to turkey gizzard myosin light chain kinase on, 479–480
in skinned swine carotid media, dependence of stress and velocity on, 399–409
and slowing of crossbridge cycling rate in mammalian smooth muscle, 289–300
and tension development, 132
time course of, and smooth muscle contraction, 359

Myosin, smooth muscle
-actin interactions, and tropomyosin, 159–177; see also Actin
activation state, and vanadate inhibition of crossbridges in skinned smooth muscle, 475–476
affinity of thin filament for. See Km effect
ATP-induced conformational transition in, 91–104
dephosphorylation. See Smooth muscle contractions
heavy chain, location of reactive thiols of, 100–101
and heavy meromyosin, proteolytic digestion of, 93–98
isoforms in smooth muscle, 67–77
  densitometric scans from different species, 70–71
  and effect of hypertrophy of uterine smooth muscle on band patterns, 73
  effect of filamin, 74–77
  effect of phosphorylation, 76–77
  heavy chain, in human smooth muscle from different tissues, 451
  pyrophosphate gel electrophoresis for separation of, 67–77
magnesium-ATPase activity of
  actin-activation of, and tropomyosin binding to actin, 144–149
  conversion of myosin to 10S form, and thiol reagents, 101–104
  dependence of shortening in skinned single smooth muscle cells, 487
  and force development in skinned chicken gizzard fibers, 430
  myosin phosphorylation and actin-activation of, 143
  and proteolytic digestion of smooth muscle myosin and heavy meromyosin, 93–98
  and stability of copolymers composed of dephosphorylated and phosphorylated smooth muscle myosin, 88, 90
  and stability of synthetic smooth muscle myosin filaments and phosphorylation of myosin light chain, 82–83
  stimulation by MalNEt, 101–104

Nitrendipine, and inhibition of myosin light chain kinase activity, 240
Noncontractile proteins, dephosphorylation by smooth muscle phosphatases, 211–212
Norepinephrine stimulation of vascular smooth muscle of spontaneously hypertensive rats
  and α-adrenoreceptor-mediated responses, 280–284
  and β-adrenoreceptor-mediated responses, 277–280
  molecular mechanisms of increased contractility to, 277–285
Nucleotides, and conformational change in smooth and striated muscle, 63–64; see also Cyclic nucleotides

Optical microscopy of smooth muscle cells
  and three dimensional organization of contractile elements, 6–8
  and structural changes during contractions, 18–22
Oxygen consumption, and isometric force in smooth muscle, 463–464

Papain
  digestion of arterial carotid smooth muscle cells with, contractions and, 453–454
  myosin and heavy meromyosin degradation in presence of ATP, 93–98

Pelleting assay, of time course of copolymer formation by smooth muscle myosin filaments, 85–86

pH, and actin-activated ATPase activation of smooth muscle myosin, 165–167, 174–175

Phorbol ester-induced contractions in skinned vascular smooth muscle, protein kinase C and, 219–229

Phosphatase catalytic subunit, and latch state in glycerinated porcine carotid artery, 419–422; *see also* Smooth muscle phosphatases, characterization and function

Phosphate
burst rate constant measurements, and kinetic mechanism for striated muscle actomyosin S1, 60–61
inorganic, and contraction in smooth and striated muscles, 36–37

Phosphodiesterase isozymes in canine trachealis, identity and role of, 269–273

Phosphoinositide metabolism, and norepinephrine stimulation of vascular smooth muscle, 282

Phospholipids. *See* Inositol phospholipids

Phosphorylase kinase, and actin-myosin interaction, 251

Phosphorylation
of caldesmon, 130–131
conformational changes in myosin and heavy meromyosin from chicken gizzard associated with, 91–104
and contractile activity in smooth muscle cells, 207
and effect of high myosin light chain phosphatase on latch phenomenon in smooth muscle, 387–397
effects on myosin isoforms in smooth muscle, 76–77
and kinetic mechanisms of heavy meromyosin and myosin in absence of actin, 63–64
of myosin light chain by calcium-dependent myosin light chain kinase, smooth muscle contraction and, 109–116
and norepinephrine stimulation of adenylate cyclase activity in vascular smooth muscle of spontaneously hypertensive rats, 279–280
and polycation-modulable protein phosphatases, 195–205
and regulation of smooth muscle actomyosin S1, 62
and smooth muscle actomyosin magnesium-ATPase activity, 175
and structural change in smooth muscle myosin and heavy meromyosin, 62–63
and subunit exchange between smooth muscle myosin filaments, 82–90
see also Myosin phosphorylation in smooth muscle; Smooth muscle phosphorylation inhibitors

Phosphotransferase reaction, catalyzed by myosin light chain kinases, 186

Photolysis. *See* Caged ATP

Polycation
effect on smooth muscle phosphatases, 211
-modulable phosphatases, aortic. *See* Aortic polycation-modulable protein phosphatases

Polyclonal antibodies
against chicken gizzard caldesmon, and tissue and species distribution study, 126–127
and removal of myosin light chain kinase from bovine stomach extract without affecting leiotonin activity, 113

Polylysine, effects on activities of smooth muscle phosphatases, 210, 211–212

Polypeptides, of aortic polycation-modulable protein phosphatases, 196–201

Porcine carotid artery. *See* Swine carotid artery

Portal vein, rabbit or guinea pig, saponin skinned strips. *See* Saponin

Potassium chloride
and actin-activated ATPase activity of smooth muscle myosin, 167–169
and reduced smooth muscle contractions, 365–367

Potassium depolarization inhibition, and smooth muscle contraction kinetics, 384

Pregnancy, and myosin isoforms in rat myometrium, 70–71, 73–75
Protease-activated kinase, and actin-myosin interaction, 251
Protein kinase C
 and actin-myosin interaction, 252
 activation and contraction in skinned vascular smooth muscle, 219–229
 inhibitor, 258–259
Protein kinases, and actin-myosin interaction, 251–252; *see also* Cyclic AMP-dependent protein kinase; Protein kinase C
Protein phosphatases. *See* Aortic polycation-modulable protein phosphatases
Proteolytic digestion, and removal of myosin light chain kinase from bovine stomach extract without affecting leiotonin activity, 112–113
Pyrophosphate gel electrophoresis, and detection of myosin isoforms in smooth muscle, 67–77

Rabbit
 aortic strips, inhibition of phosphorylation, 252–253
 colonic smooth muscle
  association of intracellular calcium ions release with contraction of, 485–486
  cell cultures, 465–466
 portal vein, saponin skinned strips of, in study of crossbridge transients in smooth muscle, 28–38
 smooth muscle. *See* Crossbridge(s)
 taenia coli, and slowing of crossbridge cycling rate studies, 289–300
 uterus, myosin isoforms, 70
Rat
 anococcygeus birefringence, and density of myosin filaments, 459–460
 aorta smooth muscle cells cytochalasin-like activity, 469–470
 aortic strips, inhibition of phosphorylation in, 252–253
 smooth muscle, myosin isoforms in, 70–77
  pregnancy and myometrium and, 70, 73–75
 thoracic aorta, calcium sensitivity studies, 320–322
 tissues, subcellular localization of caldesmon, 127–128
 tracheal muscle. *See* Smooth muscle contractions
 see also Spontaneously hypertensive rat vascular smooth muscle, molecular mechanisms of increased contractility to norepinephrine stimulation of
Rate constants, and kinetic mechanism for non-regulated actomyosin S1, 59–62
Relaxation
 rates in skinned smooth muscle, and ADP, 45
 from rigor in skinned smooth muscles, ATP-induced, effects of ADP on, 52–54
Rigor state
 and crossbridge structure of smooth muscle, 29–32
 relaxation from in skinned smooth muscle, effects of ADP on ATP-induced, 52–54

Saponin
 -exposed skinned isolated smooth muscle cells, structural and organizational studies, 8–17
 -skinned strips of rabbit or guinea pig portal vein, in study of crossbridge transients in smooth muscle, 28–38
SDS polyacrylamide gel electrophoresis
 of caldesmon, 123
 of myosin isoforms in smooth muscle, 68–69
 of papain digested heavy meromyosin, 96
 and relative masses of smooth muscle myosin light chain kinases, 184–188
Sedimentation patterns, of smooth muscle myosin filaments, 84–85
Sedimentation rates
 and magnesium-ATP-induced conformational change in smooth muscle myosin, 91–104

relationship of proteolytic digestibility of heavy meromyosin to, 95–98
Serotonin, and receptor-mediated activation of smooth muscle, 380, 381
Shortening
  in skinned smooth muscle cells, calcium and magnesium dependence of, 487
  velocity of skinned smooth muscle
    and calcium, 322–324
    and calmodulin, 431–433
    maximal, effects of ADP on, 49
Skeletal muscle
  myosin filaments, exchange of molecules between monomer pool and filament, 90
  role of tropomyosin in, 159
  tropomyosin binding to actin, magnesium ions and ionic strength and, 147
Skinned fibers
  chicken gizzard cells, non-calcium ion activated contraction, 437–447
  contraction, and protein kinase C activation, 219–229
  coronary artery, effects of cyclic, AMP-dependent protein kinase in, 471
  and dependence of stress and velocity on calcium ions and myosin phosphorylation, 399–409
  effects of aortic polycation-modulable phosphatases on, 204–205
  effects of calcium on mechanics and energetics of contractility, 319–329
  stiffness, and energetics of shortening in, 333–343
  vanadate inhibition of cycling and noncycling crossbridges in, 475–476
Smooth muscle
  carbachol-stimulated trachealis, inositol phospholipids synthesis in, 449–450
  crossbridge attachment model, 467
  crossbridge transients in, 27–38
    and actin and myosin filaments, 29–32
    and crossbridge structure, 29–32
    initiation by photolysis of caged ATP, 32–38
    and light chain phosphorylation, 33–34

  in rigor and relaxed states, 29–32
  existence of thin filament-linked regulatory mechanism in, 121–122
  human, myosin heavy chain isoforms in different tissues, 451
  hypoxia-induced relaxation, and energy supply to contractile apparatus, 463–464
  myosin isoforms in. *See* Myosin, smooth muscle
  rabbit colonic cell cultures, 465–466
  see also under various types of smooth muscle, e.g., Airway smooth muscle; Arterial smooth muscle; Mammalian smooth muscle; Vertebrate smooth muscle cells; *entries beginning with* Smooth muscle
Smooth muscle actomyosin. *See* Actomyosin, smooth muscle
Smooth muscle caldesmon. *See* Caldesmon, smooth muscle
Smooth muscle cell filaments, molecular structure and organization, 1–22
  actin and myosin filaments and fusiforms dense bodies of, 2–6
  and flexibility of force generating mechanism, 2
  and immunochemical labeling of fusiform dense bodies, 16–17
  optical microscopy and three dimensional organization of contractile elements, 6–8
  and structure of contractile apparatus, 11–17
  and structural changes associated with contraction, 1–2, 18–22
  and use of saponin-exposed isolated skinned cells, 8–17
  and use of single, isolated cells, 2–6
Smooth muscle cells
  force:velocity relationship and helical shortening in, 303–315
  and analysis of slack test data, 289–302, 306–307
  and arrangement of contractile apparatus within cell, 315

and crossbridge cycle kinetics,
313-314
form of, 313
and isotonic releases, 304-306, 307-308, 311-313
and video images of contraction of freely shortening cells,
309-310
and Vmax estimates, 309, 314-315
myosin phosphorylation in. *See* Myosin phosphorylation in smooth muscle
rat aorta, cytochalasin-like activity in,
469-470
see also Smooth muscle cell filaments, molecular structure and organization
Smooth muscle contractility
crossbridge properties during stretch,
347-355
effects of calcium on mechanics and energetics, 319-329
and efficiency, 328-329
and isometric force, 320-322
and shortening velocity, 322-324
stiffness and energetics of active shortening, 333-343
see also Smooth muscle contractions
Smooth muscle contractions
and aortic polycation-modulable phosphatases, 204-205
of arterial carotid smooth muscle cells prepared by digestion with papain,
453-454
association with intracellular calcium ions in colonic muscle, 485-486
and birefringence of rat anococcygeus,
459-460
calcium/calmodulin activation at low myosin phosphorylation levels,
427-435
crossbridge slowing mechanisms and,
357-374
and fatigue, 372-373
and internal resistance to shortening,
367-372
and intracellular acidosis, 373-374
and normally cycling and latch/slowly cycling crossbridges, 360-364

reduced activation with KCL as agonist, 365-367
and time course of myosin light chain phosphorylation, 359
and crossbridge transients, 27-38
energetics of
active shortening in skinned smooth muscle, stiffness and, 333, 343
effects of calcium on mechanics and,
319-329
energy supply to contractile apparatus in hypoxia-induced relaxation,
463-464
influence of ATP, ADP, AMPPNP on, 43-55
energy usage
length dependence of, 27
and slowing of crossbridge cycling rate, 289-300; *see also* Latch-bridge state
and high myosin light chain phosphatase activity, latch phenomenon and,
387-397
kinetics of, activation mode and, 377-385
and barium, 382
and electrical field stimulation, 380, 381
and inhibition of contractions,
382-384
and inhibition of response to electrical field stimulation, 382-383
and inhibition of response to potassium depolarization, 384
and measurement of contraction kinetics, 377
and membrane depolarization, 382
and re-acceleration of down-regulated contraction kinetics, 384-385
and receptor-mediated activation, 380, 381
mechanics of. *See* Mechanics of smooth muscle contractions
myosin light chain kinase and cyclic AMP-dependent kinase in vascular smooth muscle of spontaneously hypertensive and normotensive rats, 234-239
and direct myosin light chain phosphorylation inhibitors, 235

and pharmacologic modulation of vascular contractile protein interactions, 239–244
myosin phosphorylation and regulation of, 119–121
non-calcium ion activated, 437–447
protein kinase C activation and, 219–229
relaxation by myosin dephosphorylation, 415–425
-relaxation cycle, and phosphorylation of myosin light chain by calcium-dependent myosin light chain kinase, 109–116
sliding filament mechanism of, 27
and smooth muscle phosphatases, 207–216
and stiffness during active shortening, 483
and structural changes in isolated, skinned smooth muscle cells, 18–22
Smooth muscle myosin. *See* Myosin, smooth muscle
Smooth muscle myosin filaments. *See* Myosin filaments, smooth muscle
Smooth muscle phosphatases, characterization and function, 207–216
effect of polycations, 211
effect of polylysine, 210, 211–212
physical properties of, 209–210
possible mode of action, 213–215
and regulation of smooth muscle contraction, 207–216
substrate specificity of, 211–212
Smooth muscle phosphorylation inhibitors, 251–260
cyclic AMP-dependent protein kinase inhibitors, 256–258
and effects of protein kinases on actin-myosin interaction, 251–252
myosin light chain-kinase inhibitors, 252–256
protein kinase C inhibitor, 258–259
see also Phosphorylation
Sodium nitroprusside, and role of phosphodiesterase isozymes in canine trachealis, 272–273
Spontaneously hypertensive rat vascular smooth muscle, molecular mechanisms of increased contractility to norepinephrine stimulation of, 277–285
Steric blocking mechanism, and kinetic model for striated muscle actomyosin S1 activation, 62
Stiffness
and energetics of active shortening in skinned smooth muscle, 333–343
and biochemical correlates of crossbridge turnover, 341–343
and force-extension relationships of elastic components, 338–341
and homogeneity of length step, 336–338
measurements, and crossbridge properties of smooth muscle during stretch, 347–355
Stress maintenance in vascular smooth muscle, and protein kinase C, 219–229
Stretching. *See* Smooth muscle contractility
Striated muscle
actomyosin. *See* Actomyosin, smooth muscle
heavy meromyosin, calcium and ATP and activation of, 62–65
proteins, and smooth muscle actomyosin ATPase, 27
tropomyosin, and inhibition of ATPase activity, 159
Z discs, compared with contractile elements of smooth muscle cells, 8, 9
Structural mechanism of regulation of smooth muscle and striated muscle actomyosin, compared with kinetic mechanism, 59–65
Superprecipitation, as criterion for assessing in vitro smooth muscle contraction, 110
Swine carotid artery
effects of calcium on energetics of cells of, 324–325
and latch phenomenon studies of high myosin light chain phosphatase, 387–397
latch state in skinned preparations, 411–412
myosin phosphorylation and calcium ions and stress and velocity in, 399–409

prepared by digestion with papain, contraction of, 453–454
Synthetic myosin filaments, subunit exchange studies with, 81–90
Synthetic peptide substrates, in studies of smooth and skeletal muscle myosin light chains, 186–188

Temperature, and crossbridge properties of smooth muscle during stretch, 348–349
Tension development in smooth muscle
myosin phosphorylation and, 121, 132
from rigor state in presence of calcium and photolysis of caged ATP, 32–38
and vanadate, 475–476
see also Latchbridge state
Thick filaments, composition of, 183
Thin filament-associated proteins, modulation of actomyosin ATPase by, 143–155
and effect of calcium and magnesium ions on tropomyosin binding to actin, 144–149
and effect of caldesmon on actomyosin ATPase, 149–154
and effect of ionic strength on tropomyosin binding to actin, 147–149
see also Caldesmon, Smooth muscle
Thiol reagents, effects on smooth muscle myosin, 100–104
Tilting experiments, and smooth muscle cell filaments molecular structure and organization, 13, 15–16
Toad. *See Bufo marinus* isolated smooth muscle cells
Tracheal smooth muscle
canine
carbachol-stimulated, inositol phospholipids synthesis in, 449–450
mechanism of shortening-induced dependence of contractility in, 461–462
cells, myosin phosphorylation in, 188–192
and measurement of nonphosphorylated, monophosphorylated, and diphosphorylated forms of myosin, 189–192
stiffness and length relationship during active shortening, 483
see also Bovine tracheal strips, effect of muscarinic tone on response to; Canine tracheal smooth muscle
Tropomyosin
association with caldesmon, 127–130
binding to actin
effect of calcium and magnesium ions on, and actin-activation of magnesium-ATPase of myosin, 144–149
effect of ionic strength on, 147–149
binding assays, 163
and myosin binding to actin in absence of calcium, 59
stimulation of vertebrate smooth muscle actomyosin magnesium-ATPase activity, modification by myosin phosphorylation and its conformational state, 159–177
see also Km effect
Tryptic digestion, of aortic polycation-modulable protein phosphatases, 201, 202–203

Uterine smooth muscle myosin isoforms, 67–77

Vanadate, inhibition of crossbridges in skinned smooth muscle, 475–476
Vascular smooth muscle
actomyosin magnesium-ATPase activity stimulation by tropomyosin and modification by myosin phosphorylation, 159–177
and ATPase and phosphorylation assays, 162–163
and binding of smooth muscle tropomyosin to smooth muscle actin, 171–173
enzyme kinetic analysis, 176
and fresh *vs.* frozen preparations, 173
and influence of pH on actin-activated ATPase activity of smooth muscle myosin, 165–167
and myosin-actin interactions, 160

and myosin conformational changes, 173–175
and potassium chloride dependence of actin activation, 167–169
and relationship between phosphorylation and magnesium-ATPase, 175
and tropomyosin binding to actin, 176–177
tropomyosin binding assays, 163
and tropomyosin- and myosin light chain kinase-free actomyosin, 163–164
and Vmax effect and Km effect, 169–171
calcium metabolism, effects of reduced extracellular calcium on, 481
contractions, dependence on myosin phosphorylation and calcium ions, 399–409
comparisons with intact carotid media, 405–406
and dependence of unloaded shortening velocity on phosphorylation, 408
and deterioration and attainment of steady-state, 405
and regulation of stress maintenance, 406–408, 409
cyclic AMP-dependent kinase and myosin light chain kinase in, 234–235

receptor-operated calcium channels in, guanylate-cyclase dependent gating of, 455–457
of spontaneously hypertensive rats, increased contractility to norepinephrine stimulation in, 277–285
Verapamil, and inhibition of smooth muscle response to electrical field stimulation, 382–383
Vertebrate smooth muscle cells
actin and myosin filaments in, 2–4
fusiform dense bodies in, 4
see also Smooth muscle cells
Vmax effect
estimates, in force:velocity relationships in smooth muscle cells, 292–294, 309, 314–315
and stimulating effect of tropomyosin on activation of myosin ATPase by actin, 169–171
for synthetic peptide substrates of myosin light chain kinases, 187–188
see also Maximum shortening velocity in smooth muscle cells

W-7, and inhibition of smooth muscle phosphorylation, 252–256
Western blotting, and relative masses of myosin light chain kinases from smooth muscle, 184–188
Work efficiency, and slowing of crossbridge cycling rate in mammalian smooth muscle, 289–300